AF389343

Remote Sensing in Forest Fire Monitoring and Post-fire Damage Analysis

Remote Sensing in Forest Fire Monitoring and Post-fire Damage Analysis

Editors

Víctor Fernández-García
Leonor Calvo
Susana Suarez-Seoane
Elena Marcos

Basel • Beijing • Wuhan • Barcelona • Belgrade • Novi Sad • Cluj • Manchester

Editors

Víctor Fernández-García
University of Lausanne
Lausanne, Switzerland

Leonor Calvo
Universidad de León
León, Spain

Susana Suarez-Seoane
University of Oviedo
Mieres, Spain

Elena Marcos
Universidad de León
León, Spain

Editorial Office
MDPI
St. Alban-Anlage 66
4052 Basel, Switzerland

This is a reprint of articles from the Special Issue published online in the open access journal *Remote Sensing* (ISSN 2072-4292) (available at: https://www.mdpi.com/journal/remotesensing/special_issues/remote_sensing_forest_fire_monitoring_post_fire_damages_analysis).

For citation purposes, cite each article independently as indicated on the article page online and as indicated below:

Lastname, A.A.; Lastname, B.B. Article Title. *Journal Name* **Year**, *Volume Number*, Page Range.

ISBN 978-3-0365-8882-7 (Hbk)
ISBN 978-3-0365-8883-4 (PDF)
doi.org/10.3390/books978-3-0365-8883-4

Contents

About the Editors

Víctor Fernández-García

Víctor Fernández-García was awarded his MSc in Natural Hazards and PhD in Science from the University of León, both with extraordinary awards. He is currently working at the University of Lausanne, leading an international project studying the role of fire on climate change, biodiversity and local livelihoods in African biodiversity hotspots.

Leonor Calvo

Leonor Calvo was awarded her PhD in Biological Sciences from the University of León. She is currently a Professor of Ecology at the University of León (Spain) and the Leader of the Research group Applied Ecology and Remote Sensing.

Susana Suarez-Seoane

Susana Suarez-Seoane is a Senior lecturer in Ecology at the University of Oviedo and a researcher at the IMIB (Biodiversity Research Institute). Her research focuses on the multi-scale assessment of landscape patterns and processes under global change scenarios, using remote sensing techniques and species distribution models as methodological tools.

Elena Marcos

Elena Marcos was awarded her PhD in Biological Sciences from the University of León. She is currently a Senior Lecturer at the Department of Biodiversity and Environmental Management of the University of León. Her activity is focused on the analysis of the effects of disturbances, mainly forest fires and atmospheric nitrogen deposition, on the functioning of wooded and shrubby ecosystems.

Preface

Wildfires are one of the most significant disturbances worldwide, impacting both natural ecosystems and human communities. Many scientists and a portion of society believe that the frequency and intensity of fires are escalating due to shifts in land use, suppression policies, and climate change, leading to critical consequences for nature and the benefits it provides to people. Additionally, the expansion of human settlements into forested areas has altered our vulnerability to fire, further affecting the potential impacts of future events. In the context of these widespread changes, remote sensing emerges as a valuable tool with which to address the environmental and social challenges of forest fires and offer solutions. Its versatility, wealth of information, and rapid technological advancements make it indispensable for designing proactive strategies, monitoring risks, and analyzing damages over large areas, facilitating decision making and the development of pre- and post-fire management strategies.

This Special Issue aims to explore various aspects that contribute to the advancement of (I) fire driver characterization and the development of fire predictive models, (II) the assessment of burned areas (III) the assessment of burn severity and specific fire impacts, and (IV) the analysis of post-fire trajectories using remote sensing methods. The compilation comprises ten research articles addressing these four topics, utilizing a diverse range of methodologies and remote sensing platforms. The target audience for this Special Issue includes researchers, academicians, practitioners, and policymakers working in forest ecology, environmental science, disaster management, remote sensing, and geospatial technologies in general.

The success of this Special Issue can be attributed to the dedicated efforts and expertise of many individuals. First and foremost, we express our gratitude to the 45 authors for presenting their cutting-edge research and insights into remote sensing. Their exceptional contributions have enriched this Special Issue significantly. We would also like to extend our appreciation to the more than 100 reviewers who invested their time and knowledge in meticulously evaluating the submissions, thereby enhancing the quality of our Special Issue. Finally, we would like to acknowledge the editorial staff for their efficient management of the editorial processes, which facilitated the success of this Special Issue.

Víctor Fernández-García, Leonor Calvo, Susana Suarez-Seoane, and Elena Marcos
Editors

Editorial

Remote Sensing Advances in Fire Science: From Fire Predictors to Post-Fire Monitoring

Víctor Fernández-García [1,2,*], Leonor Calvo [1], Susana Suárez-Seoane [3] and Elena Marcos [1]

[1] Ecology, Department of Biodiversity and Environmental Management, Faculty of Biological and Environmental Sciences, Universidad de León, 24071 León, Spain; leonor.calvo@unileon.es (L.C.); elena.marcos@unileon.es (E.M.)

[2] Institute of Geography and Sustainability, Faculty of Geosciences and Environment, Université de Lausanne, Géopolis, CH-1015 Lausanne, Switzerland

[3] Department of Organisms and Systems Biology and Biodiversity Research Institute, CSIC-University of Oviedo, 33071 Oviedo, 33600 Mieres, Spain; s.seoane@uniovi.es

[*] Correspondence: vferg@unileon.es

Citation: Fernández-García, V.; Calvo, L.; Suárez-Seoane, S.; Marcos, E. Remote Sensing Advances in Fire Science: From Fire Predictors to Post-Fire Monitoring. *Remote Sens.* **2023**, *15*, 4930. https://doi.org/10.3390/rs15204930

Received: 20 September 2023
Accepted: 7 October 2023
Published: 12 October 2023

Fire activity has significant implications for ecological communities, biogeochemical cycles, climate, and human lives and assets. Approximately over half of the Earth's land surface is susceptible to fire, with around 3% experiencing annual burning according to coarse-resolution satellites [1], a value that is probably much higher according to recent estimates from finer satellite imagery [2]. Because of the vast extent of land burned over the world, landscape fires release approximately 23% of the global CO_2 emitted annually from fossil fuels, modify Earth's energy fluxes through changes in surface albedo, and have an enormous influence on human health and the economy [1]. Fires also have a large influence on local ecosystems, affecting the ecosystem services provided to local communities. Thus, fires are a relevant phenomenon with an enormous area of impact every year.

Because of the relevance and impact of fires across the globe, the study of this phenomenon and the assessment of its consequences cannot be addressed only by field or laboratory studies. In this sense, the exploitation of remote sensing platforms, sensors, and methods is crucial to obtain accurate and extensive spatial and spatiotemporal information on fire and its consequences. Considering fire as a natural hazard, geo-spatial studies can be organized in multiple ways, one of them being based on the temporal point on which they focus in relation to fire. Thus, we can differentiate those studies focused on a pre-fire stage: for instance, those addressing topics that can help predict fire-related risks and thus are useful for pro-active management strategies. The second stage is the moment when fires occur. At that point, remote sensing might be useful to detect active fires, or to detect burn scars as evidence of fire. The next stage is the analysis of the immediate impacts and consequences of fire. This is related, but not limited to burn severity assessments, which indicate the overall environmental change caused by fire. The assessments immediately after fire are essential for addressing post-fire emergency actions when needed. Lastly, remote sensing plays a crucial role in analyzing the evolution of burned areas over time. This can be focused on multiple elements of the ecosystem but is generally focused on soil and vegetation. The assessment of post-fire trajectories is necessary to identify the areas where post-fire recovery is not satisfactory and for the implementation of restoration strategies.

The remote sensing discipline is rapidly advancing thanks to the increasing availability of sensors, data, techniques, and processing capabilities. Thus, in this Special Issue, "Remote Sensing in Forest Fire Monitoring and Post-fire Damage Analysis", we compiled 10 studies [1–10] representing significant advances in the remote sensing of fires, with regard to the different aspects and temporal stages exposed above. In relation to the pre-fire stage, our Special Issue provides new insights into the relevant predictors of fire activity, such as live fuel moisture content [3] and surface fuel load [4], or soil moisture availability [5]. Moreover, Stoyanova et al. [5] explored the potential of land surface temperature status and dynamics as novel indicators of fire risk. In relation to the second stage, a new burned area mapping algorithm using Sentinel-2 [2] is presented, which has been revealed as the most accurate non-commercial imagery for burned area mapping. Furthermore, Park et al. [6] presented a novel approach to improve disaster responses, based on the application of deep learning to detect the multiple elements that might interact in a fire situation. Regarding the assessment of burned areas and fire impacts, our Special Issue includes the first analysis of trends in burn severity at the global scale, revealing the aggravation of fires in many forest biomes [1]. Moreover, Silva Cardoza et al. [7] proposed an improved burn severity algorithm by combining relative phenological correction with former burn severity metrics. In terms of analyzing post-fire trajectories, our Special Issue encompasses a variety of advances, providing cutting-edge information on the drivers and dynamics of post-fire regeneration [8], the performance of physical-based models to measure forest resilience to fire [9], and the identification of optimal parametrizations and wavelengths for LiDAR classifications in post-fire environments [10].

The work provided in this Special Issue contains examples of the multiple advancements in the remote sensing discipline. For instance, the presented studies demonstrate advancements using different remote sensing platforms exploiting imagery from geosynchronous orbit satellites (METEOSAT) [5], sun-synchronous polar orbit satellites (MODIS) [1,3], low-earth-orbit satellites (Sentinel-2) [2,7–9], aircrafts [4,10], or UAVs [6]. Likewise, examples are also provided of how the remote sensing and fire sciences can be advanced using different sensor types (passive [1–3,5–9] and active [4,10]) and methodological approaches (deep learning, machine learning, radiative transfer models, spectral mixture analysis, interpolation techniques, linear models, and spectral indices).

Conflicts of Interest: The authors declare no conflict of interest.

References

1. Fernández-García, V.; Alonso-González, E. Global Patterns and Dynamics of Burned Area and Burn Severity. *Remote Sens.* **2023**, *15*, 3401. [CrossRef]
2. Sali, M.; Piaser, E.; Boschetti, M.; Brivio, P.A.; Sona, G.; Bordogna, G.; Stroppiana, D. A Burned Area Mapping Algorithm for Sentinel-2 Data Based on Approximate Reasoning and Region Growing. *Remote Sens.* **2021**, *13*, 2214. [CrossRef]
3. Cunill Camprubí, À.; González-Moreno, P.; Resco de Dios, V. Live Fuel Moisture Content Mapping in the Mediterranean Basin Using Random Forests and Combining MODIS Spectral and Thermal Data. *Remote Sens.* **2022**, *14*, 3162. [CrossRef]
4. Lin, C.; Ma, S.-E.; Huang, L.-P.; Chen, C.-I.; Lin, P.-T.; Yang, Z.-K.; Lin, K.-T. Generating a Baseline Map of Surface Fuel Loading Using Stratified Random Sampling Inventory Data through Cokriging and Multiple Linear Regression Methods. *Remote Sens.* **2021**, *13*, 1561. [CrossRef]
5. Stoyanova, J.S.; Georgiev, C.G.; Neytchev, P.N. Satellite Observations of Fire Activity in Relation to Biophysical Forcing Effect of Land Surface Temperature in Mediterranean Climate. *Remote Sens.* **2022**, *14*, 1747. [CrossRef]
6. Park, M.; Tran, D.Q.; Lee, S.; Park, S. Multilabel Image Classification with Deep Transfer Learning for Decision Support on Wildfire Response. *Remote Sens.* **2021**, *13*, 3985. [CrossRef]
7. Silva-Cardoza, A.I.; Vega-Nieva, D.J.; Briseño-Reyes, J.; Briones-Herrera, C.I.; López-Serrano, P.M.; Corral-Rivas, J.J.; Parks, S.A.; Holsinger, L.M. Evaluating a New Relative Phenological Correction and the Effect of Sentinel-Based Earth Engine Compositing Approaches to Map Fire Severity and Burned Area. *Remote Sens.* **2022**, *14*, 3122. [CrossRef]
8. Avetisyan, D.; Velizarova, E.; Filchev, L. Post-Fire Forest Vegetation State Monitoring through Satellite Remote Sensing and In Situ Data. *Remote Sens.* **2022**, *14*, 6266. [CrossRef]
9. Fernández-Guisuraga, J.M.; Suárez-Seoane, S.; Quintano, C.; Fernández-Manso, A.; Calvo, L. Comparison of Physical-Based Models to Measure Forest Resilience to Fire as a Function of Burn Severity. *Remote Sens.* **2022**, *14*, 5138. [CrossRef]
10. Nelson, K.; Chasmer, L.; Hopkinson, C. Quantifying Lidar Elevation Accuracy: Parameterization and Wavelength Selection for Optimal Ground Classifications Based on Time since Fire/Disturbance. *Remote Sens.* **2022**, *14*, 5080. [CrossRef]

Article

Global Patterns and Dynamics of Burned Area and Burn Severity

Víctor Fernández-García [1,2,*] and Esteban Alonso-González [3]

1. Institute of Geography and Sustainability (IGD), Université de Lausanne, Mouline–Géopolis, 1015 Lausanne, Switzerland
2. Ecology, Department of Biodiversity and Environmental Management, Faculty of Biology and Environmental Sciences, Universidad de León, 24071 León, Spain
3. Centre d'Etudes Spatiales de la Biosphère, CESBIO, Université de Toulouse, CNES/CNRS/INRAE/IRD/UPS, 31000 Toulouse, France; esteban.alonso-gonzalez@univ-tlse3.fr
* Correspondence: vferg@unileon.es or victor.fernandezgarcia@unil.ch

Abstract: It is a widespread assumption that burned area and severity are increasing worldwide due to climate change. This issue has motivated former analysis based on satellite imagery, revealing a decreasing trend in global burned areas. However, few studies have addressed burn severity trends, rarely relating them to climate variables, and none of them at the global scale. Within this context, we characterized the spatiotemporal patterns of burned area and severity by biomes and continents and we analyzed their relationships with climate over 17 years. African flooded and non-flooded grasslands and savannas were the most fire-prone biomes on Earth, whereas taiga and tundra exhibited the highest burn severity. Our temporal analysis updated the evidence of a decreasing trend in the global burned area (-1.50% year^{-1}; $p < 0.01$) and revealed increases in the fraction of burned area affected by high severity (0.95% year^{-1}; $p < 0.05$). Likewise, the regions with significant increases in mean burn severity, and burned areas at high severity outnumbered those with significant decreases. Among them, increases in severely burned areas in the temperate broadleaf and mixed forests of South America and tropical moist broadleaf forests of Australia were particularly intense. Although the spatial patterns of burned area and severity are clearly driven by climate, we did not find climate warming to increase burned area and burn severity over time, suggesting other factors as the primary drivers of current shifts in fire regimes at the planetary scale.

Keywords: fire severity; burn severity; spatial patterns; trends; biomes; continents; climate warming

Citation: Fernández-García, V.; Alonso-González, E. Global Patterns and Dynamics of Burned Area and Burn Severity. *Remote Sens.* **2023**, *15*, 3401. https://doi.org/10.3390/rs15133401

Academic Editor: Luis A. Ruiz

Received: 31 May 2023
Revised: 27 June 2023
Accepted: 2 July 2023
Published: 4 July 2023

1. Introduction

Fire activity plays a key role in shaping ecological communities, biogeochemical cycles, climate, and human lives and assets [1,2]. More than half of the land surface on Earth is prone to fire [3], with about 3% burning annually [4]. As a result, landscape fires generate annual emissions estimated in 2.2 PgC, an equivalent of 23% of global CO_2 from fossil fuels [5], influence global albedo [6], and cause premature deaths from poor air quality, dozens of direct fatalities, and annual economic losses estimated at above US$2500 million [7]. Despite these global figures, burned area (BA) and its directly related fire regime variable, the fire frequency [8], are largely heterogeneous across space and time because of differences in their main determining factors. Among the BA determining factors there are fuel load, which is linked to primary productivity and herbivory [1,9,10]; fuel connectivity, which is enabled by non-fragmented landscapes and non-sparse vegetation [11,12]; flammability of fuels, closely linked to weather and vegetation properties [2,13,14]; and ignition sources, which might be natural or anthropogenic, the last exceeding 90% of ignition events in most terrestrial biomes [8]. In the same way, the ecological consequences of fire vary depending on the intrinsic ecosystem traits, post-fire environmental conditions, and fire characteristics, mainly burn severity (BS). BS is closely linked to fire intensity and here is defined as the immediate degree of overall environmental change caused by a fire in ecosystems, including biomass loss, vegetation mortality, and biochemical and physical impacts on soil [15]. Moreover, BS determines the post-fire processes, playing a key role in the future ecosystem pathways [15–18], and its characterization is essential for the refinement of carbon emission models [5,15,19].

Nowadays, the exploitation of some types of satellite imagery provides globally consistent BA products based on the detection of active fires and the spectral changes in surface reflectance [4,20]. Remote sensing analysis revealed grasslands and savannas as the most fire-prone biomes on Earth, mainly those located in Africa and Northern Australia. In these regions many areas burn annually or biennially, exhibiting a large quantitative difference from the rest of the biomes, where generally less than 2% of the area burns every year [4,5]. Satellite imagery also allows the characterization of spatial patterns of BS [19,21,22]. BS has been customarily assessed by spectral indices based on the near-infrared and shortwave infrared reflectance, such as the difference in the normalized burn ratio (dNBR) [16,22,23]. The dNBR accurately matches field measurements of the overall environmental change caused by a fire in multiple ecosystems (mainly calculated through composite indices combining several ecosystem metrics) and has been adopted as a standard BS metric by the EFFIS and MTBS programs of the European Union and the United States of America, respectively [19]. Despite the lack of comprehensive global spatiotemporal BS characterizations [7], current knowledge of maximum fire intensities [20], biomass burned fractions [5] and BS drivers (mainly the fuel load available to burn and fuel continuity) [24–26] anticipate a large spatial variability in BS across the globe.

The evolution of BA and BS arouses great interest in the media and in the scientific community, which frequently warned about the increase or worsening of fire activity during recent years [7], often attributing that assumed trend to climate change (see examples in [17,27–29]). In this sense, climate warming is expected to cause increases in fire weather danger in many regions [14], which is a driver of BA in a large proportion of Earth's land surface [13] and influences BS [24,25]. However, former empirical evidence of BA and BS increases are often based on constrained observations, in terms of timescales or spatial coverage [7]. In fact, global quantitative BA analysis has shown significant decreases in the global BA between 2003 and 2015, mainly concentrated in the tropical savannas of Africa, South America, and Asian steppes, albeit BA increases have been detected in many regions such as the Siberian Arctic [30] and others dominated by closed canopy forests [4], including Australia where BA increases have been fostered by climate change [31]. Likewise, little is known about BS trends at the global scale as most of the literature focuses on USA, Australia, and Mediterranean Europe, which might lead to a global "Western" biased perception [7]. For instance, it is well documented a shift from low to high severity fire regimes in southwestern US forests, caused by the implementation of fire suppression policies after the European settlement [1], and extensive spatiotemporal studies have revealed a generalized increase in high-severity fires in some parts of Western US between 1984–2015 [21]. In Australia, an increase in the proportion of BA at high severity has been detected between 2013 and 2017 [17], and in Europe, an increased prevalence of extreme wildfire events attributed to climate change and human alterations of landscapes has been reported [29].

Here, we characterize the spatial patterns and temporal trends of BA, BS, and BA by BS levels at the global scale and by biomes and continents for the period 2003–2019 (17 years). In addition, we studied the potential relationships between spatiotemporal patterns of BA, BS, and BA by severity levels with climate variables (mean annual temperature, annual temperature range, annual precipitation, and annual precipitation range) to identify the former role of climate change on fire activity. To achieve these objectives, we used NASA-MODIS-derived products at 500 m spatial resolution and ERA5 ECMWF re-analysis climatic data.

2. Materials and Methods

2.1. Data Sources

The BA and BS data were obtained from the MOSEV database [19]. MOSEV has been developed based on MODIS information from the MCD64A1 collection 6 [32], which is the standard NASA global BA product and probably the most used by the scientific community [19]; and from the Terra MOD09A1 and Aqua MYD09A1 version 6 products, which provide the highest quality surface reflectance observations in eight-day periods. MOSEV offers, among other information, monthly global data at 500 m spatial resolution from 2000 to 2019 of date of burning and BS measured by the dNBR spectral index ranging from −2000 to 2000 [19]. Burn severity in the MOSEV database refers to short-term BS, also called initial assessment of BS as the closest pre- and post-fire MOD09A1 and MYD09A1 scenes are used for calculations. The alternative to initial BS assessments, are the extended assessments that are often made one year after burning and thus are influenced by ecosystem recovery [15,16], BS in the MOSEV database is directly related to short-term BS calculated with Landsat imagery (R = 0.74 for dNBR; R = 0.42 for RdNBR) [19]. Here, we used the dNBR instead of the relativized index RdNBR, also available in MOSEV, according to (I) its better performance when using MODIS; (II) to its similar or better fit to field composite metrics of BS such as the Composite Burn Index in different types of ecosystems; and (III) to be consistent with our definition of BS, which is understood as the overall degree of environmental change caused by fire [19]. For the present study, the period 2003–2019 was selected for consistency because Aqua products were not available for entire years before 2003. Mean temperature and precipitation data were extracted for each month of the time series from the ERA5 ECMWF reanalysis. ERA5 was downloaded from the Copernicus Climate Data Store (CDS), using the CDS Python API.

2.2. Data Preparation

All data were extracted globally and by the regions conformed by the intersection of terrestrial biomes (i.e., global vegetation units) [33] with the continent boundaries [34]. The extraction of the global and regional data from both MOSEV and ERA5 was designed in the R programming language [35]. All the routines were implemented in the supercomputing facilities of the Spanish Research Council (CSIC).

For each year, we computed globally and by regions the BA, the percentage of the land burned, the mean BS, the percentage of land burned at different BS levels, and the percentage of the BA burned at different BS levels. The number of BS levels and threshold values depend on the user's objectives [16,19,36]. The differentiation of three categories (low, moderate, and high BS) is probably the most common approach [19,37,38] and many studies have followed the pioneering proposal by Key and Benson [16] to differentiate levels. In the present study, we differentiated low, moderate, and high BS levels by the standard <270, 270–440, and >440 dNBR ranges, respectively, following the thresholds proposed in [16]. We have used the dNBR value of >440 to differentiate the high severity rather than other common higher values (e.g., 660) as those have been found to be excessive by multiple studies [37,38], and because values above 660 are scarce in the MOSEV database [19]. Moreover, we highlight that we have applied the same thresholds to all the regions for consistency; thus, same categories reflect similar overall spectral change with respect to the pre-burn situation regardless of the region.

Likewise, for each year we computed the mean annual temperature, the annual temperature range as the difference between the hottest and coldest months, the annual precipitation, and the annual precipitation range as the difference between the wettest and driest months. This raw database with the annual values of all study variables was checked for coherence before performing the statistics based on regional data. Thus, biomes corresponding to lakes, water bodies, rock, and ice were removed, as well as those regions with BA registered less than 10 years of the study period (<60% of the time).

2.3. Data Analysis

All the statistical analyses were performed in R software [35]. The spatial patterns of all fire variables (BA, percentage of the land burned, mean BS, percentage of land burned at different BS levels, percentage of the BA burned at different BS levels), and climate variables (mean annual temperature, annual temperature range, annual precipitation, and annual precipitation range) were studied by regions calculating the mean ($\pm$standard error) values for the entire study period (2003–2019). Likewise, we computed the mean global values of all fire variables.

To perform the temporal trend analyses (2003–2019; n = 17), we normalized the global and regional annual data of fire variables to the corresponding 2003 values (with values of 2003 set to 100%), to facilitate interpretation and intercomparison. Afterward, we calculated the Sen robust regression estimator with 95% confidence intervals using the zyp.sen and confint.zyp functions of the zyp package [39], and the Mann–Kendall significance using the cor.test function. The Sen robust regression estimator is a non-parametric statistical analysis computed from the median of the slopes of all lines through pairs of points, and thus it is insensitive to outliers. The non-parametric Mann–Kendall significance indicates if a variable consistently decreases or increases over time.

The influence of climate variables (independent) on each fire variable (dependent) was studied with the data by regions (n = 64; n = 63 for BA at high severity analyses because of the absence of this BS level in one region) using two complementary statistical approaches: first, the relationship of each single climate variable with each fire variable was shown by fitting the local polynomial regression (loess), and the Spearman's rank correlation coefficients and significances were calculated using the cor.test function; secondly, conditional inference regression trees were implemented to account for complex combined effects of climate variables. All mean 2003–2019 BA data were square root transformed for these statistical analyses and to facilitate outputs visualization. Regression trees were built with the ctree function of the partykit package [40], including as predictors the four climate variables together. Regression trees include the most significant partitions up to $p < 0.10$ calculated with the approximated finite-sample distribution of Monte Carlo for each node. The variance explained by the regression tree predictions against our data was also shown. The analyses were run to analyze both, the synchronous relationships between fire and climate by using the 2003–2019 means; and the influence of temporal trends, using the Sen slopes as input data.

To analyze the relationships between pairs of fire variables, as well as between pairs of climate variables, correlation matrices with the loess fit and Spearman's rank correlation coefficients and significances were computed for the synchronous spatial data (mean 2003–2019 values) and trend data (Sen slopes) by regions (n = 64) using the pairs.panels function of the psych package [41].

3. Results and Discussion

3.1. Spatial Patterns of Burned Area and Burn Severity

We observed that 4.78 ± 0.12 Mkm^2 (mean $\pm$ standard error), and around $3.26 \pm 0.08\%$ of Earth's land surface burned annually in the period 2003–2019 (Table 1), similar to NASA-MODIS MCD64A1 BA data reported for 2003–2015 [4]. The spatial patterns of absolute BA were highly heterogeneous among regions (biomes and continents) (Table A1). The largest contributors to global BA were the tropical and subtropical grasslands, savannas, and shrublands of Africa (2.79 ± 0.07 Mkm^2 year^{-1}), Australia (0.36 ± 0.03 Mkm^2 year^{-1}) and South America (0.26 ± 0.02 Mkm^2 year^{-1}), a consequence of both their proneness to fire and their vast extent.

Relativizing the BA to the total extent of each region (Figure 1; Table A1), we found that the most fire-prone regions were the flooded grasslands and savannas of Africa ($26.97 \pm 1.01\%$ year^{-1}), as well as the tropical and subtropical grasslands, savannas and shrublands of Africa ($20.07 \pm 0.49\%$ year^{-1}), and Australia ($17.01 \pm 1.32\%$ year^{-1}). There, the seasonally dry climate enables positive feedback interactions between primary productivity, fuel connectivity, and fire [1,12,42] which are essential for maintaining the open structure of these biomes [9,43]. Among grasslands and savannas, the highest potential to burn in those seasonally flooded has been attributed to greater fuel accumulation rates as consequence of a higher productivity and the low herbivory of their less palatable vegetation [10]. We found other biomes to be particularly fire-prone with more than 5% of their area burned annually (Figure 1; Table A1), including the tropical and subtropical dry broadleaf forests of Africa and montane grasslands and shrublands of Africa and Australia. Likewise, a surprisingly high BA fraction (4.38% to 3.42% year^{-1}) was detected in tropical and subtropical moist broadleaf forests of Africa and Australia as fire is not an intrinsic ecological process in rainforests [1,31]. The BA fraction was below 1.20% year^{-1} in all temperate forests and boreal biomes, except for the Australian temperate broadleaf and mixed forests (1.72% year^{-1}). Although Mediterranean biomes are considered among the most fire-prone on Earth [3], their BA ranged from 0.45% year^{-1} in Europe to 1.56% year^{-1} in North America. Our results revealed that most biomes burned more in Africa and Australia than on the other continents. In Africa, extensive burning is facilitated by the low population density and decreases in grazing by bulk-feeding herbivores [12], whereas the proneness of Australian biomes to fire is fostered by the extreme climate seasonality, the high number of dry lightning events, aboriginal fire management practices, the physiognomy of eucalypts vegetation [31,42,44], and even by the colonization of exotic grasses [45].

Table 1. Global mean values and normalized trends (value in 2003 = 100%) in burned area (BA), burn severity (BS), and BA by BS levels at the global scale for the period 2003–2019. Both, BA, and BA by BS levels are presented in absolute area values and in percentage with respect to the total extent of the global land surface, BA by BS levels are also presented in percentage with respect to the global BA. Note that BA and BA by BS levels, whether calculated as absolute area or as percentage with respect to the total extent of the region results in the same trend values. The entire land area of the world was considered in calculations, including regions with scarce or null fire occurrence. Significant trends ($p < 0.05$) are denoted by boldface and the number of asterisks denotes the level of statistical significance ($p < 0.05$, $p < 0.01$, and $p < 0.001$). SE: standard error; CL: confidence limit.

Fire Variable	Global Mean		Global Trend			
	Value (±SE)	Unit	Value	Lower 95% CL	Upper 95% CL	Unit
BA	4.78 (±0.12)	Mkm^2 year^{-1}	**−1.50 *****	**−2.02**	**−0.82**	% year^{-1}
	3.26 (±0.08)	% land year^{-1}				
BS	175.74 (±0.89)	dNBR	0.13	−0.08	0.33	% year^{-1}
BA at low severity	3.59 (±0.09)	Mkm^2 year^{-1}	**−1.62 *****	**−2.10**	**−0.94**	% year^{-1}
	2.45 (±0.06)	% land year^{-1}				
	75.06 (±0.23)	% BA year^{-1}	−0.08	−0.25	0.05	% year^{-1}
BA at moderate severity	0.99 (±0.02)	Mkm^2 year^{-1}	**−1.27 *****	**−2.20**	**−0.50**	% year^{-1}
	0.67 (±0.02)	% land year^{-1}				
	20.61 (±0.18)	% BA year^{-1}	0.08	−0.25	0.48	% year^{-1}
BA at high severity	0.21 (±0.00)	Mkm^2 year^{-1}	−0.63	−1.72	0.25	% year^{-1}
	0.14 (±0.00)	% land year^{-1}				
	4.33 (±0.10)	% BA year^{-1}	**0.95 ***	**0.04**	**1.85**	% year^{-1}

BS spatial patterns were closely linked to biomes, being more consistent among continents than BA (Figure 1, Table A1). Generally, the BS patterns detected in the present study are in line with those of fire radiative power emitted by fires detected in previous work [20], revealing at the global scale the assumed relationship between BS and fire intensity [15]. In general, we found the highest mean severities in the taiga, followed by tundra, temperate coniferous forests, and Mediterranean biomes. BS was particularly high in North American taiga, which is characterized by high-intensity stand-replacing crown fire regimes [25], whereas lower mortality (42% lower) and combustion completeness (36% lower) characterize the Eurasian boreal forests, due to different traits of the dominant species [2,46]. The high severity of tundra fires is a consequence of the burning completeness of vascular vegetation and moss cover, as well as of the upper soil organic strata accumulated for decades, which contributes to permafrost degradation [47]. It is important to note that biomes at high latitudes have the largest topsoil carbon stocks on Earth (0–30 cm) [48], and seasonally become available to severely burn. In Mediterranean biomes, high severity results from a marked crown fire regime [1,3]. The lowest severities were found in tropical coniferous forests, mangroves, and in deserts and xeric shrublands; in the first, combustion completeness is flammability-limited, and the last BS is constrained by fuel-limitations. Tropical and subtropical forests in Africa exhibited lower BS than in Asia and North America, which can be a consequence of burning at smaller fire patches [46] as fire size covariates with BS [17], and potentially because of the predominance of burning in lower tree cover ranges (25–75%) [46]. Low severity values were also found in non-flooded grassy biomes, and intermediate severities generally corresponded to temperate forests. These global patterns confirmed the tradeoffs between BA fractions and BS ($\rho = -0.25$; $p < 0.05$) (Figure 1) previously detected for fire intensity [20], at the time that supports the fuel load available to burn as the primary driver of BS [25,26,49].

The analysis of BA by different severity levels, particularly at high severity, is useful to acknowledge the territorial magnitude of fire impacts [21]. Globally, we revealed that 0.21 ± 0.00 MKm^2 ($0.14 \pm 0.00\%$) of land surface burned annually at high severity (Table 1). In absolute terms, the extent of BA affected by high severity was significantly linked to the BA extent ($\rho = 0.87$; $p < 0.01$) (Figure 1), the largest contributors to the global high severity BA are the tropical and subtropical grasslands, savannas, and shrublands of Africa, South America, and the Asian taiga (Table A1). However, relativized to the total extent of each region, we found that the regions most affected by high-severity burning were the montane grasslands and shrublands of Australia ($2.23 \pm 1.77\%$ year^{-1}), flooded grasslands and savannas of North America ($1.85 \pm 0.13\%$ year^{-1}), Africa ($1.17 \pm 0.12\%$ year^{-1}) and South America ($0.96 \pm 0.19\%$ year^{-1}), tropical and subtropical grasslands, savannas, and shrublands of Oceania ($1.60 \pm 1.59\%$ year^{-1}) and South America ($0.55 \pm 0.04\%$ year^{-1}), and Mediterranean North America ($0.52 \pm 0.11\%$ year^{-1}) (Figure 1; Table A1). In the rest of the regions, the high severity BA constituted less than 1% of their extent. This is a direct consequence of the BA in each region ($\rho = 0.71$; $p < 0.01$), being related to a lesser extent to the mean BS (Figure 1). Relativizing the BA at high severity to the total BA in each region, similar patterns to BS were detected ($\rho = 0.82$; $p < 0.01$) (Figure 1), with around 40% of BA exhibiting a high severity in all boreal forests and non-European tundra (Figure 1; Table A1).

3.2. Temporal Trends in Burned Area and Burn Severity

World BA decreased -1.50% year^{-1}; $p < 0.01$ which in absolute terms is equivalent to a decrease of 1.22 MKm^2 between 2003 and 2019 (25.5% less BA with respect to 2003) (Table 1). These results elucidate an intensification of the decline in BA compared to the results already reported for the period 2003–2015 [4]. Analyzing BA trends by regions (Figure 2; Table 2), we found significant BA decreases in the European taiga (-6.21% year^{-1}; $p = 0.04$), temperate grasslands, savannas and shrublands of Asia (-2.77% year^{-1}; $p = 0.01$), Mediterranean Europe (-2.31% year^{-1}; $p = 0.03$), tropical and subtropical dry broadleaf forests of North America (-2.11% year^{-1}; $p < 0.01$), montane grasslands and shrublands of Africa (-2.08% year^{-1}; $p < 0.01$) and Asia (-1.93% year^{-1}; $p = 0.02$), and in most of tropical Africa including grasslands, savannas and shrublands (-1.64% year^{-1}; $p < 0.01$), moist broadleaf forests (-1.57% year^{-1}; $p = 0.02$) and mangroves (-1.37% year^{-1}; $p = 0.02$). Generalized decreases in BA have been formerly attributed to a decrease in the number of ignitions and to a lesser extent, to decreases in fire size, driven by human activity [4]. In this sense, the increase in human population, cropland area, and livestock density cause decreases in fire activity in the fire-prone open biomes [4], whereas increased efficiency in fire prevention, detection, and extinction, and abandonment of fire use in agriculture contribute to the decreasing trend in other regions [7,50].

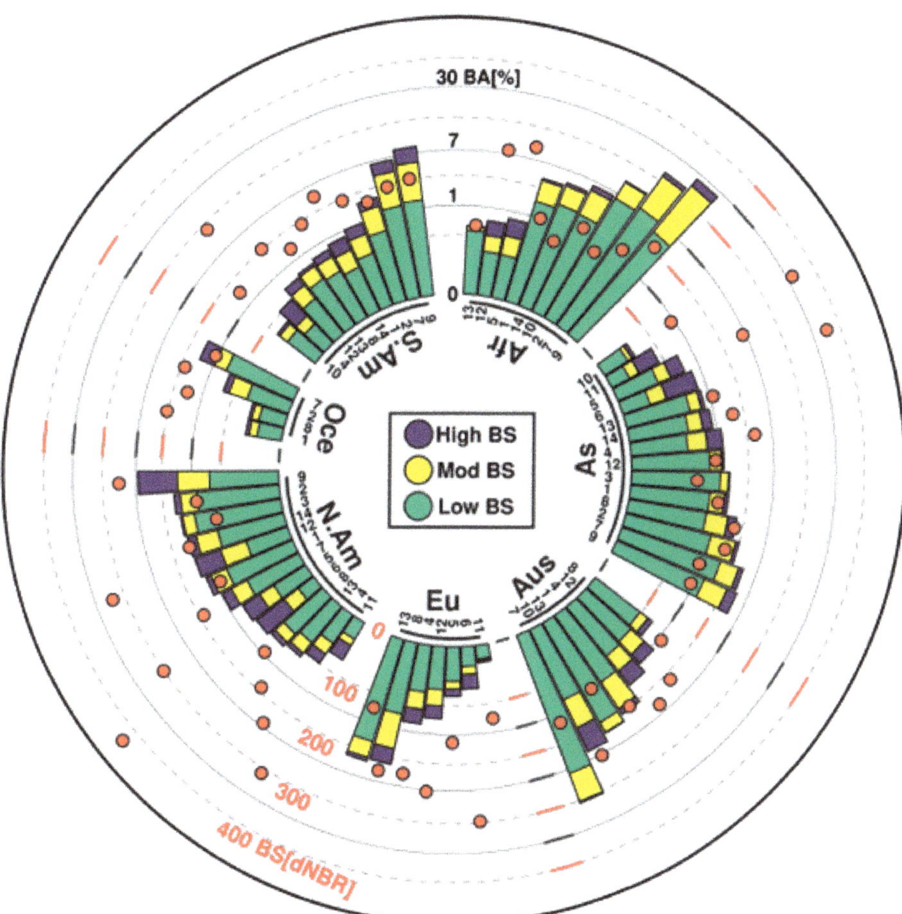

1   Tropical and subtropical moist broadleaf forest
2   Tropical and subtropical dry broadleaf forest
3   Temperate and subtropical coniferous forest
4   Temperate broadleaf and mixed forest
5   Temperate coniferous forest
6   Boreal forest/taiga
7   Tropical and subtropical grasslands, savannas and shrublands

8   Temperate grasslands, savannas and shrublands
9   Flooded grasslands and savannas
10   Montane grasslands and shrublands
11   Tundra
12   Mediterranean forests, woodlands and scrub
13   Deserts and xeric shrublands
14   Mangroves

Figure 1. Mean values of burned area (BA), burn severity (BS), and BA by BS levels by regions (biomes × continents) for the period 2003–2019. BA values are represented by square root scaled bars and expressed in percentage with respect to the total extent of the regional land surface. The percentage of the BA burned at each BS level is proportional to their fraction within the BA bars. BS is represented by red points and expressed in dNBR units ranging from −2000 to 2000. S.Am: South America, Oce: Oceania, N.Am: North America, Eu: Europe, Aus: Australia, As: Asia, Afr: Africa.

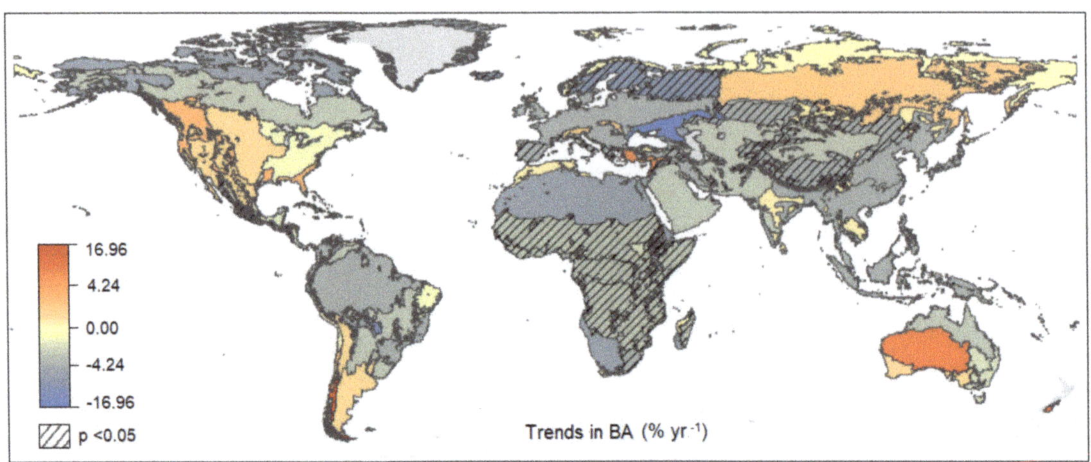

Figure 2. Normalized trends (value in 2003 = 100%) in burned area (BA) by regions (biomes × continents) for the period 2003–2019. The color ramps are square root scaled.

Globally, BS exhibited a non-significant increase (0.13% year^{-1}; $p = 0.34$) (Table 1). However, we found a large continental disparity, and regions with significant increases outnumbered those with significant decreases in BS (Figure 3, Table 2). The largest increases were found in South America, including temperate grasslands, savannas, and shrublands (2.47% year^{-1}; $p = 0.03$) and all forest biomes (1.49 to 3.56% year^{-1}; $p < 0.05$) except mangroves. Likewise, we detected significant increases in BS in tropical and subtropical moist forests of Australia (2.23% year^{-1}; $p = 0.04$), Mediterranean Africa (2.16% year^{-1}; $p < 0.01$), and Asian taiga (2.13% year^{-1}; $p = 0.03$). Land cover changes might explain BS trends in many regions, for instance, most of Mediterranean Africa, Central and Southern Chile, and Siberian taiga have experienced long-term woody encroachment and forest expansion [51], a fuel accumulation that in these regions lead to more severe fires [3,49]. Likewise, BS increases in the tropical rainforest of South America may respond to several causes. First, to increases in fuel continuity in the Eastern Brazilian coast, which is frequently burned [50]; second, to the loss of primary forests [51] that are characterized by no fire or low severity surface fires [1]; and third, to decreases in lower atmosphere moisture boosted by these forest losses that resulted in increased droughts the last years of the study period [52], making large fuel loads available to burn. Significant decreases in BS were only found in montane grasslands and shrublands of Asia (2.94% year^{-1}; $p = 0.04$) and in the European taiga (2.56% year^{-1}; $p = 0.04$), which inversely to the Asian, has experienced decreasing trends in fuel continuity and increasing moisture which have detrimental effects on BA [50] and likely on BS [24,26].

The analysis of BA by BS levels showed that decreases in global BA were mainly at the expense of decreasing low (-1.62% year^{-1}; $p < 0.01$) and moderate severity BA (-1.27% year^{-1}; $p < 0.01$) (Table 1). Globally, we found trends in BA at high severity (-0.63% year^{-1}; $p = 0.09$) to be positively related to trends in BA ($\rho = 0.67$; $p < 0.01$) as well as to trends in BS ($\rho = 0.40$; $p < 0.01$) (Figure 2). Regionally, we detected the highest increases in BA at high severity in temperate broadleaf forests of South America (40.20% year^{-1}; $p = 0.04$) (Figure 4A; Table 2), which can be a consequence of the expansion of exotic *Pinus* and *Eucalyptus* plantations with associated fuel load increases up to 40 Mg ha^{-1} between 1999 and 2006, accompanied by a large drought-driven intensification of fire activity between 2010 and 2015 [53]. We observed the BA at high severity to also aggravate in Australian tropical and subtropical moist forests (28.02% year^{-1}; $p < 0.01$), which are vulnerable to fire, as it favors alternative open biome states [44]. BA at high severity also increased in Australian grasslands, savannas, and shrublands (6.85% year^{-1}; $p = 0.01$).

In Australia, increases in fuel continuity [50] and in fire weather [31] in the last decades have been detected, which, along with the unprecedented fire events that occurred in 2019, might contribute to the detected patterns. We found significant increases in BA at high severity in the deserts and xeric shrublands and tropical and subtropical dry broadleaf forests of Asia that locally can be attributed to encroachment [49] and to increases in fuel continuity [50]. The largest decreases in high severity BA were observed in European taiga (-6.03% year^{-1}; $p = 0.01$) and in tropical and subtropical grasslands, savannas, and shrublands of North America (-5.96% year^{-1}; $p = 0.04$), a consequence of decreases in both BA and BS. Moreover, a significant global increase in the fraction of BA that is burned at high severity was detected (0.95% year^{-1}; $p = 0.04$), and regionally, trends in the fraction of BA burned at high severity were closer to BS ($\rho = 0.51$; $p < 0.01$) than to BA ($\rho = 0.11$; $p = 0.38$) (Figure 2), confirming the aggravation of impacts within burned areas in a vast proportion of South America, and in large parts of Northern Australia and Asia (Figure 4B, Table 2). In the tropical forests of Australia and the Amazon, it has been revealed that contemporary logging regimes and silvicultural practices exacerbate burn severity [54,55], suggesting decision makers and forest managers have a determinant role in past, present, and future fire regimes.

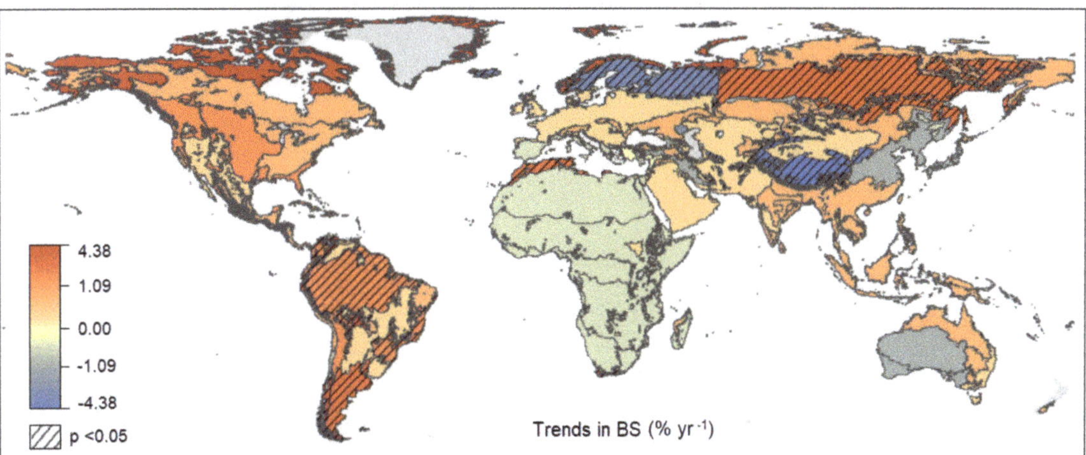

Figure 3. Normalized trends (value in 2003 = 100%) burn severity (BS) by regions (biomes × continents) for the period 2003–2019. The color ramps are square root scaled.

3.3. Relationships with Climate Variables

The climate spatial patterns (Figure 3) determined the spatial patterns of fire activity (Figures 5 and 6). Local regressions and correlation analysis (Figure 5) showed monotonic increases in BA ($\rho = 0.58$; $p < 0.01$), and monotonic decreases in BS ($\rho = -0.35$; $p < 0.01$) and in the BA at high severity ($\rho = -0.54$; $p < 0.01$) towards the warmest regions (tropics), in agreement with the results shown by the regression trees (Figure 6A–C). Moreover, the regression trees revealed mean annual temperature as the primary climate driver of regional differences in BA, BS, and BA at high severity (Figure 6A–C). Local regressions (Figure 5) and correlation analysis also showed an opposite pattern for annual temperature range, but regression trees (Figure 6A–C) suggest lower importance of annual temperature range and the other climate variables ($p > 0.05$), probably because of their strong interdependencies (Figure 4).

We found climate trends to be weakly related to all BA and BS trends (Figures 6D and 7), the detected trends being opposite to the assumption of increases in BS with climate warming. Thereby, we detected a significant inverse relationship between climate warming and the proportion of BA at high severity ($\rho = -0.26$; $p = 0.04$) (Figure 7), and both the local regression and the regression tree (Figure 6D) showed that the most stable regions in terms of proportion of BA burned at high severity were those experiencing the highest warming (>0.03 K year^{-1}). In this context, the long-term outcome of increased temperatures for fire regimes has been disclosed complex as it is caused by shifts in fire weather [31], but also by the warming effects on fuels through changes in productivity, vegetation composition [2,26], and fuel accumulation in the soil uppermost layers that are controlled by productivity–decomposition equilibriums varying at fine scales [56]. Moreover, the climate shifts would have different impacts on fire regimes depending on the former climate, productivity, and induced vegetation trajectories [24,27]. This complexity is expected to cause differential changes in fire regimes across regions, which some authors claim to be scarcely plausible until the coming decades [13,14]. Likewise, these premises highlight the difficulty of linking climate change to generalized increases in BA and BS worldwide, as different variations are expected depending on regional intrinsic characteristics.

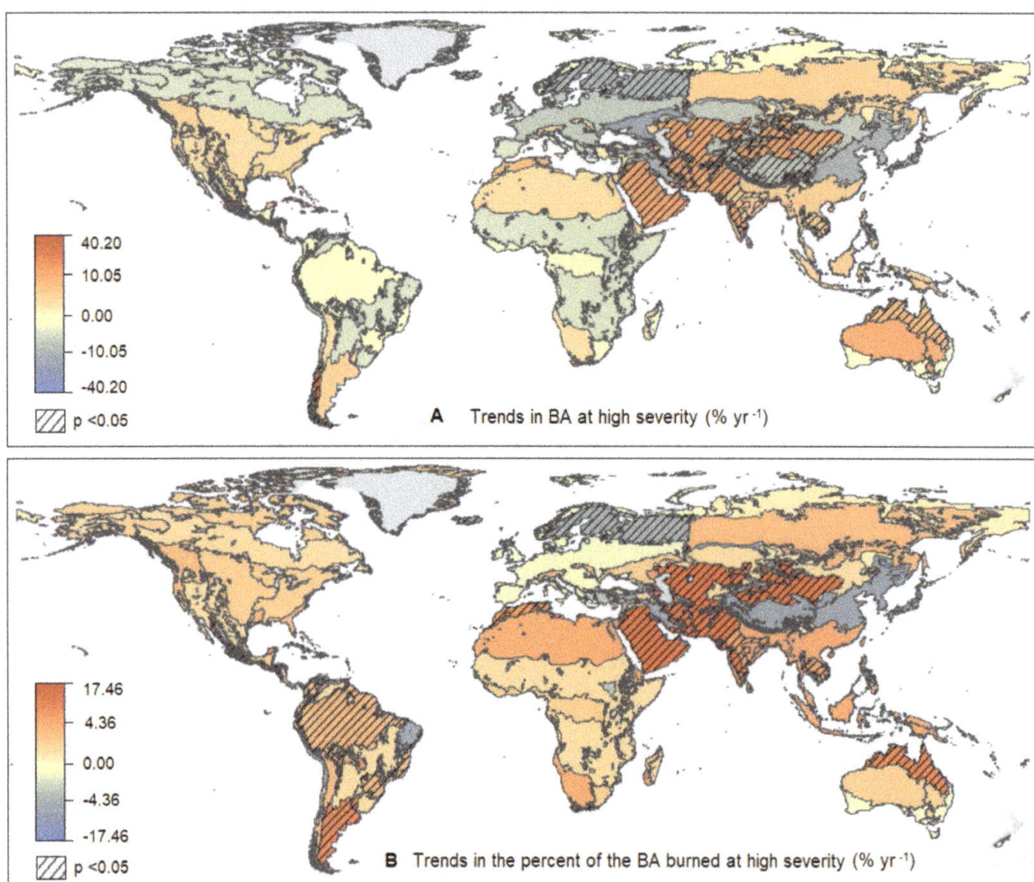

Figure 4. Normalized trends (value in 2003 = 100%) in burned area (BA) at high severity by regions (biomes × continents) for the period 2003–2019. The map in the upper part (**A**) shows the trends of BA at high severity calculated as absolute burned area at high severity (when calculated as percentage with respect to the total extent of the region the result is the same). The map in the bottom (**B**) shows the trends of BA at high severity calculated as a percentage of the burned area that burned at high severity. The color ramps are square root scaled.

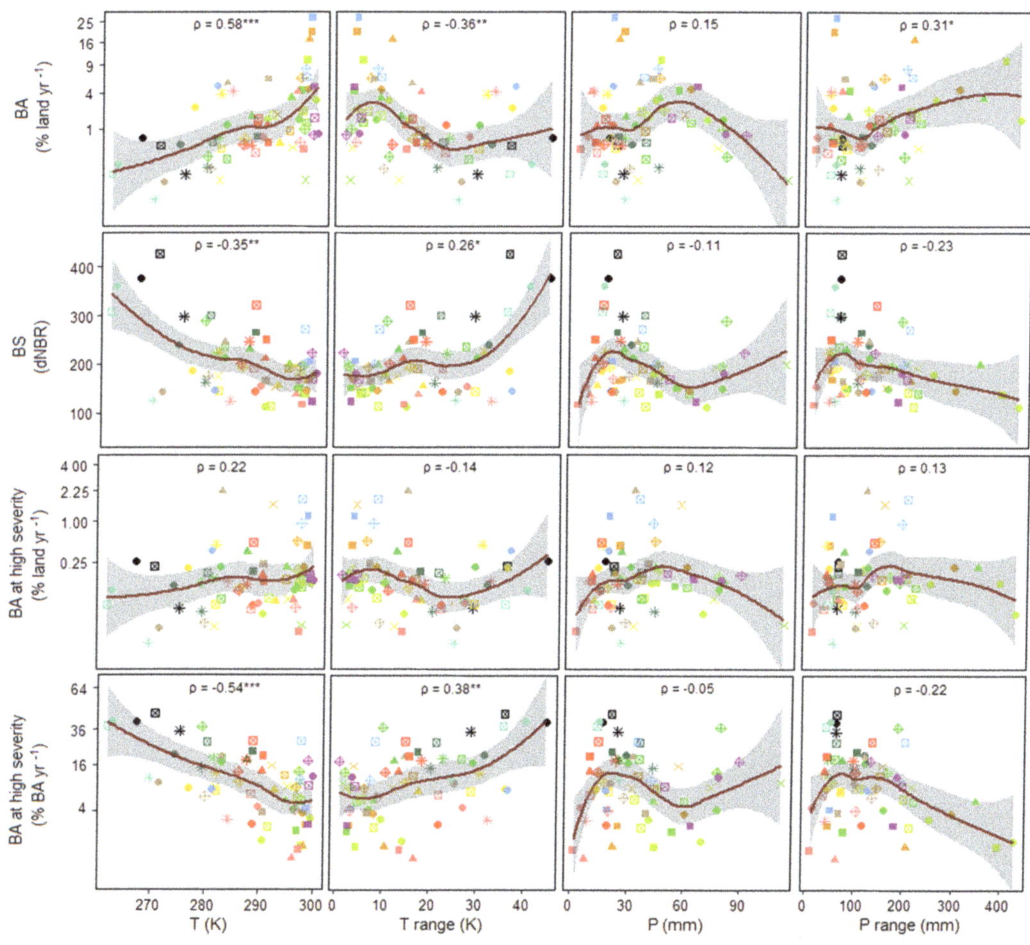

Figure 5. Scatterplots showing the synchronous relationships between the climate and fire variables for the period 2003–2019. Red lines show the local polynomial regression fitting (loess), and shaded areas ± 95% confidence intervals. Each panel also shows the Spearman's correlation coefficient (ρ), and the number of asterisks denotes the level of statistical significance ($p < 0.05$, $p < 0.01$, and $p < 0.001$). Note that the BA axes are square root transformed. BA: burned area, BS: burn severity, T: mean annual temperature: T range: annual temperature range, P: annual precipitation, P range: and annual precipitation range.

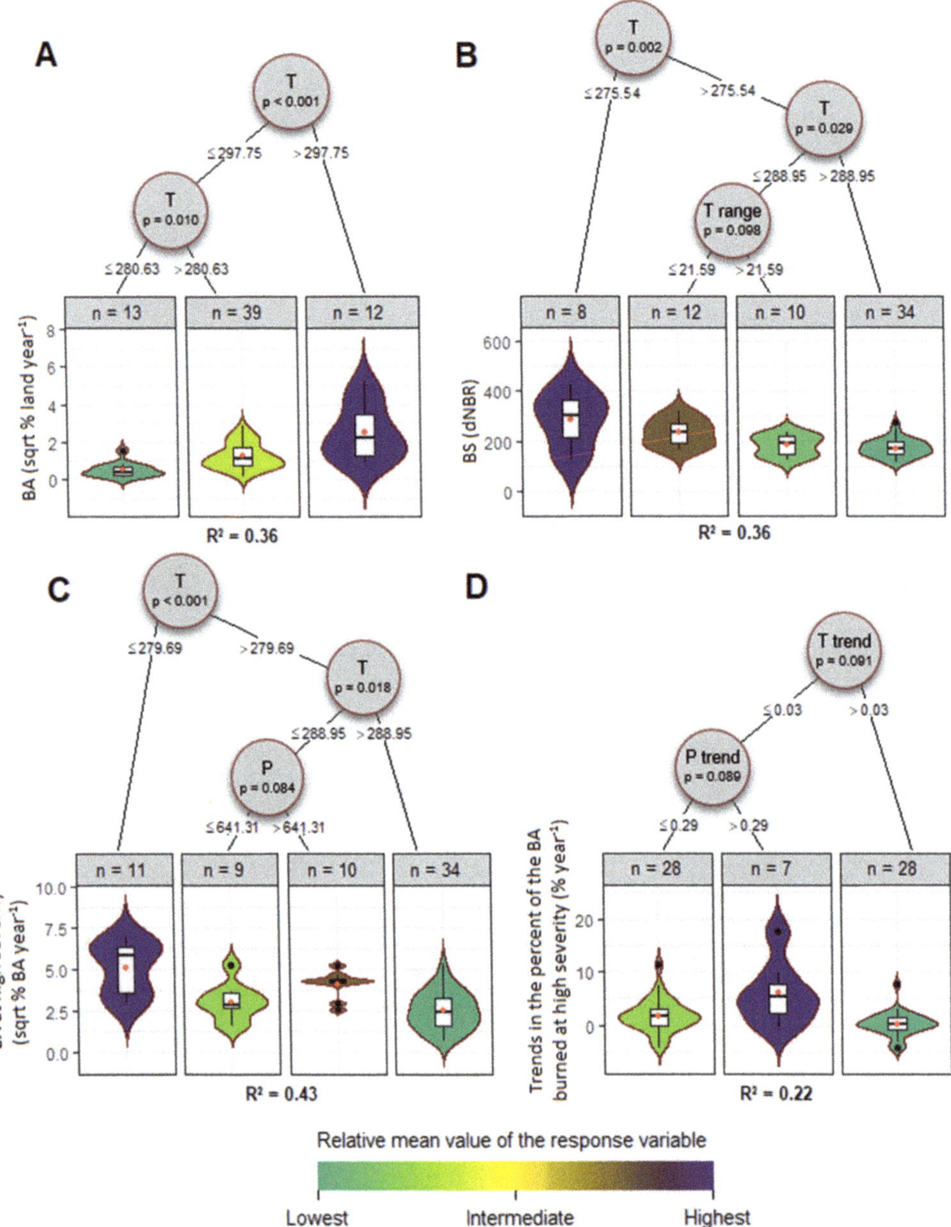

Figure 6. Conditional inference regression trees showing the most important climatic variables in determining the spatial patterns of burned area (BA) (**A**), burn severity (BS) (**B**), BA at high severity (**C**), and normalized trend (2003 = 100%) in the percentage of the BA burned at high (**D**). No further trees were able to grow for the other studied fire variables and trends at the selected confidence level for recursive binary partitioning ($p < 0.10$). R^2 values on the bottom of panels represent the variance explained by the regression tree predictions on our dataset. Red dots in the boxplots indicate the mean values. T: mean annual temperature, T range: annual temperature range, P: annual precipitation.

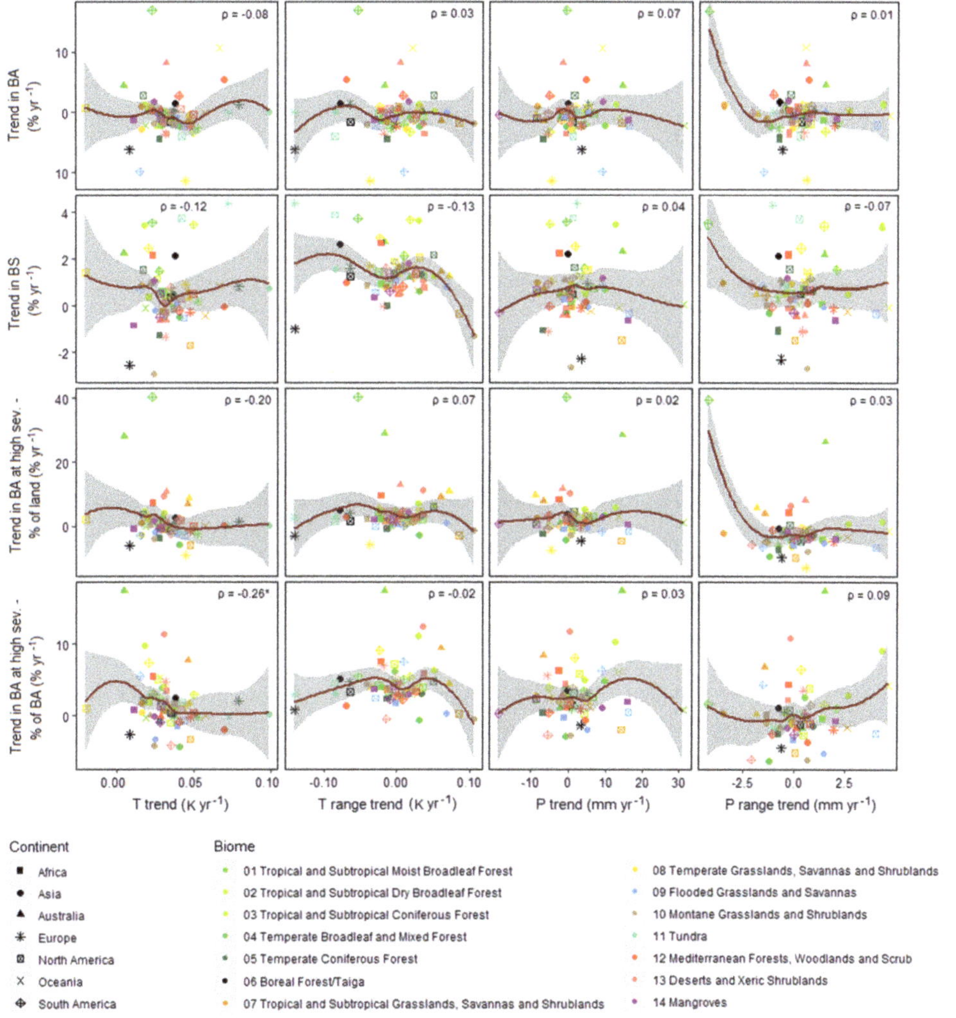

Figure 7. Scatterplots showing the relationship between the studied climate trends and normalized trends (value in 2003 = 100%) in fire variables for the period 2003–2019. Red lines show the local polynomial regression fitting (loess), and shaded areas ±95% confidence intervals. Each panel also shows the Spearman's correlation coefficient (ρ), and the number of asterisks denotes the level of statistical significance ($p < 0.05$, $p < 0.01$, and $p < 0.001$). BA: burned area, BS: burn severity, T: mean annual temperature: T range: annual temperature range, P: annual precipitation, P range: and annual precipitation range.

3.4. Implications and Final Considerations

Our study updates BA data and constitutes the first global assessment of BS spatiotemporal patterns. The information, provided by biomes and continents, is essential in revealing increasing trends in BS in several regions and increases in the fraction of BA that is burned at high severity at the planetary scale, providing scientific support for widespread assumptions. However, we did not find evidence in line with the hypothesis that climate change is increasing the mean BS worldwide. Nevertheless, several issues should be considered when interpreting our findings. First, the used BA and BS data at 500 m spatial resolution are appropriate for global fire analysis and consistent for studying trends [4,19,31], but have limitations in terms of quantifying BA absolute values. Accordingly, it has been demonstrated that the use of finer spatial resolution imagery at regional scales can result in larger BA, particularly in those regions with a predominance of small fires [57,58]. For this reason, as sensors and computational capabilities improve, we encourage future work to analyze BA and BS trends using higher spatial resolution imagery or statistically refined BA data [58], and even different methods to quantify burn severity [59]. Secondly, although there are biome–fire regime relationships, it is important to note that most biome regions have been largely modified by human activities such as agriculture, livestock, or urbanization, which also have an important role on fire regimes [2,4,8,20,60]. In this sense, we recommend exploring the implications of these drivers in determining the results presented in this study. Third, our study period is limited to 17 years because of no prior availability of both MODIS Aqua and Terra sensors. This could favor interferences of decadal ocean and atmospheric oscillations [61] in our results, but we assume little influence of El Niño Southern Oscillation (ENSO) due to 2–7 years oscillation of this phenomenon and the balanced events along our study period. Fourth, despite not finding strong evidence of climate change aggravation of BA and BS, we confirmed that climate is an essential variable influencing fire regimes worldwide [1,13,27], and further studies at finer scale map units of analysis are advisable to further disentangle the complex climate–fire–human interactions in the current context of climate change.

4. Conclusions

This study used 17 years of MODIS-derived data to analyze the spatiotemporal patterns of BA and constitutes the first BS analysis at the global scale. In addition, we showed the spatiotemporal patterns of BA and BS by regions (biomes and continents), relating them to several climate variables.

Our results updated the already known spatial patterns of BA across the globe and corroborated significant decreases in global BA. Our triple-way analyses of BS trends detected that globally (I) the mean BS exhibits non-significant increases; (II) the fraction of land affected by high BS shows non-significant decreases, as it is linked to the decreases in BA; and (III) the fraction of burned area that is affected by high BS is significantly increasing. In addition, the number of regions showing significant increases in mean BS and burned area at high BS outnumbered those with significant decreases.

We found close relationships between the spatial patterns of fire variables (BA and BS metrics) and climate variables (temperature, precipitation, and their respective interannual ranges), but our analysis by regions has not found evidence that climate warming is increasing BA nor BS, suggesting other factors as the primary drivers of change.

Given the great technical and computational advances that are currently taking place, future work may continue our analysis by using higher-resolution images, smaller analytical units, longer periods, or different methods of quantifying BS. We also encourage future research to analyze region-by-region the implications that our results may have in the field of ecology, climate regulation, and in the effectiveness and design of environmental management strategies.

Author Contributions: Conceptualization, V.F.-G. and E.A.-G.; methodology, V.F.-G. and E.A.-G.; formal analysis, V.F.-G. and E.A.-G.; investigation, V.F.-G. and E.A.-G.; writing—original draft preparation, V.F.-G.; writing—review and editing, V.F.-G. and E.A.-G.; visualization, V.F.-G. and E.A.-G. All authors have read and agreed to the published version of the manuscript.

Funding: The present study has been conducted without project funding. Víctor Fernández-García is supported by a Margarita Salas post-doctoral fellowship from the Ministry of Universities of Spain, financed with European Union-NextGenerationEU funds and granted by the University of León. Esteban Alonso-González has been funded by the Centre National d'Etudes Spatiales (CNES) postdoctoral fellowship.

Data Availability Statement: Data will be made available upon request to the corresponding author.

Conflicts of Interest: The authors declare no conflict of interest.

Appendix A

Table A1. Mean (±standard error) burned area (BA), burn severity (BS), and high severity BA by regions (biomes × continents) for the period 2003–2019. BA is expressed in mean absolute values (Mkm^2 year^{-1}) and in percentage with respect to the total extent of each region (% land year^{-1}). BS is expressed in dNBR values scaled from −2000 to 2000. High severity BA (dNBR > 440) is expressed in absolute values (Mha year^{-1}), in percentage with respect to the total extent of the land extent in each region (% land year^{-1}), and in percentage with respect to the total BA in each region (% BA year^{-1}). Standard errors indicate interannual variability. Trop.: tropical and subtropical. B: broadleaf forest. C: coniferous forest. M: mixed forest. G.: grasslands. S: savannas. Sh.: shrublands.

	Unit	Africa	Asia	Australia	Europe	N America	Oceania	S America
BA								
1 Trop. Moist B.	MKm^2 yr^{-1}	0.15 ± 0.00	0.11 ± 0.01	0.00 ± 0.00	-	0.01 ± 0.00	0.00 ± 0.00	0.09 ± 0.01
	% land yr^{-1}	4.38 ± 0.13	1.56 ± 0.10	3.42 ± 0.51	-	1.33 ± 0.20	0.03 ± 0.01	1.00 ± 0.09
2 Trop. Dry B.	MKm^2 yr^{-1}	0.02 ± 0.00	0.02 ± 0.00	-	-	0.01 ± 0.00	0.00 ± 0.00	0.01 ± 0.00
	% land yr^{-1}	10.38 ± 0.29	3.16 ± 0.12	-	-	2.10 ± 0.10	0.39 ± 0.07	2.22 ± 0.31
3 Trop. C.	MKm^2 yr^{-1}	-	0.00 ± 0.00	-	-	0.01 ± 0.00	-	-
	% land yr^{-1}	-	1.55 ± 0.21	-	-	1.81 ± 0.22	-	-
4 Temp. B. M.	MKm^2 yr^{-1}	-	0.05 ± 0.00	0.01 ± 0.00	0.03 ± 0.00	0.01 ± 0.00	-	0.00 ± 0.00
	% land yr^{-1}	-	1.12 ± 0.08	1.72 ± 0.47	0.65 ± 0.08	0.19 ± 0.02	-	0.22 ± 0.06
5 Temp. C	MKm^2 yr^{-1}	0.00 ± 0.00	0.01 ± 0.00	-	0.00 ± 0.00	0.01 ± 0.00	-	-
	% land yr^{-1}	0.69 ± 0.14	0.46 ± 0.09	-	0.10 ± 0.02	0.52 ± 0.06	-	-
6 Taiga	MKm^2 yr^{-1}	-	0.05 ± 0.01	-	0.00 ± 0.00	0.02 ± 0.00	-	-
	% land yr^{-1}	-	0.66 ± 0.09	-	0.05 ± 0.02	0.43 ± 0.05	-	-
7 Trop. G. S. Sh.	MKm^2 yr^{-1}	2.79 ± 0.07	0.00 ± 0.00	0.36 ± 0.03	-	0.00 ± 0.00	0.00 ± 0.00	0.26 ± 0.02
	% land yr^{-1}	20.07 ± 0.49	4.64 ± 0.64	17.01 ± 1.32	-	0.94 ± 0.08	1.86 ± 1.66	6.41 ± 0.45
8 Temp. G. S. Sh.	MKm^2 yr^{-1}	-	0.09 ± 0.01	0.00 ± 0.00	0.04 ± 0.01	0.01 ± 0.00	0.00 ± 0.00	0.01 ± 0.00
	% land yr^{-1}	-	2.42 ± 0.23	0.47 ± 0.08	3.87 ± 0.52	0.36 ± 0.03	0.03 ± 0.01	0.63 ± 0.11
9 Flooded G. S.	MKm^2 yr^{-1}	0.15 ± 0.02	0.01 ± 0.00	-	-	0.00 ± 0.00	-	0.02 ± 0.00
	% land yr^{-1}	26.97 ± 1.01	5.02 ± 0.48	-	-	6.57 ± 0.34	-	8.28 ± 1.13
10 Montane G. Sh.	MKm^2 yr^{-1}	0.05 ± 0.00	0.00 ± 0.00	0.00 ± 0.00	-	-	-	0.00 ± 0.00
	% land yr^{-1}	6.18 ± 0.23	0.03 ± 0.00	5.45 ± 4.07	-	-	-	0.09 ± 0.01
11 Tundra	MKm^2 yr^{-1}	-	0.00 ± 0.00	-	0.00 ± 0.00	0.00 ± 0.00	-	-
	% land yr^{-1}	-	0.14 ± 0.04	-	0.00 ± 0.00	0.06 ± 0.03	-	-
12 Mediterranean	MKm^2 yr^{-1}	0.00 ± 0.00	0.00 ± 0.00	0.01 ± 0.00	0.00 ± 0.00	0.00 ± 0.00	-	0.00 ± 0.00
	% land yr^{-1}	0.51 ± 0.04	1.21 ± 0.14	1.17 ± 0.18	0.45 ± 0.07	1.56 ± 0.26	-	0.46 ± 0.06
13 Deserts. Xeric Sh.	MKm^2 yr^{-1}	0.03 ± 0.01	0.08 ± 0.01	0.15 ± 0.03	0.00 ± 0.00	0.01 ± 0.00	-	0.01 ± 0.00
	% land yr^{-1}	0.35 ± 0.07	0.71 ± 0.06	4.25 ± 0.98	4.25 ± 0.94	0.29 ± 0.04	-	0.53 ± 0.04
14 Mangroves	MKm^2 yr^{-1}	0.00 ± 0.00	0.00 ± 0.00	-	-	0.00 ± 0.00	-	0.00 ± 0.00
	% land yr^{-1}	4.88 ± 0.13	0.84 ± 0.10	-	-	1.65 ± 0.09	-	0.76 ± 0.08

Table A1. *Cont.*

	Unit	Africa	Asia	Australia	Europe	N America	Oceania	S America
BS								
1 Trop. Moist B.	dNBR	150.65 ± 1.49	150.91 ± 2.93	199.82 ± 9.18		141.08 ± 3.97	201.31 ± 13.76	196.83 ± 4.68
2 Trop. Dry B.	dNBR	138.68 ± 5.78	179.15 ± 2.92			142.95 ± 3.60	193.17 ± 14.83	179.63 ± 9.25
3 Trop. C.	dNBR		112.58 ± 6.00			113.98 ± 4.44		
4 Temp. B. M.	dNBR		213.68 ± 10.67	229.70 ± 19.03	222.13 ± 13.68	237.19 ± 9.16		289.03 ± 20.45
5 Temp. C	dNBR	264.91 ± 16.53	239.20 ± 13.01		162.83 ± 22.14	300.09 ± 12.52		
6 Taiga	dNBR		376.14 ± 11.82		297.83 ± 20.35	426.25 ± 6.74		
7 Trop. G. S. Sh.	dNBR	168.41 ± 1.63	145.44 ± 7.54	164.73 ± 3.08		192.54 ± 9.15	152.60 ± 30.66	194.48 ± 2.63
8 Temp. G. S. Sh.	dNBR		185.88 ± 4.81	155.84 ± 10.56	225.06 ± 6.29	145.66 ± 12.05	175.51 ± 26.51	193.70 ± 9.47
9 Flooded G. S.	dNBR	206.48 ± 4.07	145.39 ± 9.61			272.19 ± 6.04		200.39 ± 10.41
10 Montane G. Sh.	dNBR	164.75 ± 2.60	144.27 ± 12.55	245.96 ± 42.50				176.17 ± 5.14
11 Tundra	dNBR		360.39 ± 17.10		125.97 ± 34.46	308.62 ± 26.09		
12 Mediterranean	dNBR	250.60 ± 10.29	143.40 ± 5.01	211.56 ± 12.10	247.64 ± 11.60	321.24 ± 18.28		200.95 ± 11.39
13 Deserts. Xeric Sh.	dNBR	116.77 ± 3.61	155.36 ± 5.28	137.86 ± 5.89	124.47 ± 12.75	192.51 ± 7.38		169.70 ± 4.26
14 Mangroves	dNBR	122.55 ± 3.74	180.83 ± 10.75			170.07 ± 5.46		223.69 ± 9.41
High severity BA								
1 Trop. Moist B.	Mha yr^{-1}	0.27 ± 0.01	0.45 ± 0.05	0.01 ± 0.00		0.05 ± 0.01	0.00 ± 0.00	1.10 ± 0.09
	% land yr^{-1}	0.08 ± 0.00	0.07 ± 0.01	0.20 ± 0.05		0.07 ± 0.01	0.00 ± 0.00	0.13 ± 0.01
	% BA yr^{-1}	1.80 ± 0.05	4.21 ± 0.34			5.58 ± 0.27	10.12 ± 2.04	13.28 ± 0.42
2 Trop. Dry B.	Mha yr^{-1}	0.02 ± 0.00	0.15 ± 0.01			0.06 ± 0.00	0.00 ± 0.00	0.12 ± 0.02
	% land yr^{-1}	0.09 ± 0.01	0.10 ± 0.01			0.12 ± 0.01	0.04 ± 0.01	0.18 ± 0.03
	% BA yr^{-1}	0.87 ± 0.10	2.99 ± 0.22			6.00 ± 0.37	8.69 ± 2.59	8.54 ± 0.54
3 Trop. C.	Mha yr^{-1}		0.00 ± 0.00			0.02 ± 0.00		
	% land yr^{-1}		0.01 ± 0.00			0.04 ± 0.01		
	% BA yr^{-1}		1.17 ± 0.31			2.15 ± 0.19		
4 Temp. B. M.	Mha yr^{-1}		0.90 ± 0.10	0.20 ± 0.08	0.54 ± 0.07	0.11 ± 0.02		0.04 ± 0.01
	% land yr^{-1}		0.21 ± 0.02	0.37 ± 0.14	0.12 ± 0.02	0.04 ± 0.01		0.10 ± 0.04
	% BA yr^{-1}		16.78 ± 1.34	17.86 ± 2.32	18.70 ± 1.21	18.62 ± 1.57		37.39 ± 3.57
5 Temp. C	Mha yr^{-1}	0.00 ± 0.00	0.12 ± 0.02		0.00 ± 0.00	0.36 ± 0.06		
	% land yr^{-1}	0.16 ± 0.04	0.08 ± 0.02		0.02 ± 0.00	0.16 ± 0.03		
	% BA yr^{-1}	21.75 ± 2.23	20.48 ± 1.83		14.73 ± 2.30	27.29 ± 1.84		
6 Taiga	Mha yr^{-1}		1.99 ± 0.21		0.05 ± 0.01	1.07 ± 0.13		
	% land yr^{-1}		0.26 ± 0.03		0.02 ± 0.01	0.21 ± 0.03		
	% BA yr^{-1}		41.35 ± 1.81		34.20 ± 2.77	48.25 ± 1.03		

Table A1. *Cont.*

	Unit	Africa	Asia	Australia	Europe	N America	Oceania	S America
7 Trop. G. S. Sh.	Mha yr^{-1}	6.46 ± 0.03	0.01 ± 0.00	0.31 ± 0.03		0.01 ± 0.00	0.00 ± 0.00	2.16 ± 0.17
	% land yr^{-1}	0.46 ± 0.02	0.21 ± 0.05	0.15 ± 0.01		0.12 ± 0.01	1.60 ± 1.59	0.54 ± 0.04
	% BA yr^{-1}	2.32 ± 0.11	3.66 ± 0.54	0.89 ± 0.07		12.10 ± 1.26	15.51 ± 10.41	8.51 ± 0.24
8 Temp. G. S. Sh.	Mha yr^{-1}		0.73 ± 0.07	0.02 ± 0.01	0.47 ± 0.06	0.08 ± 0.01	0.00 ± 0.00	0.12 ± 0.03
	% land yr^{-1}		0.19 ± 0.02	0.03 ± 0.01	0.48 ± 0.06	0.03 ± 0.00	0.00 ± 0.00	0.07 ± 0.02
	% BA yr^{-1}		8.23 ± 0.45	7.66 ± 1.47	12.82 ± 0.52	7.68 ± 1.15	8.49 ± 3.79	10.00 ± 1.09
9 Flooded G. S.	Mha yr^{-1}	0.66 ± 0.07	0.10 ± 0.01			0.05 ± 0.00		0.23 ± 0.05
	% yr^{-1}	1.17 ± 0.12	0.39 ± 0.06			1.85 ± 0.13		0.96 ± 0.19
	% BA yr^{-1}	4.26 ± 0.33	8.07 ± 0.87			27.92 ± 1.07		10.70 ± 1.05
10 Montane G. Sh.	Mha yr^{-1}	0.20 ± 0.01	0.01 ± 0.00	0.03 ± 0.02				0.00 ± 0.00
	% land yr^{-1}	0.23 ± 0.01	0.00 ± 0.00	2.23 ± 1.77				0.01 ± 0.00
	% BA yr^{-1}	3.77 ± 0.19	9.84 ± 1.39	19.48 ± 5.83				6.41 ± 0.55
11 Tundra	Mha yr^{-1}		0.22 ± 0.07		0.00 ± 0.00	0.11 ± 0.05		
	% land yr^{-1}		0.07 ± 0.02		0.00 ± 0.00	0.03 ± 0.01		
	% BA yr^{-1}		42.53 ± 2.38		11.09 ± 3.15	38.31 ± 3.16		
12 Mediterranean	Mha yr^{-1}	0.09 ± 0.01	0.01 ± 0.00	0.11 ± 0.02	0.08 ± 0.02	0.06 ± 0.01		0.01 ± 0.00
	% yr^{-1}	0.10 ± 0.01	0.02 ± 0.00	0.14 ± 0.03	0.09 ± 0.02	0.52 ± 0.11		0.06 ± 0.02
	% BA yr^{-1}	19.63 ± 1.63	2.40 ± 0.37	13.53 ± 1.97	17.79 ± 1.82	27.71 ± 3.22		13.33 ± 1.97
13 Deserts. Xeric Sh.	Mha yr^{-1}	0.02 ± 0.00	0.31 ± 0.03	0.12 ± 0.08	0.01 ± 0.00	0.04 ± 0.01		0.02 ± 0.00
	% land yr^{-1}	0.00 ± 0.00	0.03 ± 0.00	0.04 ± 0.02	0.11 ± 0.03	0.02 ± 0.00		0.02 ± 0.00
	% BA yr^{-1}	0.74 ± 0.12	4.65 ± 0.64	0.45 ± 0.14	2.92 ± 0.81	6.89 ± 0.90		4.14 ± 0.34
14 Mangroves	Mha yr^{-1}	0.01 ± 0.00	0.01 ± 0.00			0.01 ± 0.00		0.01 ± 0.00
	% land yr^{-1}	0.11 ± 0.01	0.11 ± 0.02			0.15 ± 0.02		0.14 ± 0.02
	% BA yr^{-1}	2.32 ± 0.13	11.61 ± 1.31			8.99 ± 0.68		17.59 ± 1.33

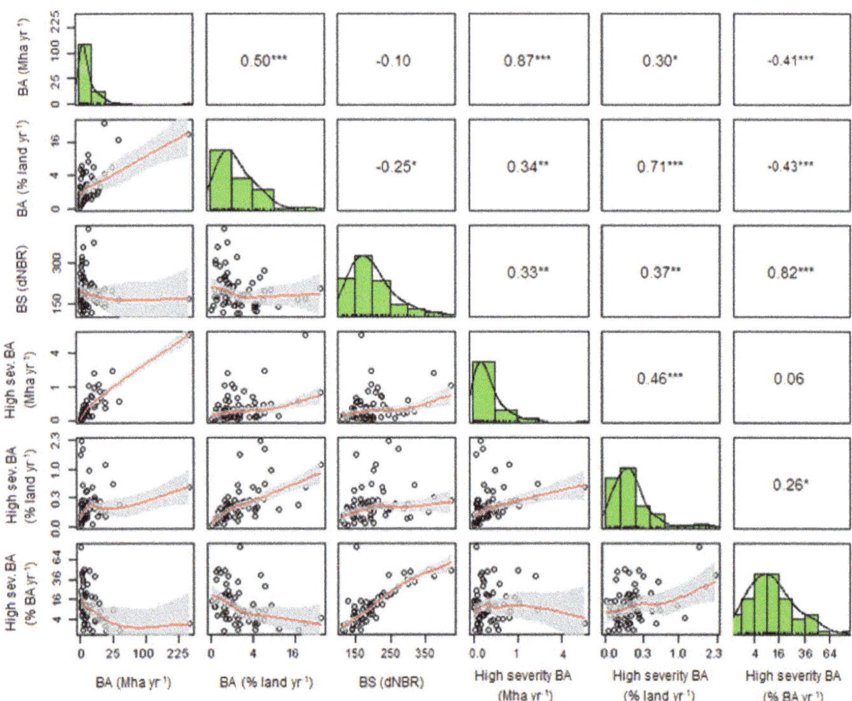

Figure 1. Synchronous relationships between the spatial patterns of mean annual burned area (BA) expressed in mean absolute values (Mkm^2 yr^{-1}) and in percentage with respect to the total extent of each region (% land yr^{-1}), burn severity (BS) expressed in dNBR units ranging from −2000 to 2000, and BA at high severity expressed in absolute values (Mha yr^{-1}), in percentage with respect to the total extent of the land extent in each region (% land yr^{-1}), and in percentage with respect the total BA in each region (% BA yr^{-1}) for the period 2003–2019. Red lines in the scatterplots show the local polynomial regression fitting (loess) and shaded areas ±95% confidence intervals. Values in the panels are Spearman's correlation coefficients (ρ) between pairs of variables, and the number of asterisks denotes the level of statistical significance ($p < 0.05$, $p < 0.01$, and $p < 0.001$). Note that all axis for BA variables were square root scaled.

Table 2. Mean trends in annual burned area (BA), in burn severity (BS), and in high severity BA (dNBR > 440) by regions (biomes × continents) for the period 2003–2019. Note that trends are normalized (value in 2003 = 100%) and expressed in percentage change per year, and therefore both, trends in mean absolute values of BA (originally measured in Mha year^{-1}), and in BA with respect to the total extent of global land extent (originally measured in % land year^{-1}) are the same. Slopes were calculated using the Sen slope estimator and significances with the Mann–Kendall test. Significant decreases ($p < 0.05$) are denoted by green boldface, significant increases by red boldface, and the number of asterisks denotes the level of statistical significance ($p < 0.05$, $p < 0.01$, and $p < 0.001$). Trop.: tropical and subtropical. B: broadleaf forest. C: coniferous forest. M: mixed forest. G.: grasslands. S: savannas. Sh.: shrublands.

	Original Units	Africa	Asia	Australia	Europe	N America	Oceania	S America
BA								
1 Trop. Moist B.		**−1.57** *	−2.20	4.43		−0.78	−2.22	−2.02
2 Trop. Dry B.		0.17	0.16			**−2.11** **	−0.70	−3.33
3 Trop. C.			1.20			−0.25		
4 Temp. B. M.			−2.43	−0.85	−2.72	−0.18		16.96

Table 2. *Cont.*

	Original Units	Africa	Asia	Australia	Europe	N America	Oceania	S America
5 Temp. C		−4.54	0.16		1.31	2.77		
6 Taiga			1.45		−6.21 *	−1.70		
7 Trop. G. S. Sh.		−1.64 ***	1.00	−1.47		−1.83	0.00	−1.38
8 Temp. G. S. Sh.			−2.77 *	−0.49	−11.40	0.74	10.75	0.88
9 Flooded G. S.		−0.92	0.07			−2.26		−9.89
10 Montane G. Sh.		−2.08 ***	−1.93 *	0.00				0.86
11 Tundra			−0.02		0.00	−4.03		
12 Mediterranean		0.15	5.43	0.47	−2.31 *	−0.36		2.78
13 Deserts. Xeric Sh.		−3.65	−1.82	8.21	−3.59	0.56		−0.18
14 Mangroves		−1.37 *	1.76			−0.98		−0.44
BS								
1 Trop. Moist B.		−0.08	0.57	2.23 *		0.42	−0.10	1.49 *
2 Trop. Dry B.		0.36	0.36			1.15	−0.23	3.47 *
3 Trop. C.			3.43			0.56		
4 Temp. B. M.			−0.49	0.32	0.19	0.60		3.56 *
5 Temp. C		−1.29	0.49		0.81	1.53		
6 Taiga			2.13 *		−2.56 *	0.35		
7 Trop. G. S. Sh.		−0.26	0.87	0.59		−1.72	−0.43	0.29
8 Temp. G. S. Sh.			0.81	0.34	0.90	1.42	0.87	2.47 *
9 Flooded G. S.		0.11	−0.22			−0.52		0.71
10 Montane G. Sh.		−0.18	−2.94 *	0.95				1.05
11 Tundra			0.73		4.38	3.75		
12 Mediterranean		2.16 **	−0.01	−0.64	−0.27	0.51		−0.21
13 Deserts. Xeric Sh.		−0.30	0.13	−0.62	−1.33	0.29		0.80
14 Mangroves		−0.85	1.07			0.53		−0.49
High severity BA								
1 Trop. Moist B.	Mha yr^{-1} % land yr^{-1}	−0.68	2.31	28.02 **		−0.76	0.00	0.07
	% BA yr^{-1}	1.13	3.90	17.46 ***		1.18	−0.33	2.46 ***
2 Trop. Dry B.	Mha yr^{-1} % land yr^{-1}	1.98	4.01 *			1.35	1.85	−1.32
	% BA yr^{-1}	1.83	4.05 **			5.03 *	5.38	2.92
3 Trop. C.	Mha yr^{-1} % land yr^{-1}		4.58			−0.35		
	% BA yr^{-1}		9.75			−1.02		
4 Temp. B. M.	Mha yr^{-1} % land yr^{-1}		−5.69	−0.14	−2.57	1.95		40.20 *
	% BA yr^{-1}		−4.30	−0.08	−0.17	1.55		2.97
5 Temp. C	Mha yr^{-1} % land yr^{-1}	−3.80	0.43		1.39	3.54		
	% BA yr^{-1}	−0.81	0.21		2.02	2.17		
6 Taiga	Mha yr^{-1} % land yr^{-1}		2.56		−6.03 *	−1.29		
	% BA yr^{-1}		2.47		−2.64 *	0.38		
7 Trop. G. S. Sh.	Mha yr^{-1} % land yr^{-1}	−1.11	1.22	6.85 *		−5.96 *	0.00	−1.31
	% BA yr^{-1}	0.68	1.73	7.72 ***		−3.30 *	0.00	0.79
8 Temp. G. S. Sh.	Mha yr^{-1} % land yr^{-1}		−2.69	8.65	−8.97	2.11	NA	2.68
	% BA yr^{-1}		0.48	1.69	1.37	1.06	NA	7.38 ***
9 Flooded G. S.	Mha yr^{-1} % land yr^{-1}	−2.37	−1.75			−2.97 *		−2.88
	% BA yr^{-1}	−1.53	−3.29			−0.69		5.48
10 Montane G. Sh.	Mha yr^{-1} % land yr^{-1}	−0.53	−4.10 *	0.00				2.20
	% BA yr^{-1}	1.60	−4.16	0.00				2.88
11 Tundra	Mha yr^{-1} % land yr^{-1}		0.05		0.00	−1.36		
	% BA yr^{-1}		0.12		0.00	1.92		

Table 2. *Cont.*

	Original Units	Africa	Asia	Australia	Europe	N America	Oceania	S America
12 Mediterranean	Mha yr^{-1} / % land yr^{-1}	7.13	0.20	−0.05	−1.24	0.15		1.99
	% BA yr^{-1}	5.41 *	−1.88	−0.10	−0.23	0.47		−0.82
13 Deserts. Xeric Sh.	Mha yr^{-1} / % land yr^{-1}	0.98	9.25 **	10.79	1.64	1.85		−2.14
	% BA yr^{-1}	3.47	11.34 *	1.90	4.75	0.83		−4.12
14 Mangroves	Mha yr^{-1} / % land yr^{-1}	−0.76	1.67			−0.59		−0.55
	% BA yr^{-1}	0.79	−0.03			1.32		−0.90

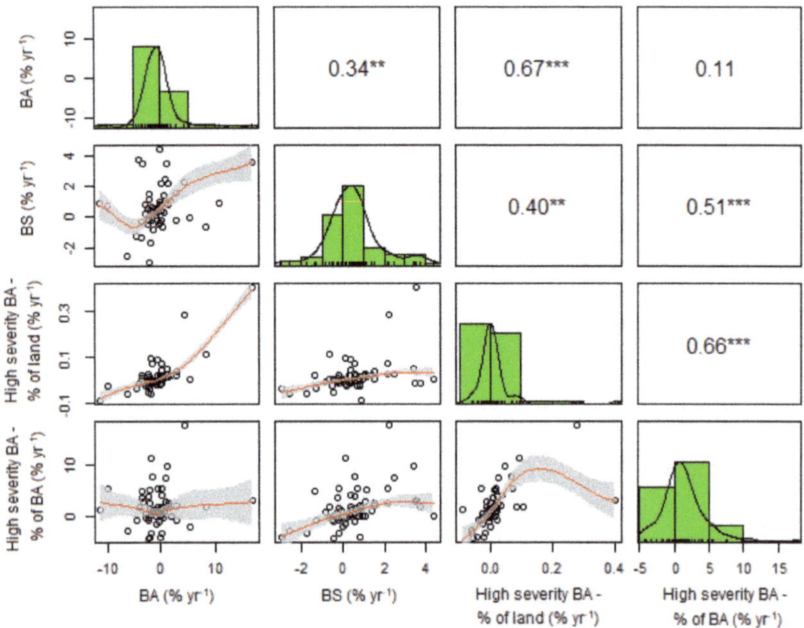

Figure 2. Relationships between the normalized trends (value in 2003 = 100%) in annual burned area (BA), burn severity (BS), and BA at high severity. Note that BA and BA by BS levels whether calculated as absolute area BA or as percentage with respect to the total extent of the region lead to same trend values. Red lines in the scatterplots show the local polynomial regression fitting (loess) fit and shaded areas ±95% confidence intervals. Values in the panels are Spearman's correlation coefficients (ρ) between pairs of variables, and the number of asterisks denotes the level of statistical significance ($p < 0.05$, $p < 0.01$, and $p < 0.001$).

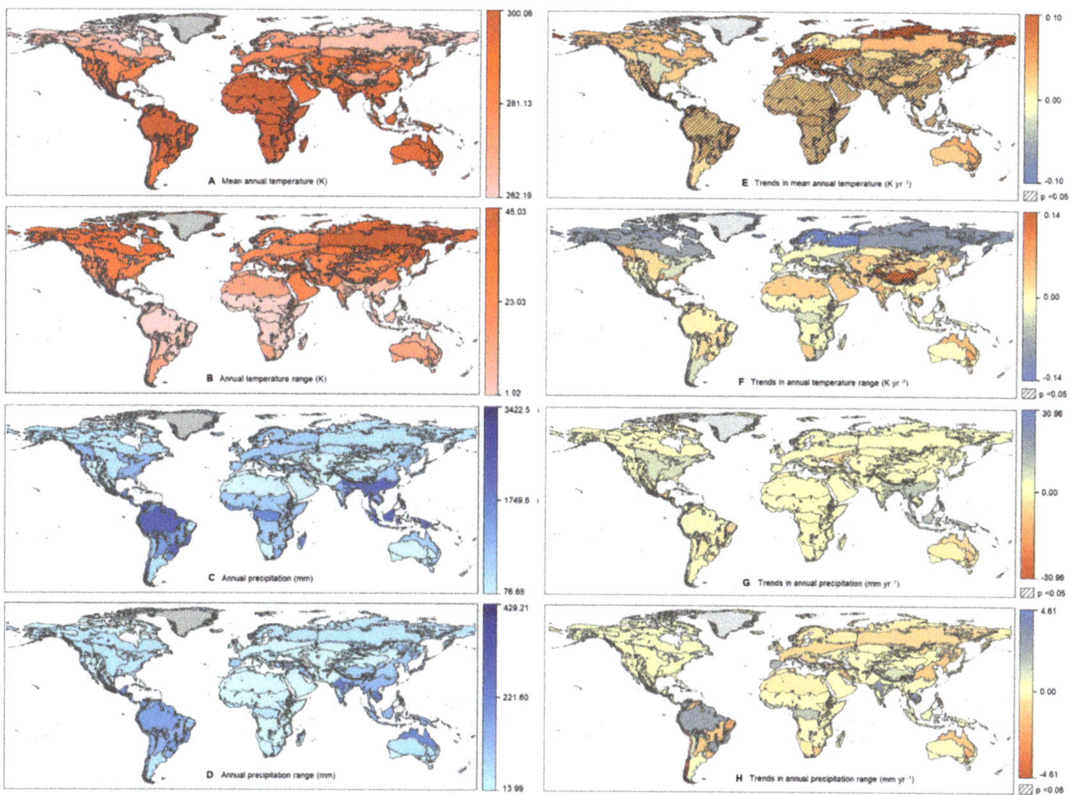

Figure 3. Spatial patterns (**A–D**) and trends (**E–H**) in mean annual temperature, annual temperature range, annual precipitation, and annual precipitation range for the period 2003–2019. Trends were calculated using the Sen estimator and significances with the Mann–Kendall test.

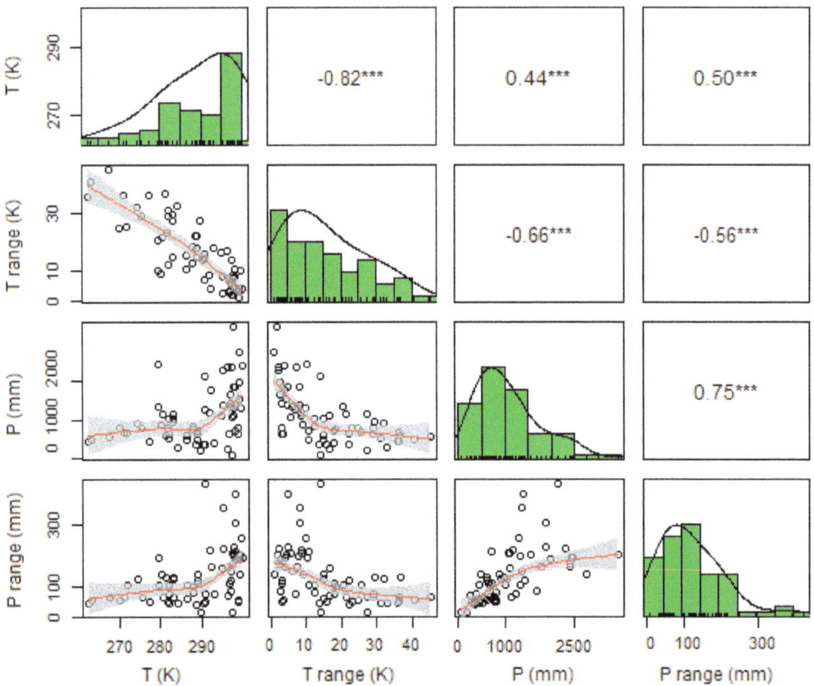

Figure 4. Synchronous relationships between the spatial patterns of mean annual temperature (T), annual temperature range (T range), annual precipitation (P) and annual precipitation range (P range) for the period 2003-2019. Red lines in the scatterplots show the local polynomial regression fitting (loess) and shaded areas ±95% confidence intervals. Values in the panels are the Spearman's correlation coefficients (ρ) between pairs of variables. the number of asterisks denotes the level of statistical significance ($p < 0.05$, $p < 0.01$, and $p < 0.001$).

References

1. Bowman, D.M.J.S.; Balch, J.K.; Artaxo, P.; Bond, W.J.; Carlson, J.M.; Cochrane, M.A.; D'Antonio, C.M.; DeFries, R.S.; Doyle, J.C.; Harrison, S.P.; et al. Fire in the Earth System. *Science* **2009**, *324*, 481–484. [CrossRef] [PubMed]
2. Archibald, S.; Lehmann, C.E.R.; Belcher, C.M.; Bond, W.J.; A Bradstock, R.; Daniau, A.-L.; Dexter, K.G.; Forrestel, E.J.; Greve, M.; He, T.; et al. Biological and geophysical feedbacks with fire in the Earth system. *Environ. Res. Lett.* **2018**, *13*, 033003. [CrossRef]
3. Keeley, J.E.; Bond, W.J.; Bradstock, R.A.; Pausas, J.G.; Rundel, P.W. *Fire in Mediterranean Ecosystems. Ecology, Evolution and Management*; Cambridge University Press: Cambridge, UK, 2012; p. 515.
4. Andela, N.; Morton, D.C.; Giglio, L.; Chen, Y.; van der Werf, G.R.; Kasibhatla, P.S.; DeFries, R.S.; Collatz, G.J.; Hantson, S.; Kloster, S.; et al. A human-driven decline in global burned area. *Science* **2017**, *356*, 1356–1362. [CrossRef] [PubMed]
5. van der Werf, G.R.; Randerson, J.T.; Giglio, L.; van Leeuwen, T.T.; Chen, Y.; Rogers, B.M.; Mu, M.; van Marle, M.J.E.; Morton, D.C.; Collatz, G.J.; et al. Global fire emissions estimates during 1997–2016. *Earth Syst. Sci. Data* **2017**, *9*, 697–720. [CrossRef]
6. Liu, Z.; Ballantyne, A.P.; Cooper, L.A. Biophysical feedback of global forest fires on surface temperature. *Nat. Commun.* **2019**, *10*, 214. [CrossRef]
7. Doerr, S.H.; Santín, C. Global trends in wildfire and its impacts: Perceptions versus realities in a changing world. *Phil. Trans. R. Soc. B* **2016**, *371*, 20150345. [CrossRef]
8. Chuvieco, E.; Pettinari, M.L.; Koutsias, N.; Forkel, M.; Hantson, S.; Turco, M. Human and climate drivers of global biomass burning variability. *Sci. Total Environ.* **2021**, *779*, 146361. [CrossRef]
9. Bond, W.J.; Woodward, F.I.; Midgley, G.F. The global distribution of ecosystems in a world without fire. *New Phytol.* **2005**, *165*, 525–538. [CrossRef]
10. Heinl, M.; Neuenschwander, A.; Silva, J.; Vanderpost, C. Interactions between fire and flooding in a southern African floodplain system (Okavango Delta, Botswana). *Lands. Ecol.* **2006**, *21*, 699–709. [CrossRef]
11. Archibald, S.; Staver, A.C.; Levin, S.A. Evolution of human-driven fire regimes in Africa. *Proc. Natl. Acad. Sci. USA* **2012**, *109*, 847–852. [CrossRef]

12. Archibald, S. Managing the human component of fire regimes: Lessons from Africa. *Phil. Trans. R. Soc. B* **2016**, *371*, 20150346. [CrossRef]

13. Bedia, J.; Herrera, S.; Gutiérrez, J.M.; Benali, A.; Brands, S.; Mota, B.; Moreno, J.M. Global patterns in the sensitivity of burned area to fire-weather: Implications for climate change. *Agr. For. Meteorol.* **2015**, *214–215*, 369–379. [CrossRef]

14. Abatzoglou, J.T.; Williams, A.P.; Barbero, R. Global emergence of anthropogenic climate change in fire weather indices. *Geophys. Res. Lett.* **2019**, *46*, 326–336. [CrossRef]

15. Keeley, J.E. Fire intensity, fire severity and burn severity: A brief review and suggested usage. *Int. J. Wildland Fire* **2009**, *18*, 116–126. [CrossRef]

16. Key, C.H.; Benson, N.C. Landscape assessment (LA) sampling and analysis methods. In *FIREMON: Fire Effects Monitoring and Inventory System*; General Technical Report, RMRS-GTR-164; Lutes, D.C., Keane, R.E., Caratti, J., Key, C.H., Benson, C., Sutherland, S., Gangi, L.J., Eds.; USDA Forest Service: Fort Collins, CO, USA, 2006; p. 400.

17. Tran, B.N.; Tanase, M.A.; Bennett, L.T.; Aponte, C. High-severity wildfires in temperate Australian forests have increased in extent and aggregation in recent decades. *PLoS ONE* **2020**, *15*, e0242484. [CrossRef]

18. Fernández-García, V.; Marcos, E.; Fulé, P.Z.; Reyes, O.; Santana, V.M.; Calvo, L. Fire regimes shape diversity and traits of vegetation under different climatic conditions. *Sci. Total Environ.* **2020**, *716*, 137137. [CrossRef]

19. Alonso-González, E.; Fernández-García, V. MOSEV: A global burn severity database from MODIS (2000–2020). *Earth Syst. Sci. Data* **2021**, *13*, 1925–1938. [CrossRef]

20. Archibald, S.; Lehmann, C.E.R.; Gómez-Dans, J.L.; Bradstock, R.A. Defining pyromes and global syndromes of fire regimes. *Proc. Natl. Acad. Sci. USA* **2013**, *110*, 6442–6447. [CrossRef]

21. Singleton, M.P.; Thode, A.E.; Sánchez-Meador, A.J.; Iniguez, J.M. Increasing trends in high-severity fire in the southwestern USA from 1984 to 2015. *For. Ecol. Manag.* **2019**, *433*, 709–719. [CrossRef]

22. Fernández-García, V.; Santamarta, M.; Fernández-Manso, A.; Quintano, C.; Marcos, E.; Calvo, L. Burn severity metrics in fire-prone pine ecosystems along a climatic gradient using Landsat imagery. *Remote Sens. Environ.* **2018**, *206*, 205–217. [CrossRef]

23. García-Llamas, P.; Suárez-Seoane, S.; Fernández-Guisuraga, J.M.; Fernández-García, V.; Fernández-Manso, A.; Quintano, C.; Taboada, A.; Marcos, E.; Calvo, L. Evaluation and comparison of Landsat 8, Sentinel-2 and Deimos-1 remote sensing indices for assessing burn severity in Mediterranean fire-prone ecosystems. *Int. J. Appl. Earth Obs. Geoinf.* **2019**, *80*, 137–144. [CrossRef]

24. Parks, S.A.; Miller, C.; Abatzoglou, J.T.; Holsinger, L.M.; Parisien, M.-A.; Dobrowski, S.Z. How will climate change affect wildland fire severity in the western US? *Environ. Res. Lett.* **2016**, *11*, 035002. [CrossRef]

25. Whitman, E.; Parisien, M.-A.; Thompson, D.K.; Hall, R.J.; Skakun, R.S.; Flannigan, M.D. Variability and drivers of burn severity in the northwestern Canadian boreal forest. *Ecosphere* **2018**, *9*, e02128. [CrossRef]

26. Fernández-García, V.; Beltrán-Marcos, D.; Fernández-Guisuraga, J.M.; Marcos, E.; Calvo, L. Predicting potential wildfire severity across Southern Europe with global data sources. *Sci. Total Environ.* **2022**, *829*, 154729. [CrossRef] [PubMed]

27. Moritz, M.A.; Parisien, M.-A.; Batllori, E.; Krawchuk, M.A.; Van Dorn, J.; Ganz, D.J.; Hayhoe, K. Climate change and disruptions to global fire activity. *Ecosphere* **2012**, *3*, 1–22. [CrossRef]

28. Turetsky, M.R.; Kane, E.S.; Harden, J.W.; Ottmar, R.D.; Manies, K.L.; Hoy, E.; Kasischke, E.S. Recent acceleration of biomass burning and carbon losses in Alaskan forests and peatlands. *Nat. Geosci.* **2011**, *4*, 27–31. [CrossRef]

29. Moreira, F.; Ascoli, D.; Safford, H.; A Adams, M.; Moreno, J.M.; Pereira, J.M.C.; Catry, F.X.; Armesto, J.; Bond, W.; E González, M.; et al. Wildfire management in Mediterranean-type regions: Paradigm change needed. *Environ. Res. Lett.* **2020**, *15*, 011001. [CrossRef]

30. Kharuk, V.I.; Dvinskaya, M.L.; Im, S.T.; Golyukov, A.S.; Smith, K.T. Wildfires in the Siberian Arctic. *Fire* **2022**, *5*, 106. [CrossRef]

31. Canadell, J.G.; Meyer, C.P.; Cook, G.D.; Dowdy, A.; Briggs, P.R.; Knauer, J.; Pepler, A.; Haverd, V. Multi-decadal increase of forest burned area in Australia is linked to climate change. *Nat. Commun.* **2021**, *12*, 6921. [CrossRef]

32. Giglio, L.; Boschetti, L.; Roy, D.P.; Humber, M.L.; Justice, O.C. The Collection 6 MODIS burned area mapping algorithm and product. *Remote Sens. Environ.* **2018**, *217*, 72–85. [CrossRef]

33. Olson, D.M.; Dinerstein, E.; Wikramanayake, E.D.; Burgess, N.D.; Powell, G.V.N.; Underwood, E.C.; D'amico, J.; Itoua, I.; Strand, H.; Morrison, J.; et al. Terrestrial ecoregions of the world: A new map of life on Earth. *Bioscience* **2001**, *51*, 933–938. [CrossRef]

34. Esri, CMI, CIA, Global Mapping International; U.S. Central Intelligence Agency-The World Factbook, World Continents. Available online: https://www.arcgis.com/home/item.html?id=a3cb207855b348a297ab85261743351d (accessed on 1 December 2021).

35. R Core Team. R: A Language and Environment for Statistical Computing. R Foundation for Statistical Computing. Available online: https://www.R-project.org/ (accessed on 1 December 2021).

36. Fernández-García, V.; Quintano, C.; Taboada, A.; Marcos, E.; Calvo, L.; Fernández-Manso, A. Remote Sensing Applied to the Study of Fire Regime Attributes and Their Influence on Post-Fire Greenness Recovery in Pine Ecosystems. *Remote Sens.* **2018**, *10*, 733. [CrossRef]

37. Miller, J.D.; Thode, A.E. Quantifying burn severity in a heterogeneous landscape with a relative version of the delta normalized burn ratio (dNBR). *Remote Sens. Environ.* **2007**, *109*, 66–80. [CrossRef]

38. Botella-Martínez, M.A.; Fernández-Manso, A. Estudio de la severidad post-incendio en la Comunidad Valenciana comparando los índices dNBR, RdNBR y RBR a partir de imágenes Landsat 8. *Rev. Teledetección* **2017**, *49*, 33–47. [CrossRef]

39. Bronaugh, D.; Werner, A. zyp: Zhang + Yue-Pilon Trends Package. R package version 0.10-1.1. Available online: https://CRAN.R-project.org/package=zyp (accessed on 1 December 2021).

40. Hothorn, T.; Zeileis, A. partykit: A Modular Toolkit for Recursive Partytioning in R. *J. Mach. Learn. Res.* **2015**, *16*, 3905–3909. Available online: https://www.jmlr.org/papers/v16/hothorn15a.html (accessed on 1 December 2021).

41. Revelle, W. psych: Procedures for Personality and Psychological Research Version 2.2.3. Available online: https://CRAN.R-project.org/package=psych (accessed on 31 May 2022).

42. Staver, A.; Archibald, S.; Levin, S.A. The Global Extent and Determinants of Savanna and Forest as Alternative Biome States. *Science* **2011**, *334*, 230–232. [CrossRef]

43. Damasceno-Junior, G.A.; Pereira, A.D.M.; Oldeland, J.; Parolin, A.P. Fire, Flood and Pantanal Vegetation. In *Flora and Vegetation of the Pantanal Wetland. Plant and Vegetation*; Damasceno-Junior, G.A., Parolin, A.P., Eds.; Springer: Cham, Switzerland, 2021; pp. 661–668. [CrossRef]

44. Bradstock, R.A.; Williams, J.E.; Gill, M.A. *Flammable Australia: The Fire Regimes and Biodiversity of a Continent*; Cambridge University Press: Cambridge, UK, 2002; p. 462.

45. Walker, Z.C.; Morgan, J.W. Perennial pasture grass invasion changes fire behaviour and recruitment potential of a native forb in a temperate Australian grassland. *Biol. Invasions* **2022**, *24*, 1755–1765. [CrossRef]

46. Van Wees, D.; van der Werf, G.R.; Randerson, J.T.; Andela, N.; Chen, Y.; Morton, D.C. The role of fire in global forest loss dynamics. *Glob. Change Biol.* **2021**, *27*, 2377–2391. [CrossRef]

47. Rocha, A.; Shaver, G.R. Postfire energy exchange in arctic tundra: The importance and climatic implications of burn severity. *Glob. Change Biol.* **2011**, *17*, 2831–2841. [CrossRef]

48. Shangguan, W.; Dai, Y.; Duan, Q.; Liu, B.; Yuan, H. A global soil data set for earth system modeling. *J. Adv. Model. Earth Syst.* **2014**, *6*, 249–263. [CrossRef]

49. Walker, X.J.; Rogers, B.M.; Veraverbeke, S.; Johnstone, J.F.; Baltzer, J.L.; Barrett, K.; Bourgeau-Chavez, L.; Day, N.J.; de Groot, W.J.; Dieleman, C.M.; et al. Fuel availability not fire weather controls boreal wildfire severity and carbon emissions. *Nat. Clim. Change* **2020**, *10*, 1130–1136. [CrossRef]

50. Kelley, D.I.; Bistinas, I.; Whitley, R.; Burton, C.; Marthews, T.R.; Dong, N. How contemporary bioclimatic and human controls change global fire regimes. *Nat. Clim. Change* **2019**, *9*, 690–696. [CrossRef]

51. Song, X.-P.; Hansen, M.C.; Stehman, S.V.; Potapov, P.V.; Tyukavina, A.; Vermote, E.F.; Townshend, J.R. Global land change from 1982 to 2016. *Nature* **2018**, *560*, 639–643. [CrossRef] [PubMed]

52. Xu, X.; Jua, G.; Zhang, X.; Riley, W.J.; Xue, Y. Climate regime shift and forest loss amplify fire in Amazonian Forests. *Glob. Change Biol.* **2020**, *26*, 5874–5885. [CrossRef] [PubMed]

53. Gonzalez, M.E.; Gómez-González, S.; Lara, A.; Garreaud, R.; Díaz-Hormazábal, I. Megadrought and its influence on the fire regime in central and south-central Chile. *Ecosphere* **2018**, *26*, 5874–5885. [CrossRef]

54. Carvalho de Andrade, D.; Ruschel, A.R.; Schwartz, G.; de Carvalho, J.O.P.; Humphries, S.; Gama, J.R.V. Forest resilience to fire in eastern Amazon depends on the intensity of pre-fire disturbance. *For. Ecol. Manag.* **2020**, *472*, 118258. [CrossRef]

55. Lindenmayer, D.B.; Kooyman, R.M.; Taylor, C.; Ward, M.; Watson, J.E.M. Recent Australian wildfires made worse by logging and associated forest management. *Nat. Ecol. Evol.* **2020**, *4*, 898–900. [CrossRef]

56. Smith, P. An overview of the permanence of soil organic carbon stocks: Influence of direct human-induced, indirect and natural effects. *Eur. J. Soil Sci.* **2005**, *56*, 673–680. [CrossRef]

57. Ramo, R.; Roteta, E.; Bistinas, I.; van Wees, D.; Bastarrika, A.; Chuvieco, E.; van der Werf, G.R. African burned area and fire carbon emissions are strongly impacted by small fires undetected by coarse resolution satellite data. *Proc. Natl. Acad. Sci. USA* **2021**, *118*, e2011160118. [CrossRef]

58. Fernández-García, V.; Kull, C.A. Refining historical burned area data from satellite observations. *Int. J. Appl. Earth Obs. Geoinf.* **2023**, *120*, 103350. [CrossRef]

59. Quintano, C.; Calvo, L.; Fernández-Manso, A.; Suárez-Seoane, S.; Fernandes, P.M.; Fernández-Guisuraga, J.M. First evaluation of fire severity retrieval from PRISMA hyperspectral data. *Remote Sens. Environ.* **2023**, *295*, 113670. [CrossRef]

60. Fernández-García, C.; Beltrán-Marcos, D.; Calvo, L. Building patterns and fuel features drive wildfire severity in wildland-urban interfaces in Southern Europe. *Landsc. Urban Plan.* **2023**, *231*, 104646. [CrossRef]

61. Wang, Y.; Huang, P. Potential fire risks in South America under anthropogenic forcing hidden by the Atlantic Multidecadal Oscillation. *Nat. Commun.* **2022**, *13*, 2437. [CrossRef]

remote sensing

Article

Post-Fire Forest Vegetation State Monitoring through Satellite Remote Sensing and In Situ Data

Daniela Avetisyan [1,*], Emiliya Velizarova [2] and Lachezar Filchev [1]

[1] Space Research and Technology Institute of the Bulgarian Academy of Sciences, Acad. G. Bonchev Str. Bl. 1, 1113 Sofia, Bulgaria
[2] Ministry of Environment and Water, Maria Luisa Blvd. 22, 1000 Sofia, Bulgaria
*   Correspondence: davetisyan@space.bas.bg

Abstract: Wildfires have significant environmental and socio-economic impacts, affecting ecosystems and people worldwide. Over the coming decades, it is expected that the intensity and impact of wildfires will grow depending on the variability of climate parameters. Although Bulgaria is not situated within the geographical borders of the Mediterranean region, which is one of the most vulnerable regions to the impacts of temperature extremes, the climate is strongly influenced by it. Forests are amongst the most vulnerable ecosystems affected by wildfires. They are insufficiently adapted to fire, and the monitoring of fire impacts and post-fire recovery processes is of utmost importance for suggesting actions to mitigate the risk and impact of that catastrophic event. This paper investigated the forest vegetation recovery process after a wildfire in the Ardino region, southeast Bulgaria from the period between 2016 and 2021. The study aimed to present a monitoring approach for the estimation of the post-fire vegetation state with an emphasis on fire-affected territory mapping, evaluation of vegetation damage, fire and burn severity estimation, and assessment of their influence on vegetation recovery. The study used satellite remotely sensed imagery and respective indices of greenness, moisture, and fire severity from Sentinel-2. It utilized the potential of the landscape approach in monitoring processes occurring in fire-affected forest ecosystems. Ancillary data about pre-fire vegetation state and slope inclinations were used to supplement our analysis for a better understanding of the fire regime and post-fire vegetation damages. Slope aspects were used to estimate and compare their impact on the ecosystems' post-fire recovery capacity. Soil data were involved in the interpretation of the results.

Keywords: fire impact; post-fire forest recovery; forest landscapes; vegetation indices; orthogonal transformation; Sentinel-2

Citation: Avetisyan, D.; Velizarova, E.; Filchev, L. Post-Fire Forest Vegetation State Monitoring through Satellite Remote Sensing and In Situ Data. *Remote Sens.* 2022, 14, 6266. https://doi.org/10.3390/rs14246266

Academic Editors: Elena Marcos, Leonor Calvo, Susana Suarez-Seoane and Víctor Fernández-García

Received: 21 October 2022
Accepted: 7 December 2022
Published: 10 December 2022

Publisher's Note: MDPI stays neutral with regard to jurisdictional claims in published maps and institutional affiliations.

1. Introduction

Forest disturbance cycles are associated with exacerbating responses to climate change [1]. Forest fires have been more frequent and severe in recent decades, especially in areas that have experienced climate change pressures for an extended period of time [2]. Due to climate change, high-temperature anomalies continue to occur, which leads to frequent forest fires [3]. The International Panel on Climate Change (IPCC) puts the Mediterranean and its adjacent lands as amongst the most vulnerable regions to the effects of global warming worldwide [4]. The models issued by IPCC agreed on a clear trend of the thermal regime based on a scenario from 1980–2000. An increase in average surface temperatures, ranging between 2.2 °C and 5.1 °C, for the period 2080–2100 was prognosed. For the same period, the models indicated pronounced rainfall regime changes showing that precipitation over lands might decrease by about 4% to 27%. Studies performed by the Department of Meteorology at the National Institute of Meteorology and Hydrology at the Bulgarian Academy of Sciences (NIMH-BAS) predicted an increase in the annual air temperature by more than 1.8 °C for the coming decades in Bulgaria. This fact increases the risk of forest fire frequency and intensity [5]. During the last two decades in Bulgaria, as well as in other countries of the Mediterranean region, many wildfires occurred and had a significant economic, political, social, and ecological impact [6–9].

In recent years, high resolution (HR) and very high resolution (VHR) optical remote sensing has become widespread concerning monitoring needs, and these strategies provide affordable multitemporal and multispectral pictures of the considered phenomena at different scales. Satellite sensors allow measurement of the impact of fires by comparing pre- and post-fire information. Applications of remote sensing technology related to fire ecology, including fire risk mapping, fuel mapping, burn severity assessment, and post-fire vegetation recovery, are widely discussed and accepted [10–15]. These technologies provide a low-cost, multi-temporal means for conducting local, regional, and global-scale fire ecology research. Moreover, the development of new technologies and techniques resulted in their rapid evolution, thus increasing the accuracy and efficiency of earth observation studies and applications [16,17]. Space and airborne sensors have been used to map burned areas, quantify the impact of fire on vegetation over large areas, and characterize post-fire ecological effects [18–20]. Emphasis has been given to the roles of multispectral sensors, lidar, and emerging Unmanned Aerial System (UAS) technologies [14]. Depending on the purpose of post-fire vegetation recovery observation and study, the assessment is performed based on groups of methods, such as image classification, vegetation indices (VIs), spectral mixture analysis (SMA) [21,22], etc. Remote sensing imagery offers an opportunity for obtaining land use and land cover information through image interpretation and classification. Spectral responses are used in image classification to identify the healthy vegetation in individual pixels [14]. Spectra-based classification approaches are conceptually simple and easy to implement [23]. The type or condition of surface features and their dynamics can be assessed by multi-temporal imaging. This type of analysis is fundamental in remote sensing and is typically called change detection [24]. VIs are the most commonly used method for assessments of vegetation state, including vegetation recovery after natural or anthropogenic disturbances [17,25,26].

Post-fire-related conditions are important for forest vegetation recovery. In this context, mapping of the burned area, representing the burn severity, is a standard technique for monitoring the post-fire effects and forest recovery patterns [17,27–29]. It was found that the differenced Normalized Burn Ratio (dNBR) and its relative form (relative differenced Normalized Burn Ratio (RdNBR)) derived from Landsat data correlate with field measurements of burn severity [30]. NBR is also used for monitoring post-fire regeneration over burned areas in ecosystems. Results showed that as vegetation regenerates, the differences between the burns and the reference area for the vegetation index decrease with time [31]. Detailed studies showed that the NIR-based vegetation indices are most appropriate for accurately assessing vegetation recovery [32]. The NDVI is found to be the most used for post-fire recovery studies, as it could be calculated alone without additional field data collection [26,33,34]. However, due to reaching saturation levels before the point where an ecosystem fully recovers its maximum biomass after disturbance, the forest recovery rate could be overestimated when using NDVI [35,36]. For estimations of variations in chlorophyll content and its changes in vegetation after a fire event, the Modified Chlorophyll Absorption Ratio Index (MCARI2) is used [37]. A quantitative analysis of forest degradation resulting from forest fires is performed by introducing the Normalized Differential Greenness Index (NDGI) [38]. The remotely sensed Moisture Stress Index (MSI) is used for canopy stress analysis and is suitable for monitoring coniferous forests and assessing specific damages that cannot be detected using NIR/R vegetation indices [39]. Spectral indices are also used to estimate other ecological parameters related to vegetation recovery. Such parameters are the Leaf Area Index (LAI) [40], the Forest Recovery Index (FRI), and Fractional Vegetation Cover (FVC) [29].

However, differences in fire severity provoke contrasting plant cover and floristic composition when ecosystems recover after forest fires. A multitude of factors such as climate, initial plant mortality, soil characteristics, the topography of the region, and vegetation composition determine the rate of recovery [41,42]. Moreover, vegetation response to fire and post-fire recovery processes differ in the various biogeographical regions and depend on vegetation type and pre-fire vegetation state [43]. For that reason, post-fire vegetation recovery is a complicated process that cannot be assessed by the application of a single, unified spectral index. Despite the advantages of spectral indices for monitoring post-fire vegetation recovery, there is still no single spectral index suitable for assessment of post-fire disturbances or vegetation recovery processes in every ecosystem, scale, and time-lag condition [44]. Even the NBR and its derivative indices developed specifically for fire-affected areas show varying accuracy under different conditions. The NBR and dNBR are considered advantageous for immediate post-fire monitoring, but their accuracy decreases with increasing temporal distance from the fire event and with the progress of vegetation recovery [45]. On the other hand, the Disturbance Index (DI) [46] is effective in monitoring forest disturbances of different origins and their temporal dynamics [47,48]. Due to the involvement of a larger range of spectral information, the DI is considered more accurate in assessing the recovery of undergrowth in forest ecosystems compared to standard monitoring methods using VIs [19,47]. The DI is based on a linear orthogonal transformation of multispectral satellite images [49,50], which increases its ability to differentiate the three main components: soil, vegetation, and moisture [51]. As a result of a fire, these three components are altered to the greatest extent. The DI more precisely separates the unvegetated spectral signatures closely linked to the stand-replacing disturbance from all other forest signatures [46]. This feature makes the DI particularly suitable for monitoring the dynamics of post-fire vegetation recovery.

This paper dissects the forest vegetation recovery stages during a period of six years after a wildfire in the Ardino region in Bulgaria. The study aims to present the potential for exploiting remotely sensed imagery and respective indices of greenness, moisture, and fire severity from Sentinel-2 to support post-fire observation and forest management with an emphasis on fire-affected territory mapping, vegetation damage assessment, fire and burn severity assessment, and their influence on the ability of vegetation to recover. The study used two groups of spectral indices for monitoring the area affected by the wildfire. The first group encompassed VIs using individual spectral bands for their calculation, and the second group included indices utilizing a larger range of spectral information through the orthogonalization of multispectral data. The Normalized Difference Vegetation Index (NDVI), the Modified Chlorophyll Absorption Ratio Index (MCARI2), and the Moisture Stress Index (MSI) are the indices that belong to the first group, and the Normalized Differential Greenness Index (NDGI), the Normalized Differential Wetness Index (NDWNI), and the Disturbance Index (DI) are the indices from the second group. The study took into account the landscape characteristics of the area influencing the processes occurring in fire-affected forest ecosystems and their post-fire recovery dynamics. Ancillary data about pre-fire vegetation state and slope inclinations were used to supplement our analysis for a better understanding of the fire regime and post-fire vegetation damages. Slope aspects were used to estimate and compare their impact on the ecosystems' resilience, vulnerability, and post-fire recovery capacity. Soil data were involved in the interpretation of the results.

2. Materials and Methods

2.1. Study Area

The study area was situated in the southeastern part of the Rhodope Mountains, near Ardino town. The X and Y coordinates of the centroid were calculated to be 25°6′7″E and 41°34′30″N. A significant fire took place on 29 July 2016 in the study area (Figure 1). 2016 was the year with the highest wind speed during the summer months of the study period (2016–2021). The average wind speed in the summer of 2016 ranged between 4.7 m/s (in September) and 6.1 m/s (in August), and the average maximum wind speed was between 7.4 m/s (in September) and 8.1 m/s (in July) [52]. Generally, based on hourly weather simulations over the past 30 years in the study area, the maximum wind speed was observed in March (reaching up to 20 m/s average value) and the minimum windspeed was in September (starting from 3 m/s average value). Days with wind speeds above 12 m/s predominated between February and September, and days with wind speeds above 5 m/s predominated between October and January. The wind direction was mainly from the north and northwest [53].

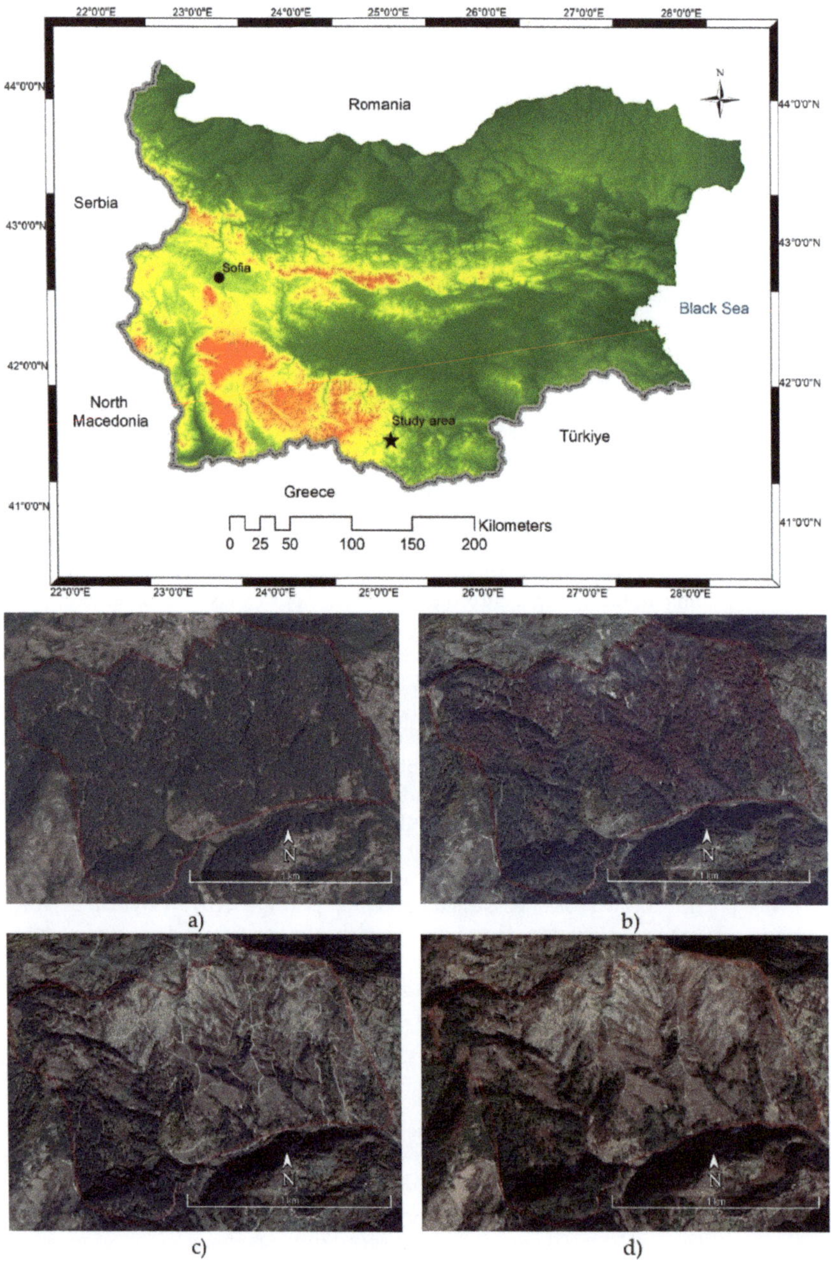

Figure 1. Location of the studied area and land cover change in the pre-fire 2013 (**a**) and post-fire 2016 (**b**), 2018 (**c**), and 2020 (**d**) years. Source: Google Earth Pro—Airbus and Maxar Technologies images.

The area affected by the forest fire was 100 ha. The main tree species were Scots pine (*Pinus sylvestris* L.) and Black pine (*Pinus nigra* Arn.). The endemic vegetation in the region refers to the Thracian province of the European deciduous forest area with the main tree species being *Quercus frainetto and Q.cerris*. On karst terrains, *Q. frainetto*, *Q. cerris*, and *Q.pubescens*, mixed with *Carpinus orientalis, Fraxinus ornus, Syringa vulgaris, Cotinus coggygria*, and *Ostrya carpinifolia* can be found [54]. However, because of erosion processes in the 1950s and the expansion of bare lands, massive afforestation with coniferous species has been performed.

Lithogenic diversity was represented by pre-Paleozoic and Paleozoic metamorphic rocks and phyllitoids covered by a Paleogene volcanogenic–sedimentary complex [54]. The terrain was hilly with steep slopes (>15°). It influenced the fire regime and is also considered a condition for the development of water erosion. Approximately 70% of the area contained slopes above 15°. The mean value of the slopes was 16° and the maximum was 27°. The slopes predominantly had east, southeast, and south exposures and define warm and dry conditions for vegetation development. The soils were shallow Lithosols mixed with Rendzinas [54].

The study area has a Continental Mediterranean climate with hot summers and mild winters. The minimum amount of precipitation is in summer, and the maximum amount is in winter. Over the last 40 years, a clear trend towards an increase in both average air temperature and precipitation sum has been observed. The average air temperature change showed a linear trend with an increase of 2.2 °C, starting from 10.2 °C in 1979 and reaching up to 12.4 °C in 2021. Summer is getting hotter, with temperatures in July and August consistently higher for the past 16 years. On the other hand, the precipitation sums in the summer months decreased, especially in August, showing a persistent negative tendency during the last 14 years. The winter is getting warmer too. February has been distinguished by sustained higher average air temperatures since 2012. The precipitation sum has also increased during the winters. Overall, there has been an increase (by 126 mm) in precipitation sum over the past 40 years. This increase is primarily due to increased winter precipitations [53]. The indicated climate changes have led to the transition of the studied area from a Warm-summer Mediterranean climate (Csb), according to the Köppen climate classification, to a Hot-summer Mediterranean climate (Csa). In addition, the slopes in the area, which predominantly had east, southeast, and south exposures, determined warm and dry conditions for the development of vegetation. The observed trends significantly increase the risk of fires in the area, loss of biological diversity, and degradation of ecosystems.

2.2. Characteristics of Climatic Anomalies Observed during the Period of 2016–2021

Figure 2 shows the mean air temperature and precipitation sum anomalies for the period of 2016–2021 on an annual basis and for the summer months (July, August, and September). As a reference, the period between 1981 and 2010 was used.

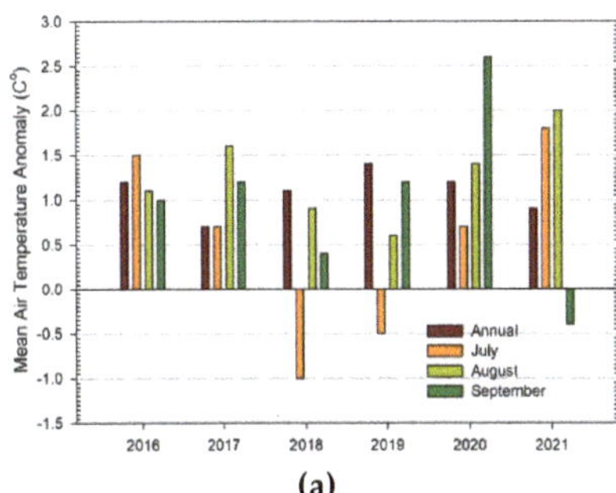

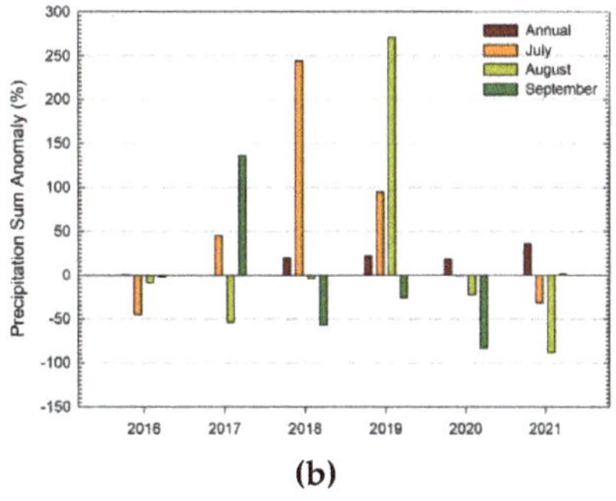

Figure 2. Mean air temperature (**a**) and precipitation sum (**b**) anomalies for the period of 2016–2021 on an annual basis and for the summer months (July, August, and September).

In terms of mean air temperature, 2019 was the year with the highest positive anomaly (+1.4 °C). Amongst the summer months, August was with positive anomalies only. The highest anomalous value (+1.6 °C) was recorded in 2017. The highest anomalous value for the summers between 2016 and 2021 was recorded in September 2020 (+2.6 °C). Three of the years during the period of 2016–2021 were characterized by positive anomalies in all three summer months. Overall for the three summer months, 2020 was with the highest positive anomaly (+4.7°C). The lowest positive anomaly, overall for the three summer months, was recorded in 2018 (+0.3°C). The mean air temperature in July 2018 was 1 °C lower in comparison with the reference period (Figure 2a) [52].

Higher sums of annual precipitation were observed during the period of 2016–2021. The wettest year was 2021, which had a 36% higher precipitation sum in comparison with the reference period. There was no definite trend in the distribution of precipitation during the summer months. All three summer months of 2016 and 2020 had lower precipitation. In 2020, the total precipitation for the summer was 106% lower in comparison with the reference period. In September 2020 only, the precipitation sum was lower by 83%. The wettest summer was in 2019. In that summer, the total amount of precipitations was 366% higher in comparison with the reference period (Figure 2b) [52].

2.3. Data

2.3.1. In Situ Data

In situ data included climatic data, soil data, and field observations.

The climatic data were based on measurements at the Kardzhali meteorological station (331 m.a.s.l) situated 21 km from Ardino [53] and on hourly weather simulations with 30 km spatial resolution over the past 30 years [52]. The in situ climatic data included mean air temperature and precipitation sum for a period between 1979 and 2021 as well as data about the mean air temperature and precipitation sum anomalies for the period of 2016–2021 on an annual basis and for the summer months (July, August, and September) [53]. In addition, ERA5 model data, which combined satellite and in situ historical observation, were used to outline how climate change has already affected the Ardino region in the last 40 years [52].

The soil data included soil types [54], soil chemical composition, and organic matter content, which were measured in a pine-dominated mixed woodland (790 m.a.s.l) situated near Ardino in 2015 [55]. Soil field data is part of LUCAS 2015 Topsoil datasets, which is freely available through the European Soil Data Centre (ESDAC) of the Joint Research Centre [55]. Because the soil data refer to the pre-fire period and the soil samples were not taken from the affected area specifically, they were used only for result interpretation as auxiliary data.

Interactive three-dimensional panoramas were used as a means of field observations. They were acquired via Google Street View technology in the summer of 2021. These were used to generate high-quality photographs of eight locations affected by the fire [56].

2.3.2. Satellite Data

Satellite data acquired from Sentinel-2A and Sentinel-2B multispectral sensors of the European Space Agency Program for Earth Observation "Copernicus" [57] were used to assess post-fire vegetation recovery. The temporal resolution of every individual Sentinel-2 satellite is ten days, and their combined resolution is five days. More detailed information about the spectral and spatial resolution of the Sentinel-2 satellites can be found in Table A1 [57].

The Sentinel-2 image acquired on 10 July 2016 was representative for the period before the fire event (29 July 2016), and the images acquired on 24 August 2017 and 2018, 14 August 2019, 02 September 2020, and 23 August 2021 were used for assessment of the forest vegetation state after the fire.

High-resolution forest layers (HRLs), which are freely available through the Copernicus Land Monitoring Service [58], were used in the validation process. The layers included

Tree Cover Density (TCD) and Forest Type (FTY) products. The TCD product represents the level of tree cover density in a range from 0–100% for 2012, 2015, and 2018 reference years. The Forest Type product represents the dominant leaf type with a Minimum Map Unit (MMU) of 0.5 ha. Both products are pixel-based, and the minimum mapping width is 20 m. The forests' HRLs in 2015 were verified for the territory of Bulgaria using in situ data [59]. The verification procedure was performed on three levels, including the highly recommended quantitative verification. According to the results obtained by Tepeliev et al. [59], the HRLs used in the present study are generally correctly mapped for the territory of Bulgaria. Hence, we assumed that they could be used as independent reference data in the validation procedure.

A Digital Elevation Model (DEM) with a spatial resolution of 25 m was used to obtain slopes and slopes' aspects. The dataset is freely available through the Copernicus Land Monitoring Service [60].

2.4. Methods

The proposed approach for monitoring post-fire vegetation state and estimating its dynamics included the following basic steps. First, spectral indices for the period between 2016 and 2021 were calculated to assess forest vegetation recovery dynamics. Second, statistical regression analyses using three of the spectral indices as variables were performed. The indices involved in the linear regression analyses were DI, MCARI2, and MSI. These indices are representative of key post-fire characteristics of the affected territories. The DI was used for the assessment of disturbance of forest ecosystems, burn severity, and vegetation damage. MCARI2 is representative for vegetation regrowth, and MSI shows stress in ecosystems caused by moisture deficiency. The third step considered differences in landscapes and the conditions for forest recovery they create, which is predetermined by the impact of slope exposures and their influence on the heat–moisture ratio. The assessment of the slope exposure factor was based on a differentiated evaluation of indices dynamics and their interpretation. The final step consisted of the validation of obtained results using statistical regression analyses involving the forest HRLs as independent reference data. Interactive three-dimensional panoramas from Google Street View were also used in the validation process. The interactive panoramas were used to generate photographs in different directions in order to demonstrate the state of various ecosystems. X and Y coordinates and altitude of the point locations of the photographs were extracted. The obtained point locations were georeferenced and digitized to be used in overlay analyses with the indices' rasters. The eight locations affected by the fire that were observed via this technology were linked to the obtained results as a means for field observation. For each location, indices values were extracted by taking into account the observed perspective and the distance of the objects from the point of capture.

Spectral Indices Selected for Assessment of Post-Fire Vegetation Recovery

The spectral indices presented in Table 1 were selected and calculated to assess the post-fire vegetation state.

The most well-known and used vegetation index for quantifying green vegetation in the near-infrared wavelength region and chlorophyll absorption in the red wavelength region is the NDVI. The NDVI strongly correlates with climate variations and their impact on plant growth. That makes this index especially suitable for estimations of climate-related vegetation changes. Moreover, changes in NDVI values correlate with the de Martonne aridity index [19].

The MCARI has been proposed to estimate variations in chlorophyll content and its concentration changes. Unlike MCARI, the newly designed MCARI2 is less sensitive to chlorophyll concentration variations but has a high linear relationship with near-infrared canopy reflectance and high linearity with the green LAI. The LAI is an important variable used to estimate the biophysical processes of different vegetation types and predict their growth and productivity [21,62]. Both the NDVI and MCARI2 range between −1 and +1.

The highest values indicate dense and "healthy" vegetation, and the lowest values indicate dead plants or inanimate objects.

Table 1. Spectral indices calculated in this study.

Index	Abbreviation	Formula
Normalized Difference Vegetation Index [61]	NDVI	$NDVI = \frac{\rho NIR - \rho RED}{\rho NIR + \rho RED}$ (1)
Modified Chlorophyll Adsorption Ratio Index [62]	MCARI2	$MCARI_2 = \frac{1.5[2.5(\rho_{800} - \rho_{670}) - 1.3(\rho_{800} - \rho_{550})]}{\sqrt{(2\rho_{800} + 1)^2 - (6\rho_{800} - 5\sqrt{\rho_{670}}) - 0.5}}$ (2)
Moisture Stress Index [63]	MSI	$MSI = \frac{SWIR1}{NIR}$ (3)
Normalized Differential Wetness Index [38,64,65]	NDWNI	$NDWNI = \frac{W_n(t_2) - W_n(t_1)}{|W_n(t_2)| + |W_n(t_1)|}$, (4) $W_n(t) = \frac{W(t) - E\{W(t)\}}{St.Dev.[W(t)]}$, (5)
Normalized Differential Greenness Index [38]	NDGI	$NDGI = \frac{GR_n(t_2) - GR_n(t_1)}{|GR_n(t_2)| + |GR_n(t_1)|}$, (6) $GR_n(t) = \frac{GR(t) - E\{GR(t)\}}{St.Dev.[GR(t)]}$, (7)
Disturbance Index [46]	DI	$DI = nBR - (nGR + nW)$ (8)

The remotely sensed MSI is used for canopy stress analysis, productivity prediction, and biophysical modeling. It detects plant water stress for these plants only, which are able to tolerate low leaf water content through cellular adjustments. In the current study, the MSI is especially suitable for monitoring coniferous forests and assessing specific damages that cannot be detected using NIR/R vegetation indices [39]. Considering coniferous vegetation, the differences in the MSI between damaged and undamaged stands are not necessarily related to differences in the LAI. The MSI ranges from 0 to more than 3. Higher values indicate greater moisture stress.

Additionally, a quantitative analysis of forest degradation resulting from forest fires was performed by introducing the NDGI and the NDWNI. Both indices are based on satellite image orthogonalization. In the process of orthogonalization, three differentiable classes (soil brightness, greenness, and wetness axes) related to the main components of the Earth's surface (soil, vegetation, and water) are obtained. The NDGI uses the greenness component, which corresponds to vegetation's spectral reflectance characteristics (SRC), and the NDWNI uses the wetness component, which corresponds to water's SRC. These indices quantitatively estimate the slightly positive and negative values of change in the vegetation's green mass and moisture content for a given period [38,64,65]. Both indices range from −1 to +1. The positive NDGI values indicate plant growth and improvement of the vegetation state, and the negative values indicate deterioration of the vegetation state or deforestation. The positive NDWNI values indicate an increase in moisture content in ecosystems, and the negative values indicate a decrease in moisture content.

The DI has also been used to monitor disturbances by forest fires. The index values range widely, with positive values indicating disturbances. Higher values indicate more severe disturbance. The DI is modified by weighting each input component to maximize the difference between disturbed and undisturbed forest canopy. The weights reduce the effects of background variations while emphasizing the variations caused by disturbance [48]. The model for calculating the DI includes three steps: the first step is the decomposition of each of the three major Tasseled Cap components (brightness (BR), greenness (GR), and wetness (W)); the second step is to calculate the averages and standard deviations for each of the Tasseled Cap components; and the third step is to calculate the normalized values of the components. These steps are needed to normalize the radiometric changes. In the normalization, the following equations were used [46]:

$$nBR = (BR - E\{BR\}/St.Dev\ (BR)$$

(9)

$$nGR = (GR - E\{GR\}/St.Dev\ (GR) \tag{10}$$

$$nW = (W - E\{W\}/St.Dev\ (W) \tag{11}$$

where E {BR}, E {GR}, and E {W} are the average values of the brightness, greenness, and wetness, respectively. St.Dev (BR), St.Dev (GR), and St.Dev (W) are the respective standard deviations of these Tasseled Cap components. Therefore, nBR, nGR, and nW are the normalized values of brightness, greenness, and wetness, respectively. The DI is computed according to the equation presented in Table 1.

3. Results

3.1. Forest Vegetation Recovery for the Entire Study Area

Areas with negative NDGI values predominated until 2019. The NDGI values were negative even a year before the fire event, which indicates a degraded state of forest vegetation. The territories with negative NDGI values had maximum territorial spread in the pre-fire year and a year after the fire (Figure 3, Table A2). The areas in the range from −1 to −0.8 were dominant between 2016 and 2017 in the year immediately following the fire (Figure 3, Table A2). With increasing temporal distance from the fire event, the maximum territorial spread shifted to the category with slightly positive values (from 0 to 0.2). The areas with positive NDGI values, or an increase in biomass, had minimal spread in the year immediately following the fire (Figure 3, Table A2). The positive NDGI values had the highest growth between 2019 and 2020 (Figure 3, Table A2).

Surprisingly, the areas with no disturbance or negative DI values prevailed during the entire studied period (Figure 3, Table A2). However, this maximum was highest in the year before the fire event. Moreover, during the same year, the highest territorial spread of the areas with great disturbance was observed. Areas from the same category (DI > 5) were also recorded in the following 2017 and 2018 years (Figure 3, Table A2). After 2019, with increasing temporal distance from the fire event, areas falling into this category were not observed. In the year before the fire event, a large portion of the areas had DI values between 0 and 2. In 2017, the areas were almost evenly distributed in the categories with DI values between 0 and 4. In the following four years, between 2018 and 2021, the maximum spread had areas with DI values between 0 and 3 (Figure 3, Table A2).

The first post-fire year was characterized by the most pronounced stress due to moisture deficiency. In the same year, the highest territorial spread of the category with MSI values above 1.5 was also observed (Figure 3, Table A2). The MSI category with values between 0.5 and 1.5 had the largest share of the area. The maximum spread had territories with MSI between 1.1 and 1.3. In the pre-fire year, 60.5% were in the category with MSI between 0.5 and 0.7. Between 2018 and 2021, the area was concentrated in the MSI categories between 0.5 and 1.3. The maximum spread was between 0.7 and 1.3 (Figure 3, Table A2).

Regarding the NDWNI, the maximum area was mainly in the category between −0.2 and 0.2, indicating weak dynamics in the moisture content. An exception was the one-year period immediately after the fire when a significant increase in moisture content was observed (Figure 3, Table A2). In the years before the fire and between 2018 and 2021, a significantly smaller share of the area had positive NDWNI dynamics (Figure 3, Table A2).

During the year before the fire, 94% of the area had MCARI2 values above 0.6. Moreover, almost half of the area had MCARI2 values between 0.7 and 0.8. This category (0.7–0.8) also had a maximum territorial spread in the post-fire years, but the area falling into this category was significantly less (Figure 3, Table A2). In the post-fire year of 2017, the area was more evenly distributed between the individual categories. In the following years, the area was concentrated mainly in the MCARI2 category with values between 0.4 and 0.9. The share of the areas in this category gradually increased, reaching 96% in 2021 (Figure 3, Table A2).

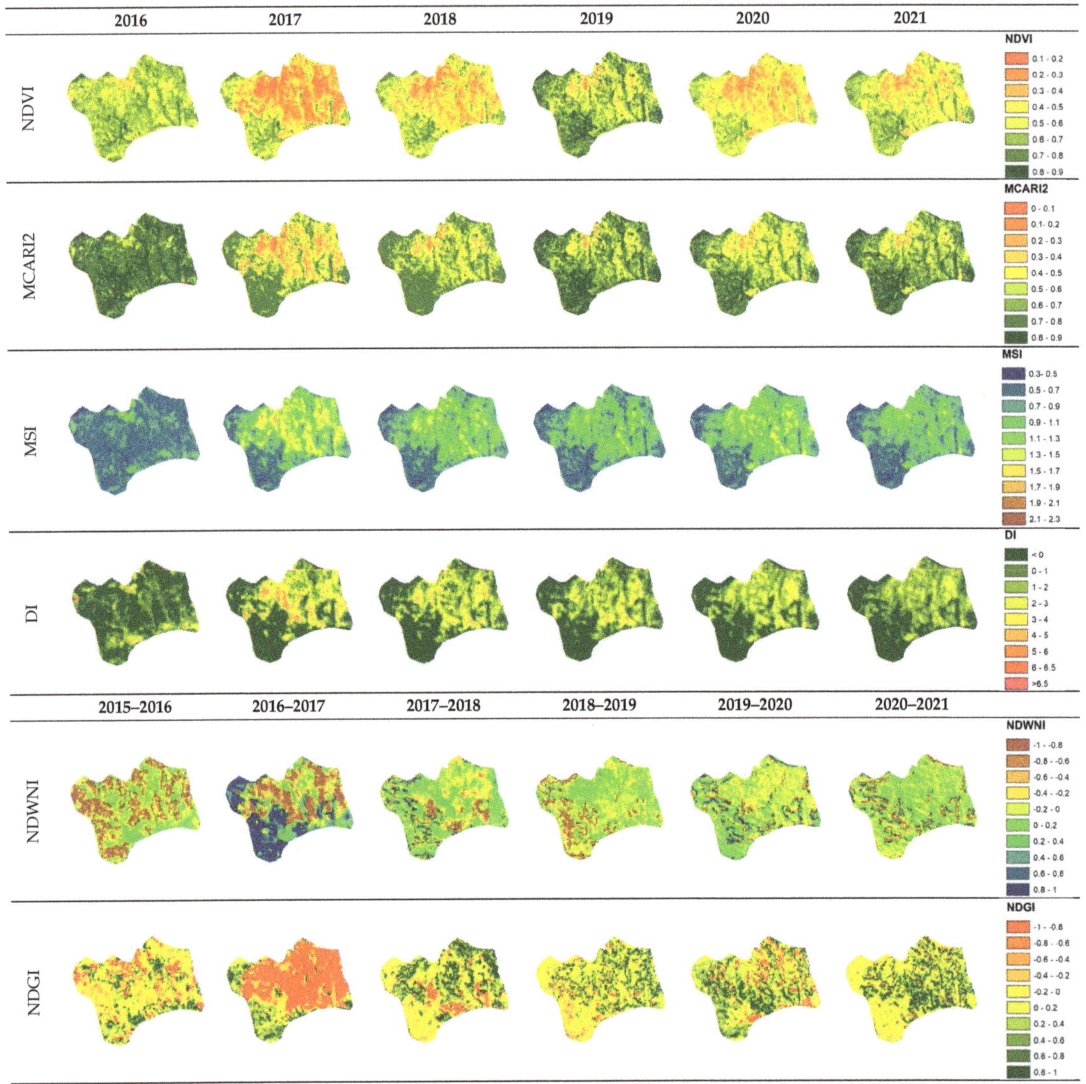

Figure 3. Dynamics of the spectral indices calculated for the period of 2016–2021.

Greater dynamics were observed in the maximum territorial spread between the individual NDVI categories. In the pre-fire year, the highest proportion of territories had NDVI values between 0.6 and 0.7, whereas in the year following the fire event, maximum spread had territories with NDVI between 0.3 and 0.4. The maximum territorial spread shifted to the categories with higher NDVI values in the years between 2018 and 2021 (Figure 3, Table A2).

3.2. Forest Vegetation Recovery in the Individual Slope Aspects

The differential analysis of the dynamics of indices values on the individual slope exposures showed that in the first three years after the fire event (2017–2019), the southwest-facing slopes had faster vegetation recovery and less moisture stress. These slopes had the highest values of the NDVI and MCARI2 and the lowest values of the MSI (Table 2). The MCARI2 had a better ability to differentiate vegetation recovery through the individual slopes. On east- and southeast-facing slopes, the NDVI showed equal totals over the three years, while the MCARI2 distinguished them. The difference in MCARI2 values between the individual slope aspects was higher than those of the NDVI and was clearer (Table 2). Regarding moisture stress in the first three years after the fire event, the MSI total values gradually increased in the following sequence: northeast-facing, southeast-facing, east-facing, and south-facing slopes (Table 2). In the last two post-fire years, the northeast-facing slopes had the highest total values of NDVI and MCARI2 and the lowest moisture stress. On northeast-facing slopes, the highest totals of NDVI and MCARI2 and the lowest totals of MSI for the entire period were recorded (Table 2).

Table 2. Mean values of the spectral indices for the different slope aspects in the studied years.

Aspect	2016	2017	2018	2019	2020	2021
			NDVI			
E	0.63	0.37	0.47	0.54	0.54	0.59
NE	0.64	0.37	0.47	0.52	0.59	0.63
S	0.61	0.36	0.44	0.51	0.46	0.53
SE	0.61	0.40	0.46	0.53	0.45	0.52
SW	0.63	0.41	0.50	0.56	0.50	0.56
			MCARI2			
E	0.77	0.46	0.61	0.67	0.68	0.73
NE	0.78	0.45	0.62	0.64	0.75	0.77
S	0.76	0.46	0.57	0.64	0.58	0.65
SE	0.74	0.50	0.59	0.65	0.57	0.63
SW	0.77	0.52	0.64	0.69	0.64	0.69
			MSI			
E	0.64	1.20	0.99	0.93	0.85	0.75
NE	0.61	1.21	0.92	0.92	0.70	0.65
S	0.69	1.22	1.07	0.96	1.06	0.93
SE	0.72	1.15	1.02	0.93	1.07	0.93
SW	0.62	1.10	0.92	0.85	0.94	0.83
			DI			
E	−0.48	1.94	0.98	1.30	−0.05	−0.05
NE	−0.74	1.66	0.46	1.19	−0.18	−0.15
S	−0.09	1.85	1.63	1.39	0.11	0.09
SE	0.58	1.53	0.14	1.27	0.12	0.10
SW	−0.80	0.86	0.36	0.48	0.00	0.01

	2015–2016	2016–2017	2017–2018	2018–2019	2019–2020	2020–2021
			NDWNI			
E	−0.46	−0.29	0.09	−0.16	0.07	−0.05
NE	−0.70	−0.27	0.27	−0.39	0.18	−0.12
S	−0.23	−0.25	−0.10	−0.02	−0.10	0.00
SE	−0.18	0.02	−0.09	−0.03	−0.03	−0.01
SW	−0.37	−0.31	0.04	−0.17	−0.06	−0.11
			NDGI			
E	−0.05	−0.81	0.38	0.10	0.26	0.18
NE	−0.16	−0.76	0.38	−0.16	0.34	0.01
S	−0.02	−0.77	0.16	0.26	0.10	0.30
SE	0.00	−0.57	0.13	0.21	0.09	0.29
SW	−0.13	−0.65	0.29	0.14	0.19	0.17

In the pre-fire year, only the vegetation on southeast-facing slopes had positive DI values (Table 2). Moreover, territories most affected by the fire were situated on slopes with such exposure in the northern part of the area. The entire forest vegetation was destroyed in those territories (Figure 1). The fire most likely started there (Figure 3). The mean DI values were positive for all slopes in the following three post-fire years. The total DI values for the individual slope exposers for 2017, 2018, and 2019 decreased in the following sequence: south-facing, east-facing, northeast-facing, southeast-facing, and southwest-facing slopes (Table 2). Despite the higher DI values recorded for the east and northeast-facing slopes in the first three post-fire years, only these slopes had negative values in the last two years of observation (Table 2). On slopes with a southern component, the DI remained positive. In contrast to the trend observed in the NDVI, MCARI2, and MSI for the entire period, the lowest DI values and the optimal vegetation state were not recorded for the northeastern but for the southwestern slopes. However, as recorded by the listed indices, the vegetation had the worst condition on south-facing slopes (Table 2).

The other two indices based on TCT (NDWNI and NDGI) can measure small changes in moisture content and green mass. A slight positive increase in moisture content in the first post-fire year was recorded only for the southeast-facing slopes. In 2018 and 2019, the northeast-facing slopes had the highest positive NDWNI dynamics. The eastern slopes also showed positive dynamics in the moisture content in these years (Table 2).

The NDGI, or green mass, showed positive dynamics after the second post-fire year (2017–2018). In the period between 2017 and 2018, the vegetation on east- and northeast-facing slopes had the highest increase in NDGI values. These were the highest NDGI mean values in the entire period of observation and for all of the slope exposures (Table 2). Moreover, during the drought in 2020, the vegetation on the northeastern and eastern slopes again showed better condition compared to those on the slopes with different exposure (Table 2). The NDWNI dynamics were also positive for the northeast- and east-facing slopes in 2020. However, for the entire period of observation, the vegetation on southeast-facing slopes had the highest total increase in NDGI values (Table 2).

3.3. Linear Regression Analyses for the Apectral Indices

The statistical regression analyses were representative for the correlation and dependency between some of the key post-fire characteristics of the affected territories in the vegetation recovery process.

The correlation between the areas affected by disturbance increased with the development of ecosystem restoration processes and improving vegetation condition. The weaker correlation between 2017 and 2018 indicated a higher difference in the state of the forest vegetation in the first and second post-fire years (Figure 4c). This trend confirmed the rapid development of the post-fire succession process between the first and second years after the wildfire. The correlation was also lower between 2019 and 2020 when the severe drought in the summer of 2020 influenced the vegetation state and increased the intra-territorial differences. The increase in correlation for the periods between 2018 and 2019 and between 2020 and 2021, in turn, showed greater similarity in the state of the vegetation. In the second post-fire year (the period between 2018 and 2019), the higher correlation indicated delaying of the ecosystem restoration process, and in the 2020–2021 period, it indicated restoring balance in the ecosystems disturbed by the severe drought. The character of the territorial spread of vegetation throughout the individual categories divided by the DI values confirmed these observations (Figure 3, Table A2). In the first post-fire year, which was distinguished by weaker correlation, the vegetation was divided between eight categories. In 2021, as vegetation recovery progressed and the highest correlation was recorded, the number of these categories decreased to five. In 2021, 67.4% of the area was concentrated in two categories. 30% of this share consisted of the territories with DI between 1 and 2. That was the greatest share of the area falling into this category of disturbance for the entire period of observation. The area in this category gradually increased between 2016 and 2021, starting at 16.4% in 2016 and reaching up to 30% in 2021 (Figure 3, Table A2).

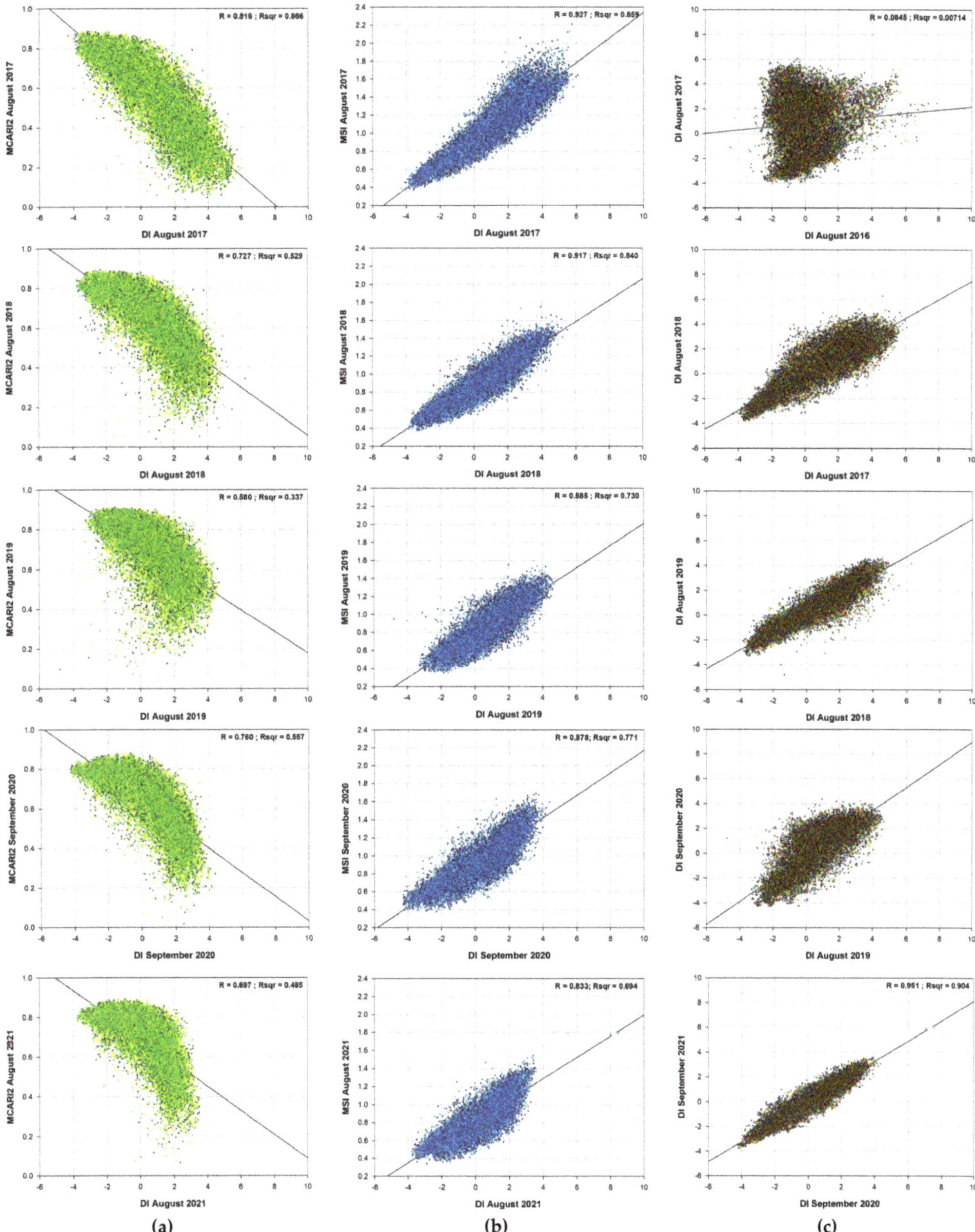

Figure 4. Linear regression analyses between DI and MCARI2 (**a**); DI and MSI (**b**); and DIs for two consecutive years (**c**) for the period between 2017 and 2021.

The correlation between MSI and DI values, in turn, showed a decreasing trend with the progression of the ecosystem restoration process (Figure 4b). In 2017 (the first post-fire year), the correlation between the disturbed ecosystems and moisture stress was highest for the entire period of observation. That indicated that for most of the forest ecosystems, higher MSI values were connected with higher DI values. In other words, the lack of moisture was connected with a higher degree of disturbance of the ecosystems (Figure 4b). The progression of ecosystem restoration processes decreased this dependency.

The correlation and the dependency between MCARI2 and DI (i.e., between leaf area and intensification of photosynthesis and the degree of disturbance of the ecosystem) showed a decreasing trend with the progression of ecosystem restoration processes for the first three post-fire years. Figure 4a shows that the areas characterized by high MCARI2 values were distinguished by lack of disturbance of the ecosystem and vice versa; areas with lower MCARI2 values (e.g., 0.2) (i.e., areas with underdeveloped vegetation and small leaf area) were distinguished by a higher degree of disturbance of the ecosystems. The progression of ecosystem restoration processes led to an increase in leaf mass and intensification of photosynthesis. The lowest MCARI2 values reached 0.4 in 2019. In these areas, the DI had decreased to below 4. The decline in the dependency between the vegetation state and the degree of disturbance of ecosystems confirmed the progression of ecosystem restoration processes. The decreasing trend in the correlation was interrupted by drought stress in 2020. In 2021, the correlations between both indices started to decline again. However, the forest recovery trend still could be observed. Starting from 2020, vegetation with DI values above 4 was not observed (Figure 4a).

3.4. Validation through HRLs

Statistical regression analyses involving the forest HRLs acquired in 2012, 2015, and 2018 as independent reference data were performed to validate the obtained results.

Generally, between 2012 and 2018, an increase in tree cover density of both broad-leaved forests and coniferous forests was observed. The largest share of the area had forests with tree cover density between 40% and 80% (Table 3). In the post-fire year of 2018, the non-forested areas, as expected, had the highest disturbance (Table 4). They had positive DI values. The DI for the forested areas in the same year was negative. The non-forested area recorded the highest moisture stress (MSI > 1) (Table 4) but also the highest increase in NDGI values. That was related to the development of vegetation succession processes in the deforested territories. The vegetation indices NDVI and MCARI2 recorded their highest values in broad-leaved forests. However, coniferous forests had the highest positive dynamics in moisture content (NDWNI) (Table 4).

Table 3. Tree-cover density (TCD) classes (%) and area of their distribution (%) within broad-leaved and coniferous forests in 2012, 2015, and 2018.

TCD (%)	Broad-Leaved Forests			Coniferous Forests		
	2012 (%)	2015 (%)	2018 (%)	2012 (%)	2015 (%)	2018 (%)
0–20	6.66	0.8	0.72	0.63	non	non
20–40	14.53	9.41	25.51	3.83	0.75	9.39
40–60	37.61	72.47	39.02	19.15	43.85	54.26
60–80	38.97	15.92	28.47	73.96	55.35	34.96
>80	2.22	1.4	6.29	2.43	0.06	1.39
Mean	43.86	52.21	56.19	47.37	55.67	60.28

The DI and MSI were the indices that showed the highest correlation with forest density (Table 5). The correlation between the DI and forest density was the highest in the coniferous forests. In the broad-leaved forests, the correlation between these two variables was slightly lower. Regarding the moisture content, both deciduous and coniferous forests showed a similar correlation. Generally, the coniferous forests were distinguished by higher dependence on the spectral indices. The difference in correlation between the forest density and NDVI was significant when comparing deciduous and coniferous forests. In deciduous forests, R was barely 0.162, whereas in coniferous forests, it was 0.529 (Table 5). The observations for the NDGI were similar. The difference between both forest types was greater than four times. The correlation between forest density and MCARI2 was also slightly higher in the coniferous compared with broad-leaved forests (Table 5).

Table 4. Mean values of the spectral indices in broad-leaved forests, coniferous forests, and in non-forested areas in 2018.

Index	Broad-Leaved Forests	Coniferous Forests	Non-Forested Areas
DI	−7.57	−18.54	15.97
MCARI2	0.79	0.76	0.57
MSI	0.71	0.68	1.06
NDGI	0.08	−0.08	0.23
NDVI	0.65	0.59	0.44
NDWNI	−0.05	0.12	−0.06

Table 5. Linear regression between forest density (independent) and spectral indices (dependent) for each of the forest types.

	Broad-Leaved Forests		Coniferous Forests		Non-Forested Areas	
Index	R	Rsqr	R	Rsqr	R	Rsqr
NDVI	0.16	0.03	0.53	0.28	0.18	0.03
MCARI2	0.46	0.21	0.51	0.26	0.18	0.03
NDGI	0.03	0	0.15	0.02	0	0
DI	0.61	0.37	0.65	0.43	0.2	0.04
MSI	0.64	0.4	0.64	0.4	0.18	0.03
NDWNI	0.06	0	0.02	0	0.09	0.01

3.5. Validation through Field Observation

Interactive three-dimensional panoramas of eight-point locations from Google Street View were also used in the validation process. The eight locations were linked to the obtained results as a means for field observation. The indices values clearly showed the trends in the manifestation of the indices in relation to the observed ecosystems' components (i.e., soil and vegetation). The lowest indices values were observed in the post-fire karst territories in point one and point eight, and the highest values were observed in the forest territories unaffected by the fire (Figure A1, Table A3).

4. Discussion

The deteriorated vegetation state and the landscape-ecological conditions (karst terrain, dry and hot summers, and warm slope exposures) caused the fire. Dry vegetation and steep terrain made the fire more intense, devastating, and hard to control. The consequences in the area of occurrence were almost complete destruction of the forest vegetation, litter cover, and soil organic layer.

The soil type and slope exposures are the main landscape-forming factors that determine differences in the processes of vegetation recovery in the study area. Unfavorable characteristics of the soils in this area, such as shallow profile, low organic matter content, acidic soil reaction, and low exchange capacity, worsened after the fire and further inhibited the recovery of vegetation, including forest. These processes manifested more significantly on the southern slopes. The removal of vegetation and litter cover as a result of a fire reduces rainfall interception, which enhances runoff and erosion rates [66]. Moreover, the burn severity of the soil surface on the south-facing slopes decreases soil carbon content and changes soil acidity [31].

The neighboring territories were also affected to a great extent. Forest management practices include a sanitary logging of burnt forest stands after a fire. For this reason, in 2018 (two years after the fire), actions to remove the burnt forest vegetation were taken. As a result, a large part of the territory was completely cut down. The recovery processes of the forest vegetation were interrupted. The manifestation of TCT-based indices (NDGI and NDWNI), which represent the dynamics in greenness and wetness after 2017–2018, was also an indication for this interruption of the recovery process (Figure 3). It was evident from a sharp decline of the areas with values that ranged between −1 and −0.8, which is typical for forested territories, and from a significant increase of the areas in the NDGI category of 0.8–1 (Figure 3, Table A2), which indicates the rapid development of grass vegetation.

The negative values of the TCT-based DI were also indicative of the landscapes' dynamic conditions in the territories affected by the fire. This index showed stable presence of areas "undisturbed" by the fire. Their share was most significant (52% of the studied area) before the fire (10 July 2016). In the years after the fire, their share remained relatively stable at about 1/3 of the territory (Figure 3, Table A2). The analysis showed that these were forest areas that remained undamaged by the fire as well as local patches of meadows in the southern zone unaffected by the fire. This category also included areas with exposed bedrock formed due to the nature of lithological type (karst) in the northern part of the study area [54]. In these areas, the fire was not a destabilizing factor for the ecosystems' condition (Figure 1).

The maximum spread of the areas in the NDWNI category (−0.2–+0.2), which had a percentage distribution similar to that observed for the negative values of DI (approximately 30%) of the studied area (Figure 3, Table A2), also confirmed the observations made in the analysis of the DI. The territories unaffected by the fire, were distinguished by the lowest dynamics in moisture content. The first post-fire year was distinguished by the highest NDWNI dynamics. It was related to rapid post-fire succession processes in a large part of the study area. However, the NDGI values, which were representative of greenness dynamics, did not indicate high dynamics in the same year, but they did indicate high dynamics in the following year after sanitary logging (Figure 3, Table A2). The increase in moisture content in the first year after the fire was not induced primarily by the better state of vegetation but was mainly a result of the meteorological condition in the summer of 2017 when the precipitation was 226% more than what it typical for the region (Figure 2b). It should be noted that in the first post-fire years, vegetation primarily comprises herbaceous species. The state of this type of vegetation is closely dependent on environmental conditions. Grasslands are less resistant to anomalies related to temperature and humidity [67,68]. Moreover, they are strongly dependent on their fluctuations [69].

The indices that were not based on orthogonalization showed the general dynamics in the vegetation state. Generally, the areas with comparatively high values of NDVI and MCARI2 (taking into consideration the character of vegetation), which corresponds with lower MSI values, had the greatest territorial spread in the pre-fire year (Figure 3, Table A2). In the first post-fire year, the areas were distributed among a large number of categories, i.e., the vegetation in the fire-affected territories was characterized by great diversity in terms of its state. There were both unaffected areas and areas with varying degrees of post-fire damage. With an increasing temporal distance from the fire event, an increasing share of the territories were concentrated in fewer and fewer categories, with a slight shift towards the categories representing a better state of vegetation. In the different post-fire years, there was a slight growth or contraction of the individual categories within the general trend of vegetation recovery (Figure 3, Table A2). However, this dynamic was the result of climatic elements in the relevant year of observation (Figure 2a,b).

Some trends stand out regarding the influence of slope exposures on the vegetation recovery process. Vegetation recovery was faster on warmer slopes in the first post-fire years (2017–2019) (Table 2). However, it should be noted that during those years, the lack of moisture was not a limiting factor for the recovery processes (Figure 2b). In all three years, the summer precipitation exceeded the norm (Figure 2b). These results are consistent with those of Wilson at al. [70], who found that vegetation had a higher recovery rate when the temperature and precipitation were higher. A rapid, "catch-up" development on the northeastern slopes was observed (Table 2) in the last two years of observation (2020 and

2021). These were dry years (the precipitation sum for the three summer months was below the norm), and the drought delayed vegetation recovery on the warmer slopes (Figure 2b). The recovery processes on the northeastern slopes for the last two years of observation were so rapid that they influenced the values indicating the total recovery for the entire period (Table 2). This rapid vegetation recovery was also recorded by the DI. The tendency of positive values (i.e., values indicating disturbance in ecosystems) was interrupted only for the slopes with northeast and east exposure (Table 2). The DI showed a significant relationship with post-fire vegetation recovery processes and that its dynamics, when assessing such a process, were correlated with various climatic and topographic factors [42]. Chen et al. [42] confirmed that the role of slope exposures in the dynamics of post-fire vegetation recovery was essential. Our results are consistent with those of Chen et al. [42]. They found that vegetation recovery on sunny sides was greater than on shady sides and that it closely related to elevation and its influence on the heat–moisture ratio [42].

The NDWNI and NDGI showed the highest positive dynamics for the north- and northeast-facing slopes. These slopes are in the western part of the study area, where due to the lower degree of slope inclinations, the forest vegetation was largely preserved from the fire (Figure 1). After the post-fire sanitary logging in 2018, small patches of these slopes were deforested. These patches remained between separate groups of trees (Figure 1). Regarding the landscape-forming factors (soil type [54], slope aspects, and inclinations), these territories had more favorable conditions for vegetation recovery.

The results of the statistical and validation analyses indicated the reliability of the methodology used to monitor and assess post-fire vegetation recovery processes. The linear regression analyses showed stronger correlations between the objects of the Earth's surface that were under favorable conditions for vegetation recovery for two consecutive years (higher R values between DI) and vice versa (weaker correlations when the territory was under post-fire or drought stress in one of two consecutive years (lower R value between DI)) (Figure 4c). The dependency between the severity of disturbance and moisture content decreased with increasing time from the fire event and with the development of the vegetation recovery process (Figure 4b). The correlation between the leaf area (indirectly represented by MCARI2) and the disturbance of the ecosystems decreased with the development of vegetation recovery. This trend of decreasing correlation was interrupted by the drought in 2020 that impacted the general state of the vegetation and resulted in withering and shrinking of tree leaves (Figure 4a).

The validation through high-resolution forest layers showed higher dependency between the forest density and the indices values for the coniferous forests (Table 5). This tendency is induced by the fact that coniferous vegetation, in contrast to deciduous vegetation, stays green during the entire year. Satellite images acquired in August or early September for each of the years were used. At the end of summer, deciduous forests undergo senescence. The process leads to reduced greenness and moisture in the tree leaves, disrupting the process of photosynthesis. In coniferous vegetation, these processes are significantly less noticeable. For that reason, the correlation between the forest density and the indices values representing the general state and functioning of vegetation were higher in the coniferous forests compared with the deciduous forests.

5. Conclusions

This study traced post-fire vegetation recovery dynamics using two groups of spectral vegetation indices and taking into consideration the local landscape factors in an area affected by a fire in southeast Bulgaria. Consistent with other studies [42,70], it was confirmed that vegetation recovery was dependent on climatic factors [70] and topography features [42]. Regarding the effectiveness of spectral vegetation indices for monitoring the post-fire vegetation state, it can be summarized that the statistical analysis and validation procedures confirmed their reliability for the assessment of restoration processes of vegetation after a fire. The NDVI, MCARI2, and MSI indicated general trends in post-fire vegetation dynamics, and the TCT-based indices (DI, NDGI, and NDWNI) were found to be suitable for more precise analyses of intra-territorial differences. The DI was advantageous for the differentiation of post-fire severity in ecosystems. The obtained results clearly showed the intra-territorial heterogeneity of post-fire vegetation recovery and the influence of local environmental factors on the dynamics of the process. The study demonstrated the need of multi-factor analysis in post-fire monitoring and could serve as a basis for further post-fire-related studies. Estimations of the impact of soil erosion would be particularly valuable. In such a study, the changes in the soil characteristics and in-depth analysis of slope steepness must be taken into account.

Author Contributions: Conceptualization, D.A., E.V. and L.F.; methodology, D.A., E.V. and L.F.; software, D.A.; validation, D.A.; formal analysis, D.A.; investigation, D.A.; resources, D.A. and L.F.; data curation, D.A.; writing—original draft preparation, D.A.; writing—review and editing, D.A.; visualization, D.A. All authors have read and agreed to the published version of the manuscript.

Funding: This research received no external funding.

Data Availability Statement: Not applicable.

Conflicts of Interest: The authors declare no conflict of interest.

Appendix A

Table A1. Spectral (in microns) and spatial (in meters) resolution of Sentinel-2 sensor.

Band	Spectral Resolution	Spatial Resolution
B1	0.443	60
B2	0.49	10
B3	0.56	10
B4	0.665	10
B5	0.705	20
B6	0.74	20
B7	0.783	20
B8	0.842	10
B8a	0.865	20
B9	0.94	60
B10	1.375	60
B11	1.61	20
B12	2.19	20

Appendix B

Table A2. Territorial spread (in %) of the individual categories divided for each of the spectral indices within the study area. This table represents the values behind the output rasters from Figure 3.

	NDVI					
Category	2016	2017	2018	2019	2020	2021
0–0.1	0.0	0.0	0.0	0.0	0.0	0.0
0.1–0.2	0.0	4.7	0.8	0.3	0.5	0.1
0.2–0.3	0.1	20.2	6.9	3.3	6.3	2.2
0.3–0.4	1.7	21.0	18.1	10.7	18.5	9.3
0.4–0.5	6.8	18.0	24.3	19.1	26.7	20.6
0.5–0.6	28.2	18.3	26.1	25.5	28.3	28.9
0.6–0.7	48.1	13.5	18.1	27.1	17.5	29.7
0.7–0.8	14.1	3.8	5.3	12.1	2.2	8.9
0.8–0.9	1.0	0.1	0.0	1.7	0.0	0.2
0.9–1	0.0	0.3	0.3	0.3	0.0	0.0

Table A2. *Cont.*

MCARI2						
Category	**2016**	**2017**	**2018**	**2019**	**2020**	**2021**
0–0.1	0.3	0.7	0.4	0.4	0.1	0.0
0.1–0.2	0.0	3.3	0.6	0.2	0.3	0.1
0.2–0.3	0.0	10.3	2.1	1.1	2.1	0.9
0.3–0.4	0.2	13.7	6.0	3.1	6.0	3.0
0.4–0.5	1.4	14.5	11.8	7.2	13.5	7.8
0.5–0.6	3.9	14.3	18.0	13.2	20.5	15.9
0.6–0.7	11.3	15.3	20.8	20.2	21.4	22.6
0.7–0.8	47.5	19.4	25.1	29.0	26.7	29.7
0.8–0.9	35.2	8.6	15.2	25.5	9.5	20.0
0.9–1	0.1	0.0	0.0	0.3	0.0	0.0

MSI						
Category	**2016**	**2017**	**2018**	**2019**	**2020**	**2021**
0.3–0.5	3.3	1.7	2.0	3.4	1.8	3.8
0.5–0.7	60.5	16.3	19.0	23.3	17.3	23.9
0.7–0.9	29.7	17.0	23.5	29.3	20.6	28.7
0.9–1.1	5.1	17.3	25.2	25.7	24.0	27.9
1.1–1.3	1.2	19.3	22.7	15.9	26.0	14.2
1.3–1.5	0.1	18.8	7.4	2.4	9.6	1.5
1.5–1.7	0.0	8.3	0.3	0.0	0.8	0.0
1.7–1.9	0.0	1.2	0.0	0.0	0.0	0.0
1.9–2.1	0.0	0.1	0.0	0.0	0.0	0.0
2.1–2.3	0.0	0.0	0.0	0.0	0.0	0.0

DI						
Category	**2016**	**2017**	**2018**	**2019**	**2020**	**2021**
<0	52.1	36.4	36.6	34.9	35.9	35.9
0–1	28.8	12.3	16.2	20.6	16.9	21.8
1–2	12.1	16.2	19.4	21.9	25.4	30.0
2–3	4.0	16.9	17.7	15.9	19.4	11.7
3–4	2.0	12.3	8.6	6.0	2.4	0.6
4–5	0.7	4.9	1.5	0.6	0.0	0.0
5–6	0.2	0.9	0.0	0.0	0.0	0.0
6–6.5	0.0	0.0	0.0	0.0	0.0	0.0
>6.5	0.0	0.0	0.0	0.0	0.0	0.0

NDGI						
Category	**2015–2016**	**2016–2017**	**2017–2018**	**2018–2019**	**2019–2020**	**2020–2021**
−1−−0.8	7.9	49.9	5.5	2.8	6.1	1.9
−0.8−−0.6	2.0	4.9	1.3	0.8	2.0	0.7
−0.6−−0.4	4.4	5.3	2.7	1.7	3.1	1.3
−0.4−−0.2	12.7	6.3	6.6	6.7	5.7	4.1
−0.2–0	29.7	8.3	24.8	23.5	12.5	20.5
0–0.2	26.6	10.8	22.2	32.3	26.1	29.2
0.2–0.4	8.8	7.8	11.0	13.1	18.3	14.1
0.4–0.6	4.2	3.9	6.3	5.7	9.1	7.8
0.6–0.8	2.4	2.0	4.1	3.2	4.4	4.5
0.8–1	1.2	0.8	15.5	10.2	12.6	15.8

NDWNI						
Category	**2015–2016**	**2016–2017**	**2017–2018**	**2018–2019**	**2019–2020**	**2020–2021**
−1−−0.8	20.4	14.8	5.8	7.8	3.9	6.3
−0.8−−0.6	5.6	6.3	2.1	2.3	1.7	2.0
−0.6−−0.4	8.4	7.2	4.8	4.4	3.8	3.2
−0.4−−0.2	13.7	8.4	12.1	10.5	12.3	7.9
−0.2–0	22.6	10.5	22.9	34.1	32.8	28.5
0–0.2	23.2	9.5	32.8	34.8	27.0	36.2
0.2–0.4	4.9	8.7	11.6	4.4	7.7	8.8
0.4–0.6	0.6	8.2	2.8	0.6	3.4	2.4
0.6–0.8	0.2	5.7	1.2	0.3	1.9	1.3
0.8–1	0.5	20.7	4.0	0.7	5.5	3.5

Appendix C

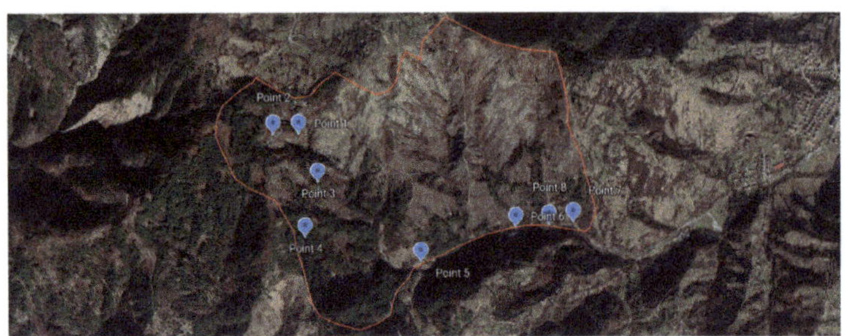

Figure A1. Point locations of field observations.

Table A3. Spectral indices values for each location. Interactive three-dimensional panoramas from Google Street view were used to generate photographs.

Index	Value	Point Location of Field Observation	Photography
NDVI	0.47		
MCARI2	0.57		
MSI	0.98	Point 1	
DI	1.27	Karst area	
NDGI	−0.53		
NDWNI	−0.09		
NDVI	0.61		
MCARI2	0.74	Point 1	
MSI	0.8	Meadows amongst coniferous	
DI	1.75	forests	
NDGI	1		
NDWNI	0.02		
NDVI	0.46		
MCARI2	0.58		
MSI	1	Point 2	
DI	0.75		
NDGI	0.29		
NDWNI	−0.01		
NDVI	0.61		
MCARI2	0.75		
MSI	0.61	Point 4	
DI	−2.52	Mixed forests	
NDGI	−0.17		
NDWNI	0.16		
NDVI	0.64		
MCARI2	0.79		
MSI	0.84	Point 5	
DI	1.06	Mixed forests	
NDGI	1		
NDWNI	0		
NDVI	0.44		
MCARI2	0.57		
MSI	1.11	Point 5	
DI	1.05	Transitional woodlands and	
NDGI	0.33	shrubs	
NDWNI	0.17		

Table A3. *Cont.*

Index	Value	Point Location of Field Observation	Photography
NDVI	0.53		
MCARI2	0.67		
MSI	1.05	Point 6	
DI	0.64	Coniferous forests	
NDGI	0.5		
NDWNI	0.47		
NDVI	0.6		
MCARI2	0.74		
MSI	0.71	Point 7	
DI	−0.46		
NDGI	0.5		
NDWNI	0.47		
NDVI	0.32		
MCARI2	0.41		
MSI	1.41	Point 8	
DI	2.91	Karst area	
NDGI	0.07		
NDWNI	0.03		

References

1. Seidl, R.; Thom, D.; Kautz, M.; Martin-Benito, D.; Peltoniemi, M.; Vacchiano, G.; Wild, J.; Ascoli, D.; Petr, M.; Honkaniemi, J.; et al. Forest disturbances under climate change. *Nat. Clim. Chang.* **2017**, *7*, 395–402. [CrossRef] [PubMed]
2. Collins, M.; Knutti, R.; Arblaster, J.; Dufresne, J.-L.; Fichefet, T.; Friedlingstein, P.; Gao, X.; Gutowski, W.J.; Johns, T.; Krinner, G.; et al. Long-term Climate Change: Projections, Commitments and Irreversibility. In *Climate Change 2013: The Physical Science Basis. Contribution of Working Group I to the Fifth Assessment Report of the Intergovernmental Panel on Climate Change*; Stocker, T.F., Qin, D., Plattner, G.-K., Tignor, M., Allen, S.K., Boschung, J., Nauels, A., Xia, Y., Bex, V., Midgley, P.M., Eds.; Cambridge University Press: Cambridge, UK; New York, NY, USA, 2013; pp. 1029–1136.
3. Wu, Z.; Li, M.; Wang, B.; Tian, Y.; Quan, Y.; Liu, J. Analysis of Factors Related to Forest Fires in Different Forest Ecosystems in China. *Forests* **2022**, *13*, 1021. [CrossRef]
4. IPCC. *Climate Change 2013: The Physical Science Basis. Contribution of Working Group I to the Fifth Assessment Report of the Intergovernmental Panel on Climate Change*; Stocker, T.F., Qin, D., Plattner, G.-K., Tignor, M., Allen, S.K., Boschung, J., Nauels, A., Xia, Y., Bex, V., Midgley, P.M., Eds.; Cambridge University Press: Cambridge, UK; New York, NY, USA, 2013; p. 1535.
5. Flannigan, M.; Amiro, B.D.; Logan, K.A.; Stocks, B.J.; Wotton, B.M. Forest fires and climate change in the 21st century. *Mitig. Adapt. Strateg. Glob. Chang.* **2005**, *11*, 847–859. [CrossRef]
6. Pausas, J.G.; Llovet, J.; Rodrigo, A.; Vallejo, R. Are wildfires a disaster in the Mediterranean basin?—A review. *Int. J. Wildland Fire* **2008**, *17*, 713–723. [CrossRef]
7. Gelabert, P.J.; Montealegre, A.L.; Lamelas, M.T.; Domingo, D. Forest structural diversity characterization in Mediterranean landscapes affected by fires using Airborne Laser Scanning data. *GIScience Remote Sens.* **2020**, *57*, 497–509. [CrossRef]
8. Moreira, F.; Ascoli, D.; Safford, H.; Adams, M.A.; Moreno, J.M.; Pereira, J.M.; Catry, F.X.; Armesto, J ; Bond, W.; González, M.E.; et al. Wildfire management in Mediterranean-type regions: Paradigm change needed. *Environ. Res. Lett.* **2020**, *15*, 011001. [CrossRef]
9. Velizarova, E.; Nedkov, R.; Avetisyan, D.; Radeva, K.; Stoyanov, A.; Georgiev, N.; Gigova, I. Application of remote sensing techniques for monitoring of the climatic parameters in forest fire vulnerable regions in Bulgaria. In Proceedings of the Seventh International Conference on Remote Sensing and Geoinformation of the Environment (RSCy2019), Paphos, Cyprus, 18–21 March 2019. [CrossRef]
10. Chuvieco, E.; Congalton, R.G. Application of remote sensing and geographic information systems to forest fire hazard mapping. *Remote Sens. Environ.* **1989**, *29*, 147–159. [CrossRef]
11. Arroyo, L.A.; Pascual, C.; Manzanera, J.A. Fire models and methods to map fuel types: The role of remote sensing. *For. Ecol. Manag.* **2008**, *256*, 1239–1252. [CrossRef]
12. Chu, T.; Guo, X. Remote Sensing Techniques in Monitoring Post-Fire Effects and Patterns of Forest Recovery in Boreal Forest Regions: A Review. *Remote Sens.* **2014**, *6*, 470–520. [CrossRef]
13. Stankova, N.; Nedkov, R. Research model of monitoring the recovery of an ecosystem after fire based on satellite and GPS data. *Ecol. Eng. Environ. Prot.* **2016**, *1*, 5–11.
14. Szpakowski, D.M.; Jensen, J.L.R. A Review of the Applications of Remote Sensing in Fire Ecology. *Remote Sens.* **2019**, *11*, 2638. [CrossRef]
15. Barrett, K.; Baxter, R.; Kukavskaya, E.; Balztera, H.; Shvetsov, E.; Buryak, L. Postfire recruitment failure in Scots pine forests of southern Siberia. *Remote Sens. Environ.* **2020**, *237*, 111539. [CrossRef]

16. Lu, D.; Weng, Q. A survey of image classification methods and techniques for improving classification performance. *Int. J. Remote Sens.* **2007**, *28*, 823–870. [CrossRef]

17. Corona, P.; Lamonaca, A.; Chirici, G. Remote sensing support for post fire forest management. *Iforest—Biogeosciences For.* **2008**, *1*, 6–12. [CrossRef]

18. Lhermitte, S.; Verbesselt, J.; Verstraeten, W.W.; Veraverbeke, S.; Coppin, P. Assessing intra-annual vegetation regrowth after fire using the pixel based regeneration index. *J. Photogramm. Remote Sens.* **2011**, *66*, 17–27. [CrossRef]

19. Velizarova, E.; Radeva, K.; Stoyanov, A.; Georgiev, N.; Gigova, I. Post-fire forest disturbance monitoring using remote sensing data and spectral indices. In Proceedings of the SPIE 11174, Seventh International Conference on Remote Sensing and Geoinformation of the Environment (RSCy2019), 111741G, Paphos, Cyprus, 27 June 2019. [CrossRef]

20. Moran, C.J.; Seielstad, C.A.; Cunningham, M.R.; Hoff, V.; Parsons, R.A.; Queen, L.; Sauerbrey, K.; Wallace, T. Deriving Fire Behavior Metrics from UAS Imagery. *Fire* **2019**, *2*, 36. [CrossRef]

21. Bannari, A.; Morin, D.; Bonn, F.; Huete, A.R. A review of vegetation indices. *Remote Sens. Rev.* **1995**, *13*, 95–120. [CrossRef]

22. Meng, R.; Denninson, P.E.; Huang, C.; Moritz, M.A.; D'Antonio, C. Effects of fire severity and post-fire climate on short term vegetation recovery of mixed-conifer and red fir forests in the Sierra Nevada Mountains of California. *Remote Sens. Environ.* **2015**, *171*, 311–325. [CrossRef]

23. Moser, G.; Serpico, S.B.; Benediktsson, J.A. Land-Cover Mapping by Markov Modeling of Spatio-contextual Information in Very-High-Resolution Remote Sensing Images. *Proc. IEEE* **2013**, *101*, 631–651. [CrossRef]

24. Salih, A.A.M.; Ganawa, E.-T.; Elmahl, A.A. Spectral mixture analysis (SMA) and change vector analysis (CVA) methods for monitoring and mapping land degradation/desertification in arid and semiarid areas (Sudan), using Landsat imagery. *Egypt. J. Remote Sens. Space Sci.* **2017**, *20*, S21–S29. [CrossRef]

25. Corona, P.; Saracino, A.; Leone, V. Plot size and shape for the early assessment of post-fire regeneration in Aleppo pine stands. *New For.* **1998**, *16*, 213–220. [CrossRef]

26. Gitas, I.; Mitri, G.; Veraverbeke, S.; Polychronaki, A. Advances in remote sensing of post-fire vegetation recovery monitoring—A review. In *Remote Sensing of Biomass—Principles and Applications*; Fatoyinbo, L., Ed.; InTech: London, UK, 2012; ISBN 978-953-51-0313-4.

27. Chirici, G.; Corona, P. An overview of passive remote sensing for post-fire monitoring. *Forest* **2005**, *2*, 282–289.

28. Nedkov, R.; Velizarova, E.; Molla, I.; Radeva, K. Application of remote sensing data for forest fires severity assessment. In *Proceedings of SPIE, Earth Resources and Environmental Remote Sensing/GIS Applications IX*; SPIE: Washington, DC, USA, 9 October 2018; Volume 10790. [CrossRef]

29. Picos, J.; Alonso, L.; Bastos, G.; Armesto, J. Event-based integrated assessment of environmental variables and wildfire severity through Sentinel-2 data. *Forests* **2019**, *10*, 1021. [CrossRef]

30. Miller, J.D.; Thode, A.E. Quantifying burn severity in a heterogeneous landscape with a relative version of the delta Normalized Burn Ratio (dNBR). *Remote Sensing of Environment* **2007**, *109*, 66–80. [CrossRef]

31. Lopez Garcia, M.J.; Caselles, V. Mapping burns and natural reforestation using Thematic Mapper data. *Geocarto International* **1991**, *6*, 31–37. [CrossRef]

32. Veraverbeke, S.; Gitas, I.; Katagis, T.; Polychronaki, A.; Somers, B.; Goossens, R. Assessing post-fire vegetation recovery using red–near infrared vegetation indices: Accounting for background and vegetation variability. *ISPRS J. Photogramm. Remote Sens.* **2012**, *68*, 28–39. [CrossRef]

33. Xofis, P.; Buckley, P.G.; Takos, I.; Mitchley, J. Long Term Post-Fire Vegetation Dynamics in North-East Mediterranean Ecosystems. *The Case of Mount Athos Greece. Fire* **2021**, *4*, 92. [CrossRef]

34. Koutsias, N.; Karamitsou, A.; Nioti, F.; Coutelieris, F. Assessment of Fire Regimes and Post-Fire Evolution of Burned Areas with the Dynamic Time Warping Method on Time Series of Satellite Images—Setting the Methodological Framework in the Peloponnese, Greece. *Remote Sens.* **2022**, *14*, 5237. [CrossRef]

35. Goetz, S.J.; Fiske, G.J.; Bunn, A.G. Using satellite time-series data sets to analyze fire disturbance and forest recovery across Canada. *Remote Sens. Environ.* **2006**, *101*, 352–365. [CrossRef]

36. Chu, T.; Guo, X.; Takeda, K. Remote sensing approach to detect post-fire vegetation regrowth in Siberian boreal larch forest. *Ecol. Indic.* **2016**, *62*, 32–46. [CrossRef]

37. Nedkov, R.; Velizarova, E.; Avetisyan, D.; Georgiev, N. Assessment of forest vegetation state through remote sensing in response to fire impact. In Proceedings of the SPIE 11524, Eighth International Conference on Remote Sensing and Geoinformation of the Environment (RSCy2020), 11524, Society of Photo-Optical Instrumentation Engineers (SPIE), Paphos, Cyprus, 16–18 March 2020. [CrossRef]

38. Nedkov, R. Normalized differential greenness index for vegetation dynamics assessment. *Comptes Rendus L'acad'emie Bulg. Des Sci.* **2017**, *70*, 1143–1146.

39. Rock, B.N.; Vogelmann, J.E.; Williams, D.L.; Vogehnann, A.F.; Hoshizaki, T. Remote detection of forest damage. *Bioscience* **1986**, *36*, 439–445. [CrossRef]

40. Zhao, X.; Chunxiang, C.; Xiliang, N.; Wei, C. Retrieval and application of leaf area index over China using HJ-1 data. *Geomat. Nat. Hazards Risk* **2017**, *8*, 478–495. [CrossRef]

41. Díaz-Delgado, R.; Lloret, F.; Pons, X. Influence of fire severity on plant regeneration by means of remote sensing imagery. *Int. J. Remote Sens.* **2003**, *24*, 1751–1763. [CrossRef]

42. Chen, X.; Chen, W.; Xu, M. Remote-Sensing Monitoring of Postfire Vegetation Dynamics in the Greater Hinggan Mountain Range Based on Long Time-Series Data: Analysis of the Effects of Six Topographic and Climatic Factors. *Remote Sens.* **2022**, *14*, 2958. [CrossRef]

43. French, N.H.F.; Kasischke, E.S.; Hall, R.J.; Murphy, K.A.; Verbyla, D.L.; Hoy, E.E.; Allen, J.L. Using Landsat Data to Assess Fire and Burn Severity in the North American Boreal Forest Region: An Overview and Summary of Results. *Int. J. Wild. Fire* **2008**, *17*, 443–462. [CrossRef]

44. Fornacca, D.; Ren, G.; Xiao, W. Evaluating the Best Spectral Indices for the Detection of Burn Scars at Several Post-Fire Dates in a Mountainous Region of Northwest Yunnan, China. *Remote Sens.* **2018**, *10*, 1196. [CrossRef]

45. Picotte, J.J.; Robertson, K. Timing constraints on remote sensing of wildland fire burned area in the southeastern US. *Remote Sens.* **2011**, *3*, 1680–1690. [CrossRef]

46. Healey, S.; Cohen, W.; Yang, Z.; Krankina, O. Comparison of Tasseled Cap-based Landsat data structures for use in forest disturbance detection. *Remote Sens. Environ.* **2005**, *97*, 301–310. [CrossRef]

47. Stankova, N.; Nedkov, R.; Ivanova, I.; Avetisyan, D. Modeling of forest ecosystems recovery after fire based on orthogonalization of multispectral satellite data. In Proceedings of the SPIE 10790, Earth Resources and Environmental Remote Sensing/GIS Applications IX, 10790, SPIE, Berlin, Germany, 10–13 September 2018. [CrossRef]

48. Thayn, J.B. Using a remotely sensed optimized Disturbance Index to detect insect defoliation in the Apostle Islands, Wisconsin, USA. *Remote Sens. Environ.* **2013**, *136*, 210–217. [CrossRef]

49. Kauth, R.; Thomas, G. The Tasseled Cap—A graphic description of the spectral—Temporal development of agricultural crops as seen by Landsat. In Proceedings of the Second Annual Symposium on Machine Processing of Remotely Sensed Data, West Lafayette, Indiana, 29 June–1 July 1976.

50. Crist, E.; Cicone, R. A physicaly-based transformation of Thematic Mapper data—The TM Tasseled Cap. *IEEE Trans. Geosci. Remote Sens.* **1984**, *22*, 256–263. [CrossRef]

51. Crist, E.; Kauth, R. The Tasseled Cap de-mystified. *Photogramm. Eng. Remote Sens.* **1986**, *52*, 81–86.

52. Климатични данни България. Available online: https://www.stringmeteo.com/synop/bg_climate.php (accessed on 19 November 2022).

53. Изменение на климата Ардино. Available online: https://www.meteoblue.com/bg/climate-change/%D0%90%D1%80%D0%B4%D0%B8%D0%BD%D0%BE_%D0%91%D1%8A%D0%BB%D0%B3%D0%B0%D1%80%D0%B8%D1%8F (accessed on 19 November 2022).

54. Velchev, A.; Penin, R.; Todorov, N.; Konteva, M. *Landscape Geography of Bulgaria*, 1st ed.; Bulvest 2000: Sofia, Bulgaria, 2011; pp. 213–235.

55. LUCAS 2015 TOPSOIL Data—ESDAC—European Commission. Available online: https://esdac.jrc.ec.europa.eu/content/lucas2015-topsoil-data (accessed on 19 November 2022).

56. "Streetview," Digital Images, Google Maps Photograph of Ardino Region, Taken 2021. Available online: https://www.google.com/streetview/explore/ (accessed on 6 December 2022).

57. Open Access Hub. Available online: https://scihub.copernicus.eu/ (accessed on 19 November 2022).

58. Forests—Copernicus Land Monitoring Service. Available online: https://land.copernicus.eu/pan-european/high-resolution-layers/forests (accessed on 19 November 2022).

59. Tepeliev, Y.; Koleva, R.; Dimitrov, V. Verification of Forest High Resolution Layers 2015: Tree Cover Density and Dominant Leaf Type in Bulgaria. *For. Ideas* **2021**, *27*, 343–353.

60. EU-DEM v1.1—Copernicus Land Monitoring Service. Available online: https://land.copernicus.eu/imagery-in-situ/eu-dem/eu-dem-v1.1 (accessed on 19 November 2022).

61. Rouse, W.; Haas, R.H.; Schell, J.A.; Deering, D.W. Monitoring vegetation systems in the Great Plains with ERTS. In Proceedings of the Third Earth Resources Technology Satellite-1 Symposium, Washington, DC, USA, 10–14 December 1973; pp. 301–317.

62. Haboudane, D.; Miller, J.R.; Pattey, E.; Zarco-Tejada, P.J.; Strachan, I.B. Hyperspectral Vegetation Indices and Novel Algorithms for Predicting Green LAI of Crop Canopies: Modeling and Validation in the Context of Precision Agriculture. *Remote Sens. Environ.* **2004**, *90*, 337–352. [CrossRef]

63. Hunt, E.R., Jr.; Rock, B.N. Detection of changes in leaf water content using Near- and Middle-Infrared reflectances. *Remote Sens Environ.* **1989**, *30*, 43–54.

64. Nedkov, R. Orthogonal Transformation of Segmented Images from the Satellite Sentinel-2. *Comptes Rendus De L'acad 'Emie Bulg. Des Sci.* **2017**, *70*, 687–692.

65. Avetisyan, D.; Nedkov, R.; Borisova, D.; Cvetanova, G. Application of spectral indices and spectral transformation methods for assessment of winter wheat state and functioning. In Proceedings of the SPIE 11149, Remote Sensing for Agriculture, Ecosystems, and Hydrology XXI, 1114929, Strasbourg, France, 21 October 2019. [CrossRef]

66. Malvar, M.C.; Prats, S.A.; Keizer, J.J. Runoff and inter-rill erosion affected by wildfire and pre-feire ploughing in eucalypt plantations of north-central Portugal. *Land Degrad. Dev.* **2016**, *27*, 1366–1378. [CrossRef]

67. Cartwright, J.M.; Littlefield, C.E.; Michalak, J.L.; Lawler, J.J.; Dobrowski, S.Z. Topographic, soil, and climate drivers of drought sensitivity in forests and shrublands of the Pacific Northwest, USA. *Sci. Rep.* **2020**, *10*, 18486. [CrossRef]

68. Avetisyan, D.; Borisova, D.; Velizarova, E. Integrated Evaluation of Vegetation Drought Stress through Satellite Remote Sensing. *Forests* **2021**, *12*, 974. [CrossRef]

69. Chen, P.; Liu, H.; Wang, Z.; Mao, D.; Liang, C.; Wen, L.; Li, Z.; Zhang, J.; Liu, D.; Zhuo, Y.; et al. Vegetation Dynamic Assessment by NDVI and Field Observations for Sustainability of China's Wulagai River Basin. *Int. J. Environ. Res. Public Health* **2021**, *18*, 2528. [CrossRef]
70. Wilson, A.M.; Latimer, A.M.; Silander, J.A. Climatic controls on ecosystem resilience: Postfire regeneration in the Cape Floristic Region of South Africa. *Proc. Natl. Acad. Sci. USA* **2015**, *112*, 9058–9063. [CrossRef] [PubMed]

remote sensing

Article

Comparison of Physical-Based Models to Measure Forest Resilience to Fire as a Function of Burn Severity

José Manuel Fernández-Guisuraga [1,2,*], Susana Suárez-Seoane [3,4], Carmen Quintano [5,6], Alfonso Fernández-Manso [7] and Leonor Calvo [1]

1 Department of Biodiversity and Environmental Management, Faculty of Biological and Environmental Sciences, University of León, 24071 León, Spain
2 Centro de Investigação e de Tecnologias Agroambientais e Biológicas, Universidade de Trás-os-Montes e Alto Douro, 5000-801 Vila Real, Portugal
3 Department of Organisms and Systems Biology (Ecology Unit), University of Oviedo, 33007 Oviedo, Spain
4 Research Institute of Biodiversity (IMIB, UO-CSIC-PA), University of Oviedo, 33600 Mieres, Spain
5 Electronic Technology Department, School of Industrial Engineering, University of Valladolid, 47011 Valladolid, Spain
6 Sustainable Forest Management Research Institute, University of Valladolid-Spanish National Institute for Agriculture and Food Research and Technology (INIA), 34004 Palencia, Spain
7 Agrarian Science and Engineering Department, School of Agricultural and Forestry Engineering, University of León, 24400 Ponferrada, Spain
* Correspondence: jofeg@unileon.es

Citation: Fernández-Guisuraga, J.M.; Suárez-Seoane, S.; Quintano, C.; Fernández-Manso, A.; Calvo, L. Comparison of Physical-Based Models to Measure Forest Resilience to Fire as a Function of Burn Severity. *Remote Sens.* 2022, 14, 5138. https://doi.org/10.3390/rs14205138

Academic Editor: Nikos Koutsias

Received: 16 September 2022
Accepted: 12 October 2022
Published: 14 October 2022

Abstract: We aimed to compare the potential of physical-based models (radiative transfer and pixel unmixing models) for evaluating the short-term resilience to fire of several shrubland communities as a function of their regenerative strategy and burn severity. The study site was located within the perimeter of a wildfire that occurred in summer 2017 in the northwestern Iberian Peninsula. A pre- and post-fire time series of Sentinel-2 satellite imagery was acquired to estimate fractional vegetation cover (FVC) from the (i) PROSAIL-D radiative transfer model inversion using the random forest algorithm, and (ii) multiple endmember spectral mixture analysis (MESMA). The FVC retrieval was validated throughout the time series by means of field data stratified by plant community type (i.e., regenerative strategy). The inversion of PROSAIL-D featured the highest overall fit for the entire time series ($R^2 > 0.75$), followed by MESMA ($R^2 > 0.64$). We estimated the resilience of shrubland communities in terms of FVC recovery using an impact-normalized resilience index and a linear model. High burn severity negatively influenced the short-term resilience of shrublands dominated by facultative seeder species. In contrast, shrublands dominated by resprouters reached pre-fire FVC values regardless of burn severity.

Keywords: fractional vegetation cover; MESMA; PROSAIL; recovery; Sentinel-2; wildfire

1. Introduction

The observed and projected increase in the severity of wildfires in the western Mediterranean Basin may hinder the resilience of plant communities to fire disturbance [1]. Wildfire severity, defined as the fire impact on the ecosystem, and operationally estimated from the amount of above- and belowground plant biomass consumed [2], is one of the key determinants of plant communities' recovery in the first post-fire periods [3]. Although field sampling methods are reliable and accurate for assessing plant community resilience to fire, its wide-scale application is limited in large-scale assessments due to its high time and economic cost [4]. In this sense, synoptic observation of the land surface using remote sensing-based techniques offers an effective way to achieve this objective.

Passive optical sensors of moderate spatial resolution, such as those onboard Landsat or Sentinel-2 satellite missions, offer the potential to accurately detect land cover changes and associated processes over extended time series [5,6]. The Landsat and Sentinel-2 multispectral imagery archive is potentially useful for assessing spatial patterns of land cover change such as post-fire forest dynamics due to the nominal 30 m/20 m spatial resolution, a consistent scale with that of most landscape-scale land cover changes [7]. In addition, their revisit period provides great coverage for multitemporal studies, especially in areas with frequent cloud cover that can deplete the availability of usable imagery [8].

To date, most studies on ecological resilience to fire have been based on the evaluation of plant community greenness recovery through spectral vegetation indices [9,10], among others. However, this product does not represent physical quantities [11], is affected by the background signal of burned areas [12], and suffers from reflectance saturation at high canopy cover in unburned areas [13]. Conversely, physical-based remote sensing techniques, such as pixel unmixing models and radiative transfer models (RTM), can be used to retrieve vegetation biophysical variables (e.g., fractional vegetation cover (FVC)) as fire resilience indicators [14].

Pixel unmixing models are based on the underlying premise that imagery pixels are constituted by several ground features that contribute to the surface reflectance captured by the optical sensor [15], being the pixel FVC, the physical fraction of vegetation cover in that pixel [16]. In this approach, FVC is directly retrieved from remote sensing data using spectral libraries or image endmembers, without initial field data needs [17]. However, endmember collection can be time-consuming in large, heterogeneous burned landscapes [11]. Among these techniques, multiple endmember spectral mixture analysis (MESMA; [18]) is an advanced method that allows the spectra of several endmembers to characterize each pixel ground component (e.g., the vegetation fraction; [19]). Remarkably, each pixel can be resolved independently with a differing number of endmembers to account for the associated terrain variability [17], for instance different vegetation species within the community. This contrasts with more conventional spectral mixture models, such as linear spectral mixture analysis, in which only one spectrum can be incorporated in each endmember [15].

RTM-based approaches simulate the physical relationships between vegetation biophysical variables, such as the FVC, and the plant community reflectance [20]. These models can be inverted using observed surface reflectance data captured by passive optical sensors for retrieving the FVC, to be implemented as a resilience indicator to fire. RTMs do not need to be parameterized with field data specific to the area of interest, which are only needed for validation of the retrieved biophysical variable [11]. Remarkably, RTM physical relationships are not site-specific, and vegetation recovery can then be monitored in burned landscapes that encompass several communities [21]. RTM inversion is usually performed using machine learning techniques, also known as hybrid inversion [22].

Although physical-based models have been used to monitor post-fire vegetation communities [11,23], to date there are no studies in the literature comparing their effectiveness to assess vegetation resilience to fire in plant communities affected by different severity levels. Therefore, the objective of this work was to compare the potentiality of two physical-based approaches (pixel unmixing and radiative transfer models) applied to passive optical data for evaluating vegetation resilience as a function of burn severity in shrubland communities with different regenerative traits. FVC was used as a resilience indicator retrieved from a time series of Sentinel-2 imagery, using hybrid RTM inversion and MESMA models.

2. Materials and Methods

2.1. Study Site Description

The study site lies within the perimeter of a wildfire that burned 9940 ha of shrub and forest plant communities in the summer of 2017 within Sierra de Cabrera (NW Spain; Figure 1). The site has a rugged topography, with prominent crests and steep slopes, and its altitude ranges from 836–1938 m. The climate is temperate Mediterranean, with a mean annual temperature of 9 °C and mean annual precipitation of 850 mm. The wildfire affected, among others, three types of shrub plant communities: shrub communities dominated by facultative seeder species (*Genista hystrix* Lange (gorse) and *Genista florida* L. (broom)) and resprouter species (*Erica australis* L. (heath)). The shrub plant communities of the site exhibit high spatial variability due to small-scale differences in post-fire recovery patterns and accumulation of burned debris as a consequence of the fire regime variability. Other plant communities affected by the fire include Pyrenean oak forests dominated by *Quercus pyrenaica* Willd, and Scots pine forests dominated by *Pinus sylvestris* L., as well as grasslands in the valley bottoms. The main land use in the site before the wildfire involved extensive livestock farming.

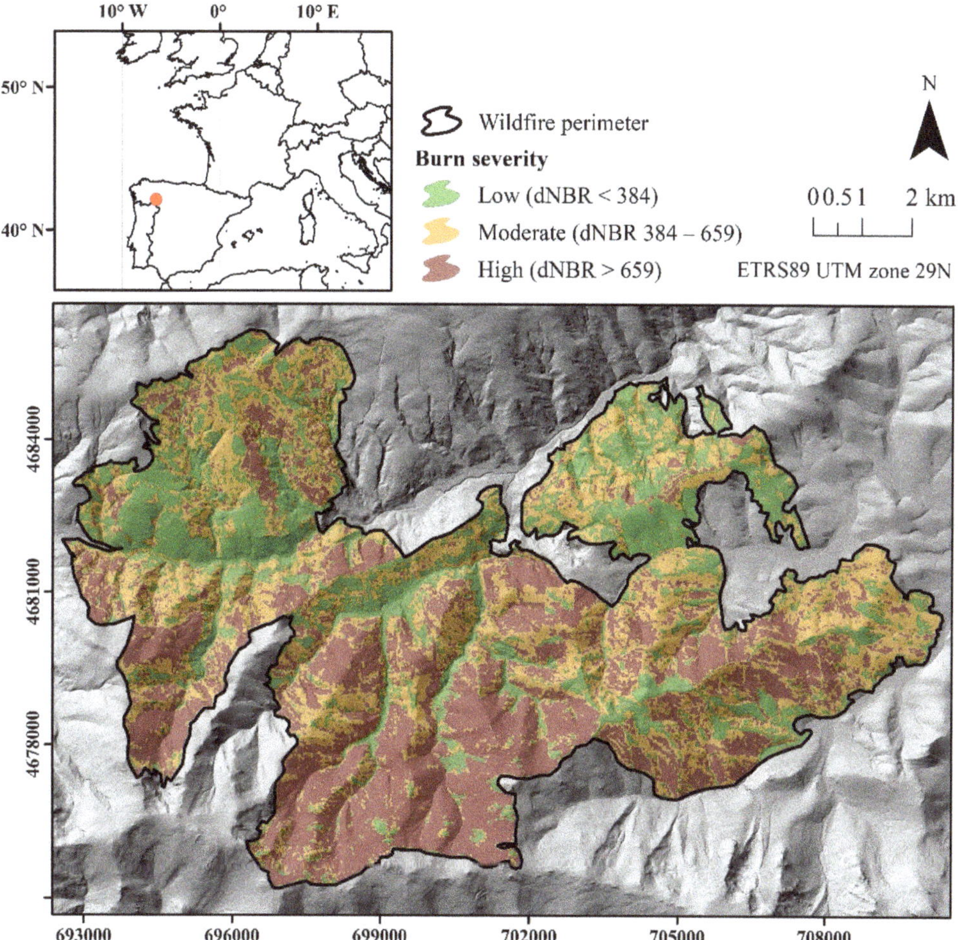

Figure 1. Wildfire perimeter and burn severity estimation from the differenced Normalized Burn Ratio (dNBR) thresholds.

Facultative seeder shrub species are able to resprout from belowground organs and to germinate during post-fire conditions; however, in our study site, their resprouting success and vigor is lower than that of obligate resprouters. The latter species rely on resprouting strategy to regenerate after wildfire because they lack a fire-resistant aerial or soil seed bank.

2.2. Sentinel-2 Imagery, Processing and Burn Severity Calculation

Sentinel-2 multispectral mission comprises two satellites (Sentinel-2A and Sentinel-2B) as part of the Copernicus program. Sentinel-2 provides thirteen bands at different spatial resolutions in the visible, near infrared and short-wave infrared regions: four bands at 10 m, six bands at 20 m and three bands at 60 m [24]. Sentinel-2 bands at a spatial resolution of 10 m were resampled to 20 m using the nearest neighbor interpolation. The bands at 60 m were discarded, as they are used for atmospheric correction and cloud detection processes and are strongly affected by atmospheric effects [20]. Two Sentinel-2 MSI Level 1C scenes were calibrated to surface reflectance (level 2A) with the Atmospheric/Topographic Correction for Satellite Imagery algorithm version 3 (ATCOR-3; [25]) by correcting for topographic and atmospheric effects. Meteorological data from the State Meteorology Agency of Spain (AEMET), the MODIS water vapor product (MOD05), and a digital terrain model (DTM) were used to set appropriate ATCOR-3 input parameters. The scenes were acquired for immediate pre-fire (13 August 2017) and post-fire (2 September 2017) scenarios to estimate burn severity through the differenced Normalized Burn Ratio (dNBR) index [26]. In this study, we selected the dNBR because it was the spectral index most related to field-based burn severity in internal testing compared to relativized indices, as well as in previous research on the site [27]. In addition, the dNBR is the primary spectral index within the European Forest Fire Information System (EFFIS) and a reference approach for initial burn severity assessment [28], which may improve the comparability of the results of our study. The dNBR was validated through the composite burn index (CBI; [29]), measured in 72 field plots of 20 m × 20 m one month after wildfire (initial burn severity assessment). Burn severity was rated in the field between 0 (unburned) and 3 (high severity). Three field burn severity categories were established based on CBI values of each plot: low (CBI < 1.25), moderate ($1.25 \leq$ CBI ≤ 2.25) and high (CBI > 2.25). These widely accepted CBI thresholds in the literature correspond to those proposed by [30]. These CBI thresholds were used to define three burn severity categories based on a dNBR thresholding approach using a linear regression model: low (dNBR < 384), moderate ($384 \leq$ dNBR ≤ 659) and high (dNBR > 659) (Figure 1). The R^2 of the linear model was equal to 0.84.

2.3. Physical-Based Models

Pixel unmixing and radiative transfer models were used to retrieve the FVC from three Sentinel-2 MSI Level 2A scenes (processed from level 1C following the methodology described in Section 2.2) acquired during the biomass peak of the study site between 2017 and 2020: (i) 1-week pre-fire (13 August 2017 (equivalent to pre-fire dNBR calculation)), (ii) 2-weeks post-fire (2 September 2017 (equivalent to post-fire dNBR calculation)), and (iii) 3-years post-fire (18 July 2020).

2.3.1. Multiple Endmember Spectral Mixture Analysis (MESMA)

Candidate endmember spectra for MESMA models were extracted from Sentinel-2 scenes (image endmembers) rather than using spectral libraries (reference endmembers) because image endmembers are acquired at the same resolution of the imagery and are influenced by the same atmospheric imagery corrections [19,31]. The first post-fire Sentinel-2 image (2-weeks post-fire) was used to collect endmembers to ensure the presence of enough non-photosynthetic material corresponding to burned vegetation. We used 500 training polygons consisting of uniform ground patches encompassing a single vegetation, soil or charred material type. Polygon size was set to encompass at least four Sentinel-2 pixels. We ensured a separation between polygons of 100 m and a uniform distribution among plant community types [32]. Training areas for soil, green vegetation (*Genista hystrix, Genista florida* and *Erica australis*) and non-photosynthetic vegetation (charred debris) were delineated using very high spatial resolution orthophotographs at a spatial resolution of 0.5 m. The Iterative endmember selection (IES) algorithm [33] was used in this study to select optimal endmembers from the candidate set and improve MESMA computational efficiency and accuracy [19]. We clustered the selected endmembers by the IES algorithm into photosynthetic vegetation (PV), non-photosynthetic vegetation (NPV), and soil spectral libraries. Next, Sentinel-2 pre- and post-fire scenes were unmixed into PV, NPV, soil, and shade fraction images. The performance of all candidate MESMA models was evaluated using the following requirements for each pixel: (i) the range between minimum and maximum fraction image value was constrained between 0 and 1; (ii) shade fraction values lower than 0.8; and (iii) maximum allowable RMSE equal to 0.025 [23,34]. The constraint in fraction images was selected on the basis of the values physically attainable in the field [32], whereas the shade constraint was chosen to maintain a reasonably high fraction value for the other endmembers [34]. For its part, the RMSE constraint is a standard in the literature [18]. Finally, fraction images were shade-normalized, being the GV shade-normalized fraction of the FVC.

IES, MESMA and shade normalization were implemented in VIPER Tools 2.0, developed by the VIPER Lab at UC Santa Barbara [35].

2.3.2. Hybrid Radiative Transfer Model (RTM) Inversion

The PROSAIL-D RTM, resulting from the coupled PROSPECT-D leaf model [36] and the 4SAIL canopy reflectance model [37], was used to simulate shrubland plant canopy reflectance. PROSPECT-D simulates leaves hemispheric reflectance and transmittance from 400–2500 nm as a function of a number of physiological and biochemical variables at the leaf level. For its part, 4SAIL simulates plant canopy reflectance based on PROSPECT-D simulations, as well as a series of variables related to canopy structure, lighting and viewing conditions [38]. The values of PROSPECT-D and 4SAIL input variables (Table 1) were obtained from satellite imagery metadata, in-depth literature review and field knowledge, considering the variability of the biophysical conditions of the shrubland communities in the study site. Then, PROSAIL-D simulations were executed in direct mode to obtain a dataset of simulated shrubland canopy reflectance and its corresponding FVC, calculated from the viewing conditions, leaf area index and leaf angle of each simulation [11]. The dataset was spectrally resampled to the Sentinel-2 band configuration using its spectral bandwidth and response function. We updated the dataset with 10% of bare soil and non-photosynthetic vegetation spectra with respect to the total simulations [39]. This amount may be representative in relation to the high variability of the vegetation biophysical conditions in highly heterogeneous plant communities [11].

Table 1. Value or range of input variables for PROSPECT-D and 4SAIL models.

Leaf Model (PROSPECT-D)	Unit	Value or Range
Structure index	unitless	1.5–2.5
Chlorophyll content	$\mu g\,cm^{-2}$	10–70
Dry matter content	$g\,cm^{-2}$	0.005–0.015
Water content	$g\,cm^{-2}$	0.005–0.015
Carotenoid content	$\mu g\,cm^{-2}$	5–40
Anthocyanin content	$\mu g\,cm^{-2}$	0–60
Brown pigment fraction	unitless	0–1
Canopy model (4SAIL)	**Unit**	**Value or range**
Leaf area index	$m^2\,m^{-2}$	0.1–3
Average leaf angle	degrees	20–90
Diffuse/direct radiation	unitless	0.1
Hot spot effect	unitless	0.001–1
Soil brightness factor	unitless	0–1
Fraction of vegetation cover	unitless	0–1
Solar zenith angle	degrees	Imagery metadata
Observation zenith angle	degrees	Imagery metadata
Sun-sensor azimuth angle	degrees	Imagery metadata

The Random Forest (RF; [40]) regression algorithm was used to model the relationship between the simulated Sentinel-2 reflectance at the shrubland canopy level and the corresponding FVC for the dataset generated by PROSAIL-D. In this study, the ntree parameter was set to 500 and the mtry parameter to one third of the number of Sentinel-2 bands, which are the default values. The calibrated RF model was then applied to the reflectance observed in the Sentinel-2 imagery for obtaining spatially explicit FVC predictions for each of the pixels in the pre- and post-fire time series imagery.

PROSAIL-D parameterization and execution in direct mode, as well as FVC retrieval through RF algorithm were performed in ARTMO [41].

2.4. FVC Retrieval Validation

In September 2017, we established 60 field plots of 20 m × 20 m in burned areas to evaluate the post-fire FVC retrieval performance through MESMA and PROSAIL-D along the post-fire time series. Likewise, 20 control plots of 20 m × 20 m were established in unburned areas to evaluate pre-fire FVC retrieval. This number of field plots has been shown to be representative of the plant communities' variability in the study site [11]. Plots were stratified by reproductive vegetation strategy (resprouters and facultative seeders). The center of each plot was georeferenced using a sub-meter accuracy GPS receiver in post-processing mode. Both control and burned plots were sampled in September 2017, with burned plots also being surveyed in July 2020. FVC was estimated in each plot as the area of the vertical projection occupied by each community stratum (i.e., herbaceous and shrub strata) in the shrubland communities, using a visual estimation method [42]. The coefficient of determination (R^2) was calculated to evaluate the retrieval performance of the FVC through MESMA and PROSAIL-D over the time series.

2.5. Data Analyses

From the spatially explicit FVC prediction maps for the time series with higher overall accuracy (i.e., PROSAIL-D RTM or MESMA pixel unmixing model), we performed a stratified random sampling of 1000 points, using reproductive vegetation strategy and burn severity categories as strata. A minimum distance of 100 m between points was ensured. For each point, the FVC value was extracted for the considered time series. We computed an impact-normalized resilience index (R_{in}) [43] that represents the recovery of the system property, in this case the FVC, with respect to the impact of the disturbance on that property:

$$R_{in} = (P_{tx} - P_{ti})/(C_0 - P_{ti})$$

where P_{tx} is the value of the system property at the time point when the resilience is evaluated after the disturbance, P_{ti} is the value of the property immediately after the disturbance, and C_0 is its control value. A value of the R_{in} index equal to 1 denotes a full recovery of the property at the considered time point. A linear regression model and subsequent Tukey's HSD test were used to evaluate the effect of regenerative vegetation strategy and burn severity (independent variables), as well as their interaction, on vegetation resilience as measured by the R_{in} index (dependent variable). All statistical analyses were performed in R.4.0.5 [44].

3. Results

The overall accuracy of the RF algorithm trained with PROSAIL-D model simulations for retrieving FVC from Sentinel-2 imagery ($R^2 = 0.75$–0.79) was substantially higher than that achieved from MESMA models ($R^2 = 0.64$–0.73), both in the immediate pre- and post-fire scenarios, as well as three years after the wildfire (Figure 2). The accuracy of the FVC retrieval for both PROSAIL-D and MESMA models was higher in the pre-fire and long-term post-fire scenarios ($R^2 > 0.69$) than in the immediate post-fire situation ($R^2 > 0.64$). The FVC estimation for the two shrubland communities considered, dominated by facultative seeder species and by resprouter species, showed no significant under- or overestimation effects over the entire range of field-measured FVC, although the estimates for the resprouters were closer tailored to the 1:1 line (Figure 2).

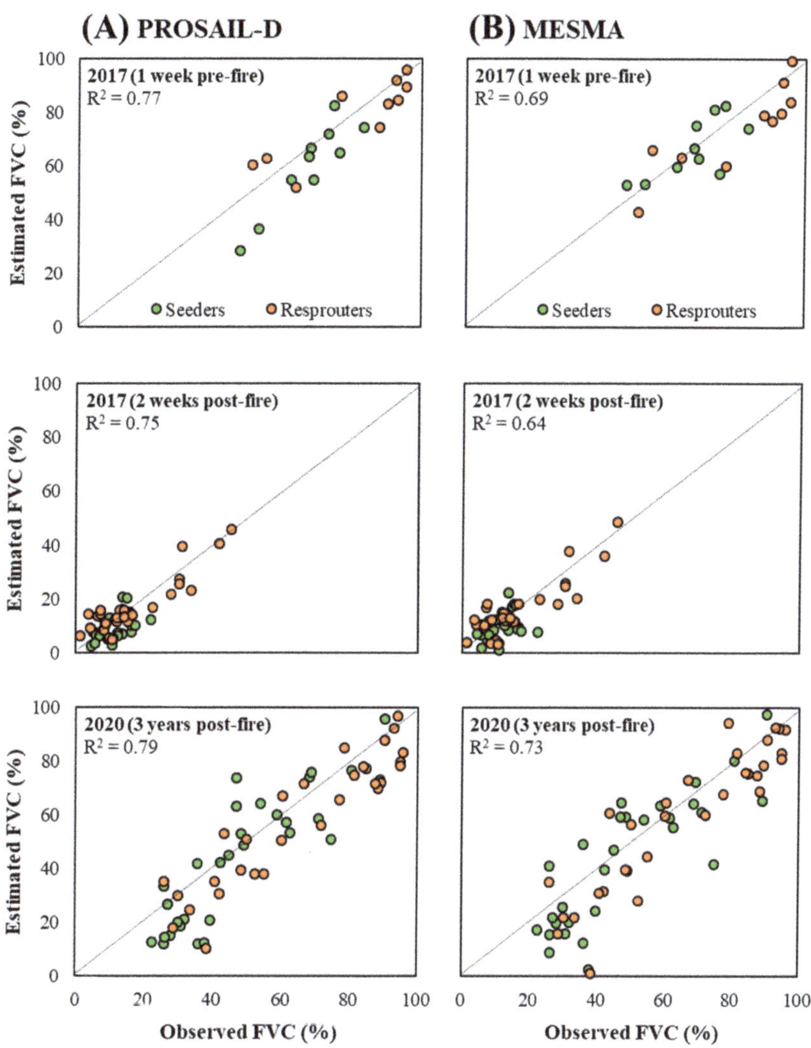

Figure 2. Relationship between the FVC measured in the field and that retrieved from Sentinel-2 imagery for the time series using PROSAIL-D (**A**) and MESMA (**B**) models.

From the PROSAIL-D spatially explicit FVC estimates, we evidenced a significant effect of vegetation reproductive strategy (p-value < 0.05) and burn severity (p-value < 0.001), as well as their interaction (p-value < 0.05), on the resilience of shrubland communities in the study site (Figure 3). High burn severity hindered the short-term resilience of shrubland communities dominated by facultative seeder species, with no FVC recovery to a pre-disturbance state being observed 3 years after wildfire (Figure 3). In contrast, communities dominated by shrub resprouter species reached pre-fire FVC values 3 years after the fire disturbance, regardless of the burn severity scenario (Figure 3). In this sense, resprouter shrubland communities featured a faster recovery rate in high burn severity scenarios than those communities dominated by facultative seeders.

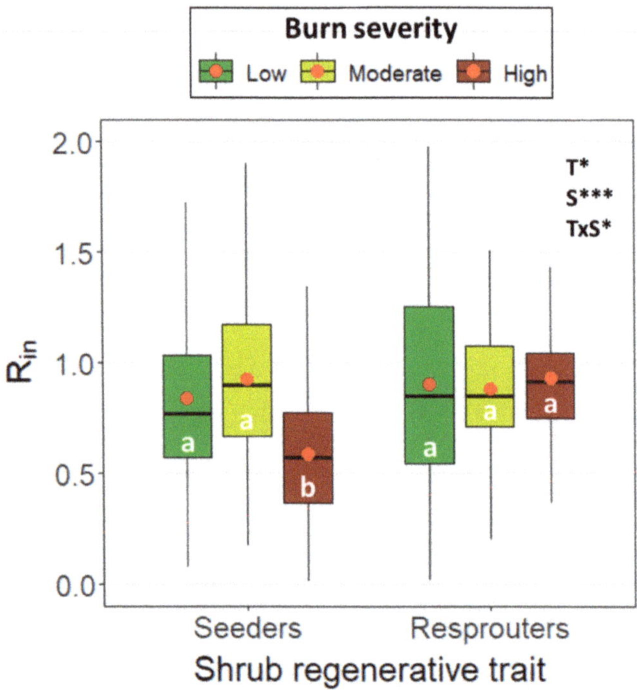

Figure 3. Boxplot showing the relationship between the impact-normalized resilience index (R_{in}) and burn severity in shrubland plant communities dominated by facultative seeders species and resprouter species. Significance of R_{in} predictors (regenerative trait: T; burn severity: S; T $\times$ S interaction) is represented as * (p-value < 0.05), ** (p-value < 0.01), and *** (p-value < 0.001). Lowercase letters denote significant differences in R_{in} at the 0.05 level between burn severity levels within each plant community.

4. Discussion

The assessment of post-fire recovery trajectories through remote sensing-based techniques is essential for (i) understanding current fire regimes in fire-prone ecosystems of the western Mediterranean Basin [3,45], (ii) providing new insights on the resilience of plant communities at several spatial scales [23,46], and (iii) supporting adaptive management strategies aimed at maintaining ecosystem functions and services endangered by changing fire regimes [47–49].

The present study demonstrated the potentiality of physical-based remote sensing approaches (i.e., radiative transfer and pixel unmixing models) for estimating FVC as an indicator of resilience to fire in several pre- and postfire scenarios of shrubland communities with different vegetation responses. Considerably accurate FVC estimates were achieved with the PROSAIL-D RTM and MESMA approaches considering (i) the complexity of biophysical parameters retrieval in shrubland communities because of the high background signal of non-photosynthetic material and bare soil exposed to the remote sensor, and (ii) the complex mixture of shrub species [11,50].

In any case, we found that hybrid inversion of the PROSAIL-D RTM using the RF algorithm outperformed MESMA models when retrieving FVC in heterogeneous shrubland plant communities. Although both approaches have a solid physical basis and feature an adequate capability to generalize the biophysical parameter retrievals [4], the delineation of spectrally pure endmembers with moderate spatial resolution imagery in pixel unmixing models can be challenging [51], especially in heterogeneous burned landscapes with fine-grained arrangement of vegetation and burned legacies. In addition, spatiotemporal changes in the vegetation's biophysical properties and background features when dealing with time series analysis may not be properly captured when delineating image endmembers in extensive burned landscapes [11]. In contrast, the PROSAIL-D model is parametrized with well-known ranges of model input variables for the plant communities under consideration, providing a strong characterization of the physical relationships between the simulated reflectance and site biophysical variability [11]. However, the use of site-specific field information to parametrize the RTM could provide higher accuracy in the retrieval of the biophysical variable of interest [52].

The higher accuracy of FVC retrieval for both PROSAIL-D and MESMA models in the pre-fire and long-term post-fire scenarios, with stronger vegetation responses than in the immediate post-fire situation, could be related to the increased influence of woody debris and bare soil on the surface reflectance at the first post-fire stages with limited canopy cover [50]. This behavior may also be related to FVC estimates closer to the 1:1 line in the case of shrublands dominated by resprouter species, which feature higher vegetation cover [53] and structural complexity [14] than the communities dominated by seeder species in the study site. In addition, a complex mixture of regenerating grass species, seedling recruitment and resprouting responses in early post-fire stages could be encompassed in a decametric Sentinel-2 pixel, increasing retrieval uncertainty [11]. Also, the non-photosynthetic material and bare soil spectra profile from expected pure pixels may not be accurately collected from decametric satellite imagery [22].

Shrubland communities dominated by both facultative seeder and resprouter species featured a high post-fire recovery ability, especially in low and moderate burn severity scenarios, in line with the results of previous field-based research [42,53–55]. However, under high burn severity scenarios, faster recovery rates were evidenced in shrubland communities dominated by resprouter species compared to those dominated by facultative seeders. In general, surviving resprouting structures confer a rapid recovery of plant biomass and a quick recolonization of the physical space occupied by the vegetation prior to the wildfire [54]. Likewise, the aerial and soil seed bank of shrub seeder species suffers considerable damage at high burn severity [56]. This behavior enabled shrubland communities dominated by resprouter species to achieve resilience in the short term after wildfire, in contrast to communities dominated by facultative seeders [57].

Our results are therefore in agreement with those obtained in previous research based exclusively on field data, which demonstrates the potential of physical-based remote sensing techniques, particularly the hybrid inversion of RTMs, to assess the resilience to fire of shrubland communities in the short term. However, both physical-based approaches (i.e., PROSAIL-D RTM and MESMA) featured several FVC retrieval uncertainties in heterogeneous fire-prone shrubland communities. First, PROSAIL-D is a turbid medium RTM, and, consequently, higher accuracies in the model inversion can be attained by using a geometric RTM to simulate canopy reflectance and transmittance in heterogeneous shrubland communities [21], but at the expense of a more complex model parameterization [58], usually with field data not available short-term after fire. Second, RTM approaches exhibit improved performance from retrieval using passive optical data at high spatial resolution, avoiding the land cover aggregation effect of mixed pixels [11]. This shortcoming would be partially solved by MESMA models, which better capture the ground spectra variability of mixed pixels. Nevertheless, non-linear mixing in sparse canopies, such as shrubland communities in the early post-fire periods, violates MESMA assumptions and has an impact on its performance [59]. However, the main limitation of PROSAIL-D RTM and MESMA lies in the impossibility to determine the post-fire recovery trajectories of specific vegetation types or growth forms within the community [31], both in terms of their composition and structure. In this sense, the fusion of remote sensing data from optical sensors processed through physical-based techniques, with that of active sensors such as synthetic aperture radar (SAR) or light detection and ranging (LiDAR), could provide valuable insights on the recovery trajectories of biophysical properties at the species level or by height strata [14,60,61].

5. Conclusions

The assessment of how shrubland communities recover from fire disturbance in fire-prone ecosystems is essential to providing new insights about the resilience of plant communities under changing fire regimes in the Mediterranean Basin. This study novelty compared the potential of radiative transfer and pixel unmixing models for evaluating the short-term resilience to fire of several shrubland communities. We found that the hybrid inversion of the PROSAIL-D RTM outperformed MESMA pixel unmixing models to retrieve FVC in heterogeneous shrubland communities. Adaptations to fire allowed shrub communities dominated by resprouter species to achieve resilience in the short-term period after wildfire, which is consistent with previous studies based exclusively on field data, and thus demonstrates the potential of physical-based remote sensing approaches in fire ecology research.

Author Contributions: Conceptualization, J.M.F.-G., S.S.-S., A.F.-M., and L.C.; formal analysis, J.M.F.-G., and C.Q.; funding acquisition, S.S.-S. and L.C.; investigation, L.C.; methodology, J.M.F.-G., C.Q., and A.F.-M.; writing—original draft, J.M.F.-G.; writing—review & editing, S.S.-S., C.Q., A.F.-M., and L.C. All authors have read and agreed to the published version of the manuscript.

Funding: This study was financially supported by the Spanish Ministry of Economy and Competitiveness and the European Regional Development Fund (ERDF), in the framework of the FIRESEVES (AGL2017-86075-C2-1-R) project; by the Regional Government of Castilla and León in the framework of the WUIFIRECYL (LE005P20) project; and by the British Ecological Society in the framework of the SR22-100154 project, where José Manuel Fernández-Guisuraga is the Principal Investigator. José Manuel Fernández-Guisuraga was supported by a Ramón Areces Foundation postdoctoral fellowship.

Data Availability Statement: The datasets analyzed for this study are available from the corresponding author on reasonable request.

Conflicts of Interest: The authors declare no conflict of interest.

References

1. Doblas-Miranda, E.; Alonso, R.; Arnan, X.; Bermejo, V.; Brotons, L.; de las Heras, J.; Estiarte, M.; Hódar, J.A.; Llorens, P.; Lloret, F.; et al. A review of the combination among global change factors in forests, shrublands and pastures of the Mediterranean Region: Beyond drought effects. *Glob. Planet. Change* **2017**, *148*, 42–54. [CrossRef]
2. Keeley, J.E. Fire intensity, fire severity and burn severity: A brief review and suggested usage. *Int. J. Wildland Fire* **2009**, *18*, 116–126. [CrossRef]
3. González-De Vega, S.; De las Heras, J.; Moya, D. Resilience of Mediterranean terrestrial ecosystems and fire severity in semiarid areas: Responses of Aleppo pine forests in the short, mid and long term. *Sci. Total Environ.* **2016**, *573*, 1171–1177. [CrossRef] [PubMed]
4. Fernández-Guisuraga, J.M.; Calvo, L.; Suárez-Seoane, S. Comparison of pixel unmixing models in the evaluation of post-fire forest resilience based on temporal series of satellite imagery at moderate and very high spatial resolution. *ISPRS J. Photogramm. Remote Sens.* **2020**, *164*, 217–228. [CrossRef]
5. Eva, H.D.; Carboni, S.; Achard, F.; Stach, N.; Durieux, L.; Faure, J.F.; Mollicone, D. Monitoring forest areas from continental to territorial levels using a sample of medium spatial resolution satellite imagery. *ISPRS Int. J. Photogramm. Remote Sens.* **2010**, *65*, 191–197. [CrossRef]
6. Nagendra, H.; Lucas, R.; Honrado, J.P.; Jongman, R.H.G.; Tarantino, C.; Adamo, M.; Mairota, P. Remote sensing for conservation monitoring: Assessing protected areas, habitat extent, habitat condition, species diversity, and threats. *Ecol. Indic.* **2013**, *33*, 45–59. [CrossRef]
7. Storey, E.A.; Stow, D.A.; O'Leary, J.F. Assessing postfire recovery of chamise chaparral using multi-temporal spectral vegetation index trajectories derived from Landsat imagery. *Remote Sens. Environ.* **2016**, *183*, 53–64. [CrossRef]
8. Stillinger, T.; Roberts, D.A.; Collar, N.M.; Dozier, J. Cloud masking for Landsat 8 and MODIS Terra over snow-covered terrain: Error analysis and spectral similarity between snow and cloud. *Water Resour. Res.* **2019**, *55*, 6169–6184. [CrossRef]

9. Viedma, O.; Melia, J.; Segarra, D.; Garcia-Haro, J. Modeling rates of ecosystem recovery after fires by using Landsat TM data. *Remote Sens. Environ.* **1997**, *61*, 383–398. [CrossRef]

10. Ireland, G.; Petropoulos, G.P. Exploring the relationships between post-fire vegetation regeneration dynamics, topography and burn severity: A case study from the Montane Cordillera Ecozones of Western Canada. *Appl. Geogr.* **2015**, *56*, 232–248. [CrossRef]

11. Fernández-Guisuraga, J.M.; Verrelst, J.; Calvo, L.; Suárez-Seoane, S. Hybrid inversion of radiative transfer models based on high spatial resolution satellite reflectance data improves fractional vegetation cover retrieval in heterogeneous ecological systems after fire. *Remote Sens. Environ.* **2021**, *255*, 112304. [CrossRef]

12. Vila, G.; Barbosa, P. Post-fire vegetation regrowth detection in the Deiva Marina region (Liguria-Italy) using Landsat TM and ETM+ data. *Ecol. Model.* **2010**, *221*, 75–84. [CrossRef]

13. Lu, D.; Chen, Q.; Wang, G.; Liu, L.; Li, G.; Moran, E. A survey of remote sensing-based aboveground biomass estimation methods in forest ecosystems. *Int. J. Digit. Earth* **2016**, *9*, 63–105. [CrossRef]

14. Fernández-Guisuraga, J.M.; Suárez-Seoane, S.; Calvo, L. Radar and multispectral remote sensing data accurately estimate vegetation vertical structure diversity as a fire resilience indicator. *Remote Sens. Ecol. Conserv.* **2022**, *in press*. [CrossRef]

15. Somers, B.; Asner, G.P.; Tits, L.; Coppin, P. Endmember variability in Spectral Mixture Analysis: A review. *Remote Sens. Environ.* **2011**, *115*, 1603–1616. [CrossRef]

16. Wang, X.; Jia, K.; Liang, S.; Li, Q.; Wei, X.; Yao, Y.; Zhang, X.; Tu, Y. Estimating Fractional Vegetation Cover from Landsat-7 ETM+ Reflectance Data Based on a Coupled Radiative Transfer and Crop Growth Model. *IEEE Trans. Geosci. Remote Sens.* **2017**, *55*, 5539–5546. [CrossRef]

17. Veraverbeke, S.; Somers, B.; Gitas, I.; Katagis, T.; Polychronaki, A.; Goossens, R. Spectral mixture analysis to assess post-fire vegetation regeneration using Landsat Thematic Mapper imagery: Accounting for soil brightness variation. *Int. J. Appl. Earth Obs. Geoinf.* **2012**, *14*, 1–11. [CrossRef]

18. Roberts, D.A.; Gardner, M.; Church, R.; Ustin, S.; Scheer, G.; Green, R.O. Mapping Chaparral in the Santa Monica Mountains Using Multiple Endmember Spectral Mixture Models. *Remote Sens. Environ.* **1998**, *65*, 267–279. [CrossRef]

19. Quintano, C.; Fernandez-Manso, A.; Roberts, D.A. Burn severity mapping from Landsat MESMA fraction images and Land Surface Temperature. *Remote Sens. Environ.* **2017**, *190*, 83–95. [CrossRef]

20. Jia, K.; Liang, S.; Gu, X.; Baret, F.; Wei, X.; Wang, X.; Yao, Y.; Yang, L.; Li, Y. Fractional vegetation cover estimation algorithm for Chinese GF-1 wide field view data. *Remote Sens. Environ.* **2016**, *177*, 184–191. [CrossRef]

21. Yebra, M.; Chuvieco, E.; Riaño, D. Estimation of live fuel moisture content from MODIS images for fire risk assessment. *Agric. For. Meteorol.* **2008**, *148*, 523–536. [CrossRef]

22. Verrelst, J.; Rivera, J.P.; Veroustraete, F.; Muñoz-Marí, J.; Clevers, J.G.P.W.; Camps-Valls, G.; Moreno, J. Experimental Sentinel-2 LAI estimation using parametric, non-parametric and physical retrieval methods—A comparison. *ISPRS J. Photogramm. Remote Sens.* **2015**, *108*, 260–272. [CrossRef]

23. Fernandez-Manso, A.; Quintano, C.; Roberts, D.A. Burn severity influence on post-fire vegetation cover resilience from Landsat MESMA fraction images time series in Mediterranean forest ecosystems. *Remote Sens. Environ.* **2016**, *184*, 112–123. [CrossRef]

24. ESA. Sentinel-2 MSI User Guide. 2022. Available online: https://sentinel.esa.int/web/sentinel/user-guides/sentinel-2-msi (accessed on 25 August 2022).

25. Richter, R.; Schläpfer, D. *Atmospheric/Topographic Correction for Satellite Imagery*; DLR Report DLR-IB 565-01/2018; ReSe Applications Schläpfer: Wessling, Germany, 2018.

26. Key, C.H. Ecological and sampling constraints on defining landscape fire severity. *Fire Ecol.* **2006**, *2*, 34–59. [CrossRef]

27. García-Llamas, P.; Suárez-Seoane, S.; Fernández-Manso, A.; Quintano, C.; Calvo, L. Evaluation of fire severity in fire prone-ecosystems of Spain under two different environmental conditions. *J. Environ. Manag.* **2020**, *271*, 110706. [CrossRef]

28. Soverel, N.O.; Perrakis, D.D.B.; Coops, N.C. Estimating burn severity from Landsat dNBR and RdNBR indices across western Canada. *Remote Sens. Environ.* **2010**, *114*, 1896–1909. [CrossRef]

29. Key, C.H.; Benson, N.C. Landscape Assessment (LA). In *FIREMON: Fire Effects Monitoring and Inventory System*; General Technical Report RMRS-GTR-164-CD; Lutes, D.C., Keane, R.E., Caratti, J.F., Key, C.H., Benson, N.C., Sutherland, S., Gangi, L.J., Eds.; Department of Agriculture, Forest Service, Rocky Mountain Research Station: Fort Collins, CO, USA, 2006; pp. 1–55.

30. Miller, J.D.; Thode, A.E. Quantifying burn severity in a heterogeneous landscape with a relative version of the delta normalized burn ratio (dNBR). *Remote Sens. Environ.* **2007**, *109*, 66–80. [CrossRef]

31. Meng, R.; Dennison, P.E.; Huang, C.; Moritz, M.A.; D'Antonio, C. Effects of fire severity and post-fire climate on short-term vegetation recovery of mixed-conifer and red fir forests in the Sierra Nevada Mountains of California. *Remote Sens. Environ.* **2015**, *171*, 311–325. [CrossRef]

32. Fernández-García, V.; Marcos, E.; Fernández-Guisuraga, J.M.; Fernández-Manso, A.; Quintano, C.; Suárez-Seoane, S.; Calvo, L. Multiple Endmember Spectral Mixture Analysis (MESMA) Applied to the Study of Habitat Diversity in the Fine-Grained Landscapes of the Cantabrian Mountains. *Remote Sens.* **2021**, *13*, 979. [CrossRef]

33. Roth, K.L.; Dennison, P.E.; Roberts, D.A. Comparing endmember selection techniques for accurate mapping of plant species and land cover using imaging spectrometer data. *Remote Sens. Environ.* **2012**, *127*, 139–152. [CrossRef]

34. Roberts, D.A.; Dennison, P.E.; Gardner, M.; Hetzel, Y.; Ustin, S.L.; Lee, C. Evaluation of the potential of hyperion for fire danger assessment by comparison to the airborne visible/infrared imaging spectrometer. *IEEE Trans. Geosci. Remote Sens.* **2003**, *41*, 1297–1310. [CrossRef]

35. Roberts, D.A.; Halligan, K.; Dennison, P.; Dudley, K.; Somers, B.; Crabbe, A. *Viper Tools User Manual*; Version 2.1; Creative Commons: San Francisco, CA, USA, 2019; p. 92.

36. Féret, J.B.; Gitelson, A.A.; Noble, S.D.; Jacquemoud, S. PROSPECT-D: Towards modeling leaf optical properties through a complete lifecycle. *Remote Sens. Environ.* **2017**, *193*, 204–215. [CrossRef]

37. Verhoef, W.; Xiao, Q.; Jia, L.; Su, Z. Unified optical-thermal four-stream radiative transfer theory for homogeneous vegetation canopies. *IEEE Trans. Geosci. Remote Sens.* **2007**, *45*, 1808–1822. [CrossRef]

38. Baret, F.; Hagolle, O.; Geiger, B.; Bicheron, P.; Miras, B.; Huc, M.; Berthelot, B.; Niño, F.; Weiss, M.; Samain, O.; et al. LAI, fAPAR and fCover CYCLOPES global products derived from VEGETATION: Part 1: Principles of the algorithm. *Remote Sens. Environ.* **2007**, *110*, 275–286. [CrossRef]

39. García-Haro, F.J.; Campos-Taberner, M.; Muñoz-Marí, J.; Laparra, V.; Camacho, F.; Sánchez-Zapero, J.; Camps-Valls, G. Derivation of global vegetation biophysical parameters from EUMETSAT Polar System. *ISPRS J. Photogramm. Remote Sens.* **2018**, *139*, 57–74. [CrossRef]

40. Breiman, L. Random forests. *Mach. Learn.* **2001**, *45*, 5–32. [CrossRef]

41. Verrelst, J.; Romijn, E.; Kooistra, L. Mapping vegetation structure in a heterogeneous river floodplain ecosystem using pointable CHRIS/PROBA data. *Remote Sens.* **2012**, *4*, 2866–2889. [CrossRef]

42. Calvo, L.; Santalla, S.; Valbuena, L.; Marcos, E.; Tárrega, R.; Luis-Calabuig, E. Post-fire natural regeneration of a Pinus pinaster forest in NW Spain. *Plant Ecol.* **2008**, *197*, 81–90. [CrossRef]

43. Ingrisch, J.; Bahn, M. Towards a Comparable Quantification of Resilience. *Trends Ecol. Evol* **2018**, *33*, 251–259. [CrossRef]

44. R Core Team. *R: A Language and Environment for Statistical Computing*; R Foundation for Statistical Computing: Vienna, Austria, 2021.

45. Fernández-Guisuraga, J.M.; Suárez-Seoane, S.; Calvo, L. Modeling Pinus pinaster forest structure after a large wildfire using remote sensing data at high spatial resolution. *For. Ecol. Manag.* **2019**, *446*, 257–271.

46. Díaz-Delgado, R.; Lloret, F.; Pons, X.; Terradas, J. Satellite evidence of decreasing resilience in Mediterranean plant communities after recurrent wildfires. *Ecology* **2002**, *83*, 2293–2303. [CrossRef]

47. Scheffer, M.; Carpenter, S.R.; Dakos, V.; van Nes, E.H. Generic Indicators of Ecological Resilience: Inferring the Chance of a Critical Transition. *Annu. Rev. Ecol. Evol. Syst.* **2015**, *46*, 145–167. [CrossRef]

48. Chergui, B.; Fahd, S.; Santos, X. Quercus suber forest and Pinus plantations show different post-fire resilience in Mediterranean north-western Africa. *Ann. For. Sci.* **2018**, *75*, 64. [CrossRef]

49. Fernández-García, V.; Quintano, C.; Taboada, A.; Marcos, E.; Calvo, L.; Fernández-Manso, A. Remote Sensing Applied to the Study of Fire Regime Attributes and Their Influence on Post-Fire Greenness Recovery in Pine Ecosystems. *Remote Sens.* **2018**, *10*, 733. [CrossRef]

50. Casas, A.; Riaño, D.; Ustin, S.L.; Dennison, P.; Salas, J. Estimation of water-related biochemical and biophysical vegetation properties using multitemporal airborne hyperspectral data and its comparison to MODIS spectral response. *Remote Sens. Environ.* **2014**, *148*, 28–41. [CrossRef]

51. Melville, B.; Fisher, A.; Lucieer, A. Ultra-high spatial resolution fractional vegetation cover from unmanned aerial multispectral imagery. *Int. J. Appl. Earth Obs. Geoinf.* **2019**, *78*, 14–24. [CrossRef]

52. Yebra, M.; Chuvieco, E. Linking ecological information and radiative transfer models to estimate fuel moisture content in the Mediterranean region of Spain: Solving the ill-posed inverse problem. *Remote Sens. Environ.* **2009**, *113*, 2403–2411. [CrossRef]

53. Huerta, S.; Fernández-García, V.; Marcos, E.; Suárez-Seoane, S.; Calvo, L. Physiological and Regenerative Plant Traits Explain Vegetation Regeneration under Different Severity Levels in Mediterranean Fire-Prone Ecosystems. *Forests* **2021**, *12*, 149. [CrossRef]

54. Calvo, L.; Santalla, S.; Marcos, E.; Valbuena, L.; Tárrega, R.; Luis, E. Regeneration after wildfire in one community dominated by obligate seeder Pinus pinaster and in another dominated by a typical resprouter Quercus pyrenaica. *For. Ecol. Manag.* **2003**, *184*, 209–223. [CrossRef]

55. Heath, J.T.; Chafer, C.J.; Bishop, T.F.A.; Ogtrop, F.F.V. Post-Fire Recovery of Eucalypt-Dominated Vegetation Communities in the Sydney Basin, Australia. *Fire Ecol.* **2016**, *12*, 53–79. [CrossRef]

56. Mamede, M.D.; de Araujo, F.S. Effects of slash and burn practices on a soil seed bank of Caatinga vegetation in Northestern Brazil. *J. Arid. Environ.* **2008**, *72*, 458–470. [CrossRef]

57. Valdecantos, A.; Baeza, M.J.; Vallejo, V.R. Vegetation Management for Promoting Ecosystem Resilience in Fire-Prone Mediterranean Shrublands. *Restor. Ecol.* **2009**, *17*, 414–421. [CrossRef]

58. Darvishzadeh, R.; Skidmore, A.; Schlerf, M.; Atzberger, C. Inversion of a radiative transfer model for estimating vegetation LAI and chlorophyll in a heterogeneous grassland. *Remote Sens. Environ.* **2008**, *112*, 2592–2604. [CrossRef]

59. Somers, B.; Cools, K.; Delalieux, S.; Stuckens, J.; Van der Zande, D.; Verstraeten, W.W.; Coppin, P. Nonlinear Hyperspectral Mixture Analysis for tree cover estimates in orchards. *Remote Sens. Environ.* **2009**, *113*, 1183–1193. [CrossRef]

60. Bergen, K.M.; Goetz, S.J.; Dubayah, R.O.; Henebry, G.M.; Hunsaker, C.T.; Imhoff, M.L.; Nelson, R.F.; Parker, G.G.; Radeloff, V.C. Remote sensing of vegetation 3-D structure for biodiversity and habitat: Review and implications for lidar and radar spaceborne missions. *J. Geophys. Res.* **2009**, *114*, G00E06. [CrossRef]
61. McCarley, T.R.; Kolden, C.A.; Vaillant, N.M.; Hudak, A.T.; Smith, A.M.S.; Wing, B.M.; Kellogg, B.S.; Kreitler, J. Multi-temporal LiDAR and Landsat quantification of fire-induced changes to forest structure. *Remote Sens. Environ.* **2017**, *191*, 419–432. [CrossRef]

remote sensing

Article

Live Fuel Moisture Content Mapping in the Mediterranean Basin Using Random Forests and Combining MODIS Spectral and Thermal Data

Àngel Cunill Camprubí [1,2,*], Pablo González-Moreno [3] and Víctor Resco de Dios [1,2,4]

[1] Joint Research Unit CTFC-AGROTECNIO-CERCA Center, 25198 Lleida, Spain; victor.resco@udl.cat
[2] Department of Crop and Forest Sciences, University of Lleida, 25198 Lleida, Spain
[3] Department of Forest Engineering, Grupo ERSAF, DendroDat Lab, Campus de Rabanales, Universidad de Córdoba, Crta. IV, km. 396, 14071 Córdoba, Spain; ir2gomop@uco.es
[4] School of Life Science and Engineering, Southwest University of Science and Technology, Mianyang 621010, China
* Correspondence: acunill86@gmail.com

Abstract: Remotely sensed vegetation indices have been widely used to estimate live fuel moisture content (LFMC). However, marked differences in vegetation structure affect the relationship between field-measured LFMC and reflectance, which limits spatial extrapolation of these indices. To overcome this limitation, we explored the potential of random forests (RF) to estimate LFMC at the subcontinental scale in the Mediterranean basin wildland. We built RF models (LFMC$_{RF}$) using a combination of MODIS spectral bands, vegetation indices, surface temperature, and the day of year as predictors. We used the Globe-LFMC and the Catalan LFMC monitoring program databases as ground-truth samples (10,374 samples). LFMC$_{RF}$ was calibrated with samples collected between 2000 and 2014 and validated with samples from 2015 to 2019, with overall root mean square errors (RMSE) of 19.9% and 16.4%, respectively, which were lower than current approaches based on radiative transfer models (RMSE ~74–78%). We used our approach to generate a public database with weekly LFMC maps across the Mediterranean basin.

Keywords: live fuel moisture content; wildfire; MODIS; spectral indices; land surface temperature; random forests

Citation: Cunill Camprubí, À.; González-Moreno, P.; Resco de Dios, V. Live Fuel Moisture Content Mapping in the Mediterranean Basin Using Random Forests and Combining MODIS Spectral and Thermal Data. *Remote Sens.* **2022**, 14, 3162. https://doi.org/10.3390/rs14133162

Academic Editor: Víctor Fernández-García

Received: 25 May 2022
Accepted: 29 June 2022
Published: 1 July 2022

Publisher's Note: MDPI stays neutral with regard to jurisdictional claims in published maps and institutional affiliations.

1. Introduction

Live fuel moisture content (LFMC), the mass of water in the foliage and small twigs relative to its total dry mass, is a key factor affecting fire potential and determining wildfire danger and activity [1,2]. Fuel moisture is directly related to the amount of energy needed to evaporate water before ignition [2,3]. Consequently, high moisture values reduce, or even inhibit, ignitability and subsequent fire spread [4].

Different studies conducted in a wide range of ecosystems have found a significant correlation between burned area and LFMC [5–7]. More specifically, these studies report that large fires only occur once fuel moisture crosses critical dryness levels. In Mediterranean regions, longer summer drought periods along with increases in temperature have been projected under climate change [8]. Such climatic changes could significantly decline LFMC and consequently enhance the length of the fire season and the rate of high intensity fires [9]. This situation could be exacerbated with intensifying fuel load accumulation and fuel connectivity as a result of rural exodus and widespread lack of land management. As a consequence, the probability and the frequency of extreme fire events is expected to increase [9]. Accurate and comprehensive spatial and temporal estimations of LFMC are thus needed to assess wildfire danger [10] and to develop early warning systems for the evolution of critical conditions [11].

Regional-scale assessment of LFMC is commonly obtained through expensive and time-consuming field inventories [12,13] or through meteorological drought indices (e.g., [14]). The latter allow spatially continuous measurements, but their validity for the Mediterranean areas has been questioned in various studies [15–17], as they do not take into account plant-specific differences and the influence of site conditions (e.g., soil water dynamics), often leading to poor predictions [3].

Remote sensing of LFMC using satellite information provides a valuable alternative to overcome the limitations of drought indices. Current approaches are mainly grouped into either physically-based simulation [18–20] or empirical methods [15,21–23]. Generally, these methods measure how water absorption and leaf properties affect reflectance in the optical spectrum [24]. Physical approaches, such as radiative transfer models (RTM), are expected to be more robust than empirical methods [10]. This is because they are based on the physical associations between leaf-canopy properties and spectral reflectance, which are independent of sensor and site conditions [19,25]. However, they are also more complex to parameterize and require additional ecological information and prior knowledge over large geographical gradients to prevent unrealistic spectra simulations [25]. In contrast, empirical approaches, which are commonly based on spectral indices (SI), are simpler and have shown similar or even better accuracies than physical models when applied locally [19,26] or across specific vegetation types [20].

Combinations of SI have been successfully employed to estimate LFMC [6,15,21,26]. In addition, some authors found stronger predictive power by including land surface temperature (LST) along with optical data to the empirical relationships [22,27–29]. The connection between LFMC and LST lies on the interaction between the plant energy balance mechanisms and its response to water stress [24]. Other recent studies implement microwave remote sensing to retrieve LFMC [30–32], but their use still has some limitations, such as data availability.

The application of empirical approaches at continental or global scales is precisely constrained by the availability of data for calibration during model development [18,21]. The biophysical and structural differences among species impact the functional relationships between LFMC and remotely sensed reflectance spectra [33,34]. Consequently, a large number of diverse sampling observations is required to reduce the effect of site dependence. Furthermore, the use of many predictive variables potentially related to LFMC may significantly improve the empirical estimations of the model [20], but also increase its complexity.

Machine learning (ML) algorithms, such as random forests (RF), are a solid alternative to physically based RTM methods or the classical regression models on which the empirical approaches are commonly based. ML algorithms are highly efficient with high dimensional data and solve the problem of model complexity by applying different functional forms in the relation between predictors and LFMC, without make explicit a priori assumptions [35]. However, using ML to estimate LFMC from remote sensing is still very recent [28,31,34,36] and has not been used in the Mediterranean basin.

Despite the importance of wildfires in the Mediterranean basin, we are currently lacking a specific method to reliably estimate LFMC at the subcontinental scale. For example, the European Forest Fires Information System (EFFIS) is using the Australian operational system [20] to estimate LFMC in the European extent, but this method has not been broadly assessed yet. Other studies have addressed LFMC modelling at local [26,37] or regional [18,25] scales and they are usually focused on specific vegetation types (e.g., grasslands or shrublands). Thus, we are still lacking a product that provides LFMC estimates for the Mediterranean basin. The only exception is the global LFMC product developed by Quan et al. [38], which is based on an RTM, and it is not yet known whether LFMC estimates could be improved through ML approaches.

The present study aims to fill this knowledge gap by developing an RF algorithm to predict LFMC within the Western Mediterranean basin using the information of the widely used Moderate Resolution Imaging Spectroradiometer (MODIS), and comparing the results with the only other method available for this area, the physically-based estimations of Quan et al. [38]. We also aim to generalize the model over a wide range of fuel types with a unique formulation by combining a forward feature selection with a spatial cross-validation and ML techniques. Finally, our ultimate goal is to develop a database of LFMC for the Mediterranean basin using available data that improves beyond currently existing products.

2. Materials and Methods

2.1. Data

2.1.1. LFMC Field Measurements

We used all the LFMC data publicly accessible within the Mediterranean basin. Most of the data available so far have been compiled in the Globe-LFMC database [39] (last accessed June 2021). The Globe-LFMC is a global compilation of 161,717 LFMC destructive field measurements of leaves and small twigs (<6 mm) from 1977 to 2018 at 1383 sampling sites with different species and characteristics in 11 fire-prone countries [39]. Most of the records in the study area come from The French Réseau hydrique database [13], but we also found a more recent LFMC time series from Catalonia (Cat-LFMC) [12]. This is a collection of 21 years (1998–2019) of biweekly field-sampled data compiled by the Catalan Forest Fire Prevention Service across nine sampling areas within this Spanish region, and focused on five species representatives of Mediterranean shrublands [12]. Cat-LFMC was added to the Globe-LFMC to extend the total number of sites and the time interval within the Mediterranean area. Both datasets have already been technically validated by correcting inconsistencies and anomalies in LFMC, as described in the relevant publications [12,13,39]. All records are properly georeferenced and inform about the species collected, the sampling protocol, land cover type, and further eco-physiological and environmental properties not used in this study. Cat-LFMC additionally includes a quality control flag, indicating possible outliers related to abrupt changes in LFMC values. These outliers were removed from the database prior to analyses.

2.1.2. MODIS Data

The MODIS MCD43A4 Collection 6 product [40] was selected as a source of above-ground spectral information, as it has shown good performance in previous studies [21,26,34]. MCD43A4 provides daily maps at 500 m spatial resolution from a 16-day composite of Nadir Bidirectional Distribution Function (NBDF)-Adjusted Reflectance for each of the 7 MODIS bands (channels 1–7, Table 1). Using a composite product may reduce the probability of cloud cover and shadows. The 'Good quality' flag from the simplified band specific quality layers (BRDF_Albedo_Band_Quality) associated with MCD43A4 was used to keep the full quality pixels of the composite.

The Terra MODIS Land Surface Temperature (LST) MOD11A2 Collection 6 product was included as a predictor of LFMC due to the impact of water availability in plant evapotranspiration and, consequently, on canopy temperature [24]. MOD11A2 is an 8-day pixel average from the MOD11A1, a daily product of LST measurements from the Terra satellite [41]. We used the daytime composite values, instead of single day measurements, because MOD11A2 had fewer data gaps (8% vs. 35%), and their effect on LFMC predictions, in terms of model RMSE, was the same (~20%, see Section S1). Daytime images cover the same period as MCD43A4, and they coincide better with the typical sample collection time, but at a 1000 m spatial resolution. They were resampled to the 500 m spatial resolution of MCD43A4 using a bilinear interpolation.

Table 1. List of potential predictors of LFMC.

Variable	Description	Wavelength (nm)	Source
NR1	Nadir Reflectance Band 1 Red	620–670	MCD43A4
NR2	Nadir Reflectance Band 2 Near infrared (NIR1)	841–876	MCD43A4
NR3	Nadir Reflectance Band 3 Blue	459–479	MCD43A4
NR4	Nadir Reflectance Band 4 Green	545–564	MCD43A4
NR5	Nadir Reflectance Band 5 Near infrared (NIR2)	1230–1250	MCD43A4
NR6	Nadir Reflectance Band 6 Shortwave infrared (SWIR1)	1628–1652	MCD43A4
NR7	Nadir Reflectance Band 7 Shortwave infrared (SWIR2)	2105–2155	MCD43A4
SI	Vegetation spectral indices: NDVI, EVI, SAVI, VARI, VIgreen, Gratio, NDII6, NDII7, NDWI, GVMI, MSI, NDTI, STI		see Table S2
LST	Land surface temperature		MOD11A2
DOY_COS DOY_SIN	Cosine and Sine of the Day of Year		

Additionally, the annual MODIS Land Cover Type (MCD12Q1) Collection 6 product [42] with the International Geosphere-Biosphere Programme (IGBP) classification scheme was used to distinguish between vegetation types in the analyses. This product replaced the land cover field included in the Globe-LFMC database, which is based on the ESA Climate Change Initiative Land Cover for the year 2015. This is because MCD12Q1 accommodates to the spatial resolution of the reflectance data and the temporal resolution of the field samples. It also was used for map production (e.g., masking water bodies and non-vegetation covers).

All MODIS images were downloaded from the NASA Land Processes Distributed Active Archive Center (LPDAAC) in the U.S. Geological Survey (USGS) Earth Resources Observation and Science Center (EROS) (https://lpdaac.usgs.gov; accessed on June 2020).

2.1.3. Landsat Data

The Landsat Collection 1 surface reflectance data included in Google Earth Engine (GEE) [43] was used to assess the MODIS subpixel spatial heterogeneity corresponding to each sampling site in the LFMC dataset. The revisit time of these satellites is 16 days, and the resolution is 30 m for the reflective bands. Similarly to Quan et al. [38], we employed Landsat 5 TM from Feb 2000 to Oct 2011 for high quality pixels, Landsat 7 ETM+ from Feb 2000 to Oct 2011 when Landsat 5 TM had poor quality pixels and also from Nov 2011 to Apr 2013, and Landsat 8 OLI from May 2013 until 2019. The use of Landsat 5 TM instead of Landsat 7 ETM+ was due to data gaps produced in the latter by failure in a sensor component [38]. Snow, cloud, and shadow pixels were removed using the Landsat internal quality band.

2.1.4. Radiative Transfer Model (RTM) Database

The global RTM-based product developed by Quan et al. [38] was used to compare the results of the ML-based approach proposed here. We chose this product because it is the only currently available database that has produced LFMC maps over the whole Mediterranean basin. It consists of a weekly collection of maps (2001–2019) generated by a physically based remote sensing model.

2.2. Methods

The following sections describe all the steps we used to estimate LFMC (Figure 1). The first section explains how we prepared the data for analyses. The second section briefly introduces the modelling approach. The last sections describe the variable selection process, the calibration and validation methods, and the software used in all steps.

2.2.1. Data Preparation

First, we cropped the Globe-LFMC dataset to the Mediterranean region (Figure 2) and used it, along with the Cat-LFMC, only for the dates with available MODIS data. Then, LFMC samples collected within the same day and site, but corresponding to different species or vegetation layers (e.g., understory and canopy), were aggregated by arithmetic means to obtain a single value per site. Nolan et al. [6] observed that average LFMC per site has a stronger correlation with spectral data than any individual vegetation layer alone. However, some studies have observed that spectral information may more closely reflect signals from the upper part of the canopy, particularly for closed forests [24]. We were interested in developing an indicator of LFMC representative of the entire canopy (upper canopy but also of the understory) because the understory often burns during a fire, which explains why we used the average LFMC value.

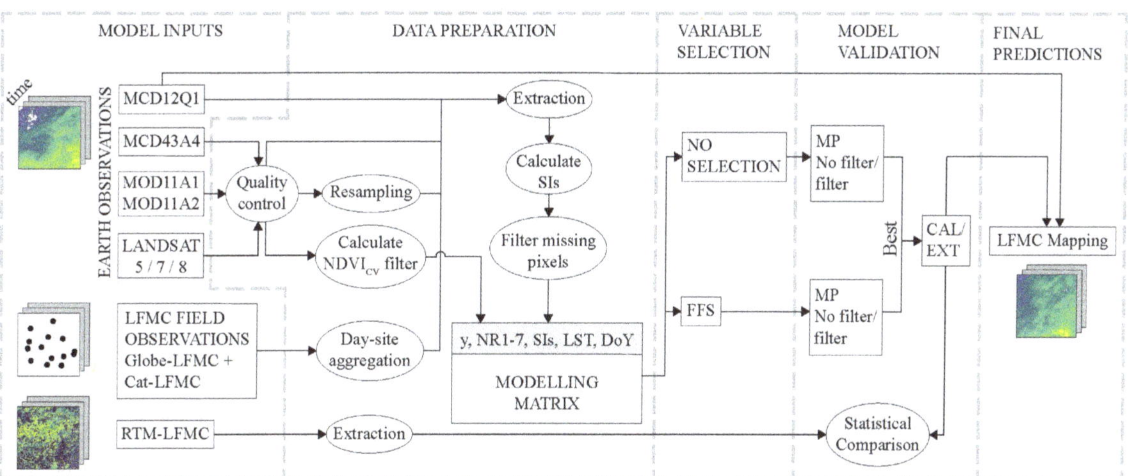

Figure 1. Overview of the pre-processing, modelling, and analysis steps.

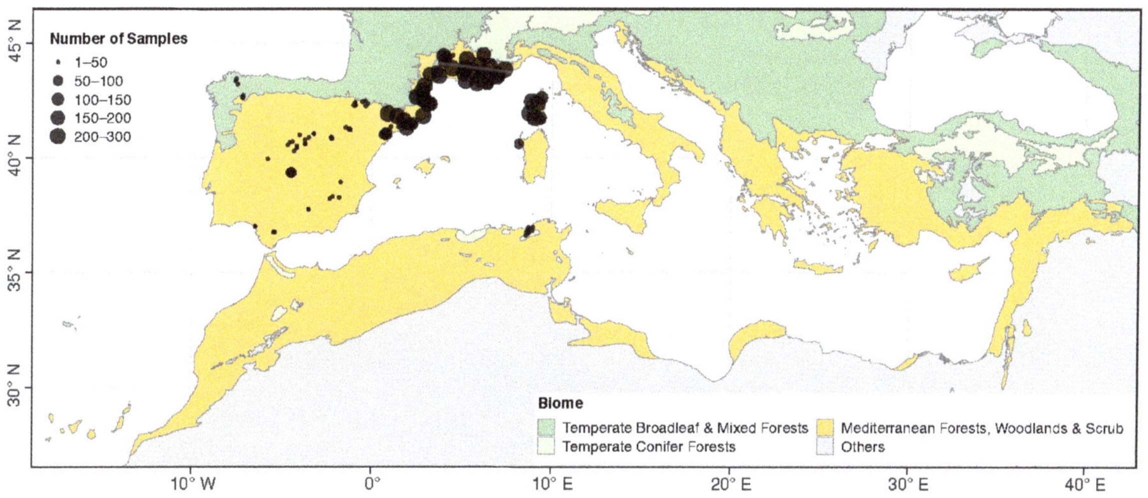

Figure 2. Distribution of sampling sites and extension of the mapping area for the database and future map productions. The background layer represents terrestrial biomes based on [44]. Gray areas were discarded from map production.

For each resultant LFMC sample, pixel values from remote sensing data were obtained by a simple pixel extraction (that is, the nearest grid cell centroid) matching their sampling date. We performed some preliminary tests observing that the simple pixel extraction method showed no significant differences (p-value = 0.9; see Section S2, Table S1) relative to conducting a focal mean (e.g., from a 3×3 window). Afterwards, various vegetation spectral indices (SI; Table S2) potentially related to LFMC were calculated by combining information from different MODIS spectral bands and used as predictors of LFMC in addition to atmospherically corrected reflectance. SI tend to reduce directional anisotropic and soil background effects and highlight specific properties of the vegetation canopy [24]. We also used land surface temperature (LST) from the LST 8-day average composite, as previously discussed (see Section S1). Finally, we added the day of year (DOY) of the ground LFMC samples as auxiliary variables to take into account the seasonal trends in LFMC [22,27]. To do so, DOY was normalized to [0, 1] and reconverted to $[-\pi, \pi]$, such that DOY 1 and DOY 366 corresponded to $-\pi$ and π, respectively. With the resulting values, we calculated the sine (DOY_SIN) and cosine (DOY_COS) to maintain the information on the periodicity as performed in Zhu et al. [34]. Consequently, DOY_SIN varied from -1 to 1 between the wettest and driest season, while DOY_COS varied from winter (coldest; -1) and summer (hottest; 1).

After defining the potential predictors described above (Table 1), we removed LFMC samples with missing data from any variable, and we discarded values outside the threshold 20–250%, which is considered the biological range of LFMC [13]. We then averaged multiple observations in the same day and MODIS-grid cell and randomly assigned the values to one of their locations to have a single daily LFMC value for a given pixel value. The resulting dataset contained a total of 10,374 LFMC field measurements between 2000 and 2019 from 118 sites located in Spain, France, Italy, and Tunisia (Figures 2, S1 and S2). These sites are mostly concentrated in the ecoregions 'Northeast Spain and Southern France Mediterranean forests' and 'Italian sclerophyllous and semi-deciduous forests' (~80%). Ecoregions with Mediterranean woodlands and coniferous, broadleaf, and mixed forest formations are also represented to a minor degree. In conjunction, mean annual temperature ranges from 6 to 20 °C and mean annual rainfall ranges from 250 to 1100 mm [44]. Site altitude ranges from 11 to 1660 m.

For model validation, Quan et al. [38] RTM data were only acquired for the sample records that coincided with the available dates of such products. We also assigned land cover information from the MCD12Q1 layers to each ground sample by matching the year of sampling with the year of the layer. Misclassified sites (e.g., croplands, permanent wetlands, and urban covers) were discarded or manually corrected based on the species collected, location, and the land cover type field included in the Globe-LFMC database. To simplify the analyses, the IGBP land cover classes present in the study were re-classified into four vegetation (or fuel) types accounting for different structural characteristics (Table S3). These new land cover classes were defined as grasslands, shrublands (closed and open shrublands), savannas (tree cover 10–60%; savannas and woody savannas), and forests (tree cover > 60%; evergreen broadleaf, evergreen needleleaf, and mixed forests).

Additionally, the NDVI coefficient of variation ($NDVI_{CV}$) derived from Landsat data were used to assess the homogeneity of vegetation 'greenness' surrounding each site coordinates, as performed in Quan et al. [38]. The authors suggest using these metrics to filter highly heterogeneous areas within a specific satellite footprint since they may not be suitable for predictive attributions [39]. Lower values correspond to more homogeneous sites. $NDVI_{CV}$ was calculated with the Landsat surface reflectance values from a 500×500 m^2 buffer that matched the MODIS cell where site coordinates were located. To do so, we adapted the GEE script publicly shared by Yebra et al. [39], such that the $NDVI_{CV}$ value was the monthly average that corresponded to the sampling date. Monthly average maximizes the quality (unmasked pixels) and the stability of the $NDVI_{CV}$ statistic. Only values with more than 80% good quality pixels (without no snow, clouds, or shadows) were retained.

2.2.2. Machine Learning Approach

Random forests (RF) was the ML algorithm chosen to empirically estimate LFMC at the Mediterranean basin because of its simplicity, its ability to deal with a large number of covariates, and because it is not necessary to have prior knowledge of the functional form of the relationships between these covariates and the response. Furthermore, the presence of outliers does not have a great influence on its performance [35].

RF is a non-parametric data-driven statistical method proposed by Breiman [45], which is based on Classification and Regression Trees (CART, also called decision trees) and bagging. Several decision trees are constructed in different bootstrap samples of the data, on which every data split (node) is forced to consider an arbitrary subset of available predictors. All individual-tree responses are then aggregated to obtain the final output predictions. The hyperparameters needed for model calibration and used in the subsequent analyses are explained in the associated Supplementary Materials (Table S4). Full details on CART, bagging, and RF can be found in Kuhn and Johnson [35].

2.2.3. Variable Selection: Forward Feature Selection

Variable selection was needed because many of the variables (or features) used as candidates to estimate LFMC were highly correlated with each other, as expected (Figure S3). This is because the SI were formed by close combinations of different spectral bands. On the other hand, predictor variables that are highly autocorrelated in space can be misinterpreted by the RF algorithm, leading to poor predictions outside the locations of the training data [46].

Here, we used the Forward Feature Selection (FFS) method proposed by Meyer et al. [46] to eliminate uninformative predictors and reduce the spatial over-fitting. First, the algorithm trains models using all possible combinations of two predictor variables and keeps those with the lowest prediction error based on a spatial cross-validation that discards entire sampling sites, as described later. Then, FFS iteratively increases the number of variables and evaluates the new model until none of the remaining variables improves the performance of the current best model. Additionally, we introduced a modification of the original method that consisted of calculating the average error over 25 different data splits. This avoided the dependence of cross-validation data splitting and aimed at stabilizing the error estimation [47].

FFS is complex and computationally intensive to execute parallel with RF parameter selection [47], and this step was performed before model calibration using a fixed set of hyperparameters (Table S4).

2.2.4. Model Selection and Performance Evaluation

In order to select the final model, we first assessed the general performance of different forms of the RF (depending on the selected predictors and whether or not the $NDVI_{CV}$ filter for heterogeneous pixels was applied) independently from a specific model calibration. We then adjusted the best performing model and evaluated its predictions.

Initial model performance assessment (MP) consisted of a bias-reduced predictive performance evaluation done using a nested 5-fold leave-location-out cross-validation (LLOCV) [47]. Nested cross-validation divides the data two times, first to develop the model and then for independently testing its performance. LLOCV means that the cross-validation folds are made of the observations left out of complete locations, assuring spatial independence [46]. More specifically, the data were divided into 5 outer folds, where one was kept for testing and the remaining were split again into 5 nested folds to iteratively train and select the optimal tuning using a standard LLOCV. Five optimal models were obtained for each outer partition, and the accuracy metrics (described in the section below) were then calculated based on the collection of predictions from all the outer folds. The same procedure was repeated 100 times with different data splits (that is, 500 independent validations), and the overall predictive power metrics were the mean of all repetitions.

Using this method, we assessed MP over 5 different model combinations with the entire set of variables, with the variables selected during the FFS, and with/without applying the $NDVI_{CV}$ filter. $NDVI_{CV}$ was treated as an additional hyperparameter and implemented in both the whole dataset (training and test) and only to the training partition. The five models consisted of: (1) all variables without filters; (2) all variables with $NDVI_{CV}$ filters on the whole dataset; (3) FFS-selected variables without filters; (4) FFS-selected variables with $NDVI_{CV}$ filters on the whole dataset; and (5) the best of all/selected variables with the $NDVI_{CV}$ filter only applied to the training partition. This method of evaluation provides an appropriate estimate of model reliability since the reported metrics are not a function of a specific model calibration, and many alternative independent datasets (outer folds) are used for testing [47]. Thus, models 1–4 allowed us to examine the effectiveness of the $NDVI_{CV}$ filter on the model performance, and the predictive improvement achieved by using only the selected features along with different parameter combinations than the fixed ones in the FFS process. With model 5 we tested how well a model optimized for homogeneous sites (defined by the selected $NDVI_{CV}$ value threshold) predicted independent sites that represent both homogeneous and heterogeneous pixels. The best alternative was employed in the subsequent calibration and validation strategies.

After selecting the best approach, we evaluated the predictions by first calibrating the model with LFMC samples from 2000 to 2014 (~80% of the total dataset) and then validated it using the samples collected in 2015–2019 (~20% of the total dataset). That is, we first determined the optimal hyperparameter values for a single model using the samples collected during 2000–2014 by training the algorithm iteratively on one-fifth of the sampling sites and tested on the remaining ones using a standard LLOCV. This process was repeated over 25 random site-resamples for each of the model candidates to stabilize the error rate and eliminate the effect of a particular data partition [47]. The model with the lowest average predictive error was selected and calibrated again to obtain predictions on the whole 5 cross-validation folds. The respective accuracy metrics (called CAL) referred to estimates within the sample period but are not independent from model calibration, as they are the outer-fold metrics in MP. We then evaluated how well the model extrapolates outside the sample period using the samples collected in 2015–2019. This validation phase (named EXT) included some new locations (3 sites) not used in CAL, which means validating future predictions also at unknown points in space.

The final model was used to compare the RF predictions against the RTM estimations produced by Quan et al. [38]. To be a fair comparison, both estimates were contrasted over the same ground-truth samples separately for the $LFMC_{RF}$ predictions inside (CAL) and outside (EXT) the training period.

The optimal hyperparameters for model calibration were chosen from an initial set of possible inputs performing a grid-search scheme [47]. We considered a wider range of possible values (Table S4) of the grid-search scheme for the MP, and then we limited the range according to the results obtained from all fitted models. For CAL, each parameter combination in the grid was iteratively assessed. In the MP, a random subset of combinations (e.g., 50) was implemented at each training process to be more computationally effective. In this case, the choice of hyperparameters was not so important since the cross-validation estimates were a generalization of the model performance.

In all cases, models were optimized to predict new locations, which is the interest of remote sensing(that is, to estimate LFMC over areas without available ground data), and it prevents spatial over-fitting [46]. For MP and CAL grid-search steps, these locations were selected using the method of Meyer et al. [46] to benefit splitting diversity. In the final model adjustment, prior to predictions, sample-site splitting was conducted by means of their coordinates and the K-means algorithm to ensure equal spatial distribution [48].

2.2.5. Validation Methods and Map Production

The predictive capabilities of the model were characterized by means of the root mean square error (RMSE), the mean absolute error (MAE), the mean bias error (MBE), and the unbiased RMSE (ubRMSE), as well as the variance explained by predictive models based on cross-validation (VEcv) [49] and the Lin's concordance correlation coefficient (CCC) [50]. RMSE, MAE, and MBE measure, respectively, squared, absolute, and mean departures between the estimated $(\hat{y}_i)$ and observed (y_i) test values of LFMC in the same units of the outcomes. RMSE was the statistic used as a criterion for parameter tuning and variable selection processes. We included the ubRMSE following Zhu et al. [34], which shows the error after removing the tendency to over- or under-predict in the model:

$$ubRMSE = \sqrt{\frac{1}{n}\sum_{1}^{n}(y_i - \hat{y}_i)^2 - \left(\frac{1}{n}\sum_{1}^{n}(y_i - \hat{y}_i)\right)^2} \tag{1}$$

Here, n is the number of observations in a validation dataset. VE_{CV} is similar to the coefficient of determination R^2, but it measures the predictive accuracy of a model by comparing observations and predictions derived from cross-validation and not the square correlation between observed and fitted values. It is defined as:

$$VE_{CV} = 1 - \frac{\sum_{1}^{n}(y_i - \hat{y}_i)^2}{\sum_{1}^{n}(y_i - \bar{y})^2} \tag{2}$$

where $\bar{y}$ is the mean of the observed values. Otherwise, CCC provides a measure of correlation relative to the line of agreement, which is expected to be unbiased with a slope of 1 and apply a penalty (Cb) if the relationship is far from this line. From CAL, EXT, and the RTM, we also obtained the slope and intercept from the linear regression between observed against predicted to assess general deviation trends.

Spatiotemporal analyses were additionally made through land cover types [31]. More specifically, we calculated general performance metrics from the CAL and EXT estimates for each land cover class, and we decomposed the mean RMSE by land cover and the month of the year to determine the temporal variability of the predictions over each vegetation functional type.

After the validations, we recalibrated the model using the whole dataset in order to consider all the available information to train the algorithm. The readjusted LFMC$_{RF}$ was then used to produce the collection of maps of the reported LFMC database.

2.2.6. Marginal Effects of the Predictors

We used partial dependence plots derived from the fitted model to evaluate the contribution of each variable to the LFMC estimations. The partial dependence function represents the average effect of a given variable on the predicted response marginalized over the effects of the rest of model inputs [51]. Mainly, we divided the distribution of values of the variable of interest into equal steps (e.g., 50). At each step, we calculated the average of all possible predictions made on the data holding the value of the step constant. Finally, we drew a line joining all average points. Resulting plots allowed for the examination of the functional relationships between the most relevant features and the LFMC estimates.

2.2.7. Software

Model building and statistical analysis were made with the statistical software R version 4.2.0 [52] and their base package for generic operations. RF was principally implemented with the R package *'ranger'* [53] but also with the *'randomForest'* library [54] to extract the partial dependence plots. The R packages *'raster'* [55] and *'sf'* [56] were used for remote sensing and spatial data manipulation, and *'doParallel'* [57] for parallel computing. An adaptation of the *stratfold3d* function of the *'sparsereg3D'* package [48] was used to make the equally spatially distributed LLOCV folds, while the spatially random splits were created with the *CreateSpacetimeFolds* function from the *'CAST'* package [58].

3. Results

3.1. Selected Variables

Results of the FFS indicated that the most important predictors of LFMC, in terms of error reduction, were the combination of LST and DOY_SIN followed by VARI, NDTI, and DOY_COS (Figure 3). These five variables alone led to an RMSE of 20.1%. We also considered NR3 and NR5 because each one represented on average an additional improvement of ~0.1% in RMSE from the previous stepwise selection, which was greater than the corresponding RMSE standard error (~0.05%) calculated from the 25 FFS realizations. Selected variables for the subsequent developments reached an overall RMSE of 19.9%.

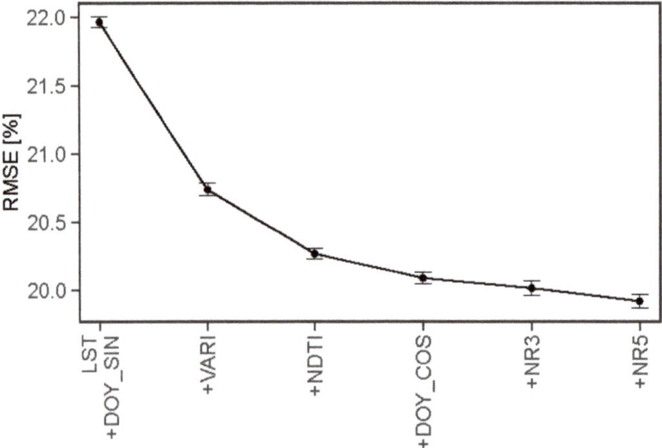

Figure 3. Selected variables derived from the combination of the Forward Feature Selection (FFS) process and the leave-location-out cross-validation (LLOCV). Black dots and vertical segments represent, respectively, the average LLOCV error and the standard error calculated from the 25 random forests computed at each FFS step.

3.2. Statistical Performance of the LFMC$_{RF}$

Calibrated and evaluated models within the general model performance assessment (MP) achieved similar results among them, with overall RMSEs (that is, from all separate iterations of each MP alternative in conjunction) ranging from 19.1% to 21.4% and VE$_{CV}$ ranging from 0.28 to 0.43. Average performance statistics (Table 2) showed that all MP alternatives tended towards a slight overprediction (MBE: 0.9–1.5%). Nonetheless, the ubRMSE values were close to the RMSE (max. difference ~0.07%), further indicating a relatively low bias of the LFMC$_{RF}$ estimates. In general, models with all the initial predictors (All$p$) showed worse performance than those with only the selected ones (Sel$p$) (Table 2). The latter benefited from the elimination of the spatially dependent variables and were used in the successive validation strategies.

Table 2. Evaluation metrics from predicted and observed values of the model performance (MP), spatial cross-validation (CAL), and the time extrapolation (EXT) assessment. Different methods based on the $NDVI_{CV}$ filter application and the complete (All$p$) or selected (Sel$p$) predictive variables. Predictions from CAL and EXT broken down by fuel type. RTM extractions and $LFMC_{RF}$ were validated on the same ground-truth observations separately if they were used in CAL or EXT.

Method	Fuel Type	Variables	Filter *	MBE (%)	MAE (%)	RMSE (%)	ubRMSE (%)	CCC	VE_{CV}	#Testing Samples/Sites
MP	All	All$p$	NF	1.10	15.70	20.57	20.54	0.53	0.32	10,374/118
	All	All$p$	F1	1.43	15.47	20.29	20.24	0.55	0.35	7633/103
	All	Sel$p$	NF	0.86	15.18	19.90	19.88	0.56	0.37	10,374/118
	All	Sel$p$	F1	1.00	15.07	19.74	19.71	0.57	0.38	7633/103
	All	Sel$p$	F2	1.06	15.18	19.92	19.89	0.57	0.39	7887/109
CAL	All	Sel$p$	NF	0.47	15.10	19.93	19.93	0.56	0.37	8983/115
	Forests			0.87	14.49	18.32	18.30	0.54	0.33	2633/27
	Savannas			1.94	15.22	19.74	19.65	0.51	0.33	4330/46
	Shrublands			−7.76	16.20	20.98	19.50	0.53	0.31	442/9
	Grasslands			−1.94	15.48	22.57	22.49	0.57	0.36	1578/43
EXT	All	Sel$p$	NF	2.75	13.05	16.35	16.12	0.69	0.52	1391/43
	Forests			7.40	13.57	16.87	15.16	0.62	0.40	456/17
	Savannas			1.63	13.18	16.46	16.38	0.69	0.55	730/22
	Shrublands			−4.62	12.08	15.27	14.56	0.72	0.54	166/3
	Grasslands			0.86	8.56	12.04	12.01	0.72	0.55	39/2
$LFMC_{RF}$ (CAL)	All	Sel$p$	NF	0.86	14.54	18.74	18.73	0.54	0.34	1152/68
RTM (CAL)	All	-	-	65.10	66.56	77.78	42.58	0.04	−10.31	1152/68
$LFMC_{RF}$ (EXT)	All	Sel$p$	NF	3.88	14.15	17.32	16.88	0.66	0.46	157/41
RTM (EXT)	All	-	-	61.87	63.10	74.41	41.33	0.07	−8.98	157/41

* NF: no filter; F1: $NDVI_{CV}$ filter applied to the entire dataset (training and test); F2: filter applied only to the training data.

Application of the $NDVI_{CV}$ filter did not show significant effects on the general model performance (Table 2; Figures S4 and S5). For example, applying the optimal filter in Sel$p$ to the entire dataset (F1) led to a small improvement in RMSE (<2%) and VE_{CV} (~0.01), but also to an increase of MBE (~0.15%) with respect to no filter application. Moreover, comparing MP with no filter and with the filter only applied to the training data (F2) resulted in increases in RMSE and MBE by 0.02% and 0.2%, respectively. In addition, the application of the filter led to the elimination of 26–28% of the dataset. It is worth noting that only a very small percentage of the data (2–4%) was deleted with $NDVI_{CV}$ application because they were above the optimal filter threshold (0.3–0.35). The rest of the data was removed because of missing rows in $NDVI_{CV}$, which were derived from poor-quality pixels in Landsat products. The model with no filter was thus used in subsequent analyses.

3.3. Prediction Assessment and Intercomparison

Accuracy metrics from the calibrated model (CAL) were consistent with the general performance (MP) of the $LFMC_{RF}$ (Table 2). These results were expected because CAL was developed with 80% of the data employed in MP, but they proved that the adjusted model was not overfitted to the particular data or by the current hyperparameter optimization (Table S4). In contrast, the EXT validation showed smaller RMSE (~3.5%) and higher VE_{CV} (~0.15) than CAL, probably due to differences in the validation samples.

We did not observe any significant bias in the $LFMC_{RF}$ estimations, as the y-intercepts and slopes were close to 0 and 1 in the fitted line between measured and predicted values of LFMC, respectively (Figure 4a,d). However, the residuals between predictions and observations revealed a linear pattern along the range of LFMC in both CAL and EXT (Figure S6). For example, the model highly underestimated values above 120% (MBE CAL = −33.97%

and MBE EXT = −22.58%) and overestimated values below 30% (MBE CAL = 45.7%; no data in EXT). This explained the aforementioned better outcomes from EXT, because the range of the actual LFMC for testing (31–209%) excluded values where the model performed worst. Within the LFMC values (30–120%) where live fuels transition from flammable to non-flammable, the model attained a smaller RMSE (MAE) of 16.75% (13.35%) for CAL and 15.10% (12.19%) for EXT relative to the overall performance of the corresponding estimates, with a small propensity to overestimate (MBE of 3.30% and 5.24% for CAL and EXT, respectively). It is worth noting that 92% of the data was within the range of 30–120%, and data below 30% may have represented curing.

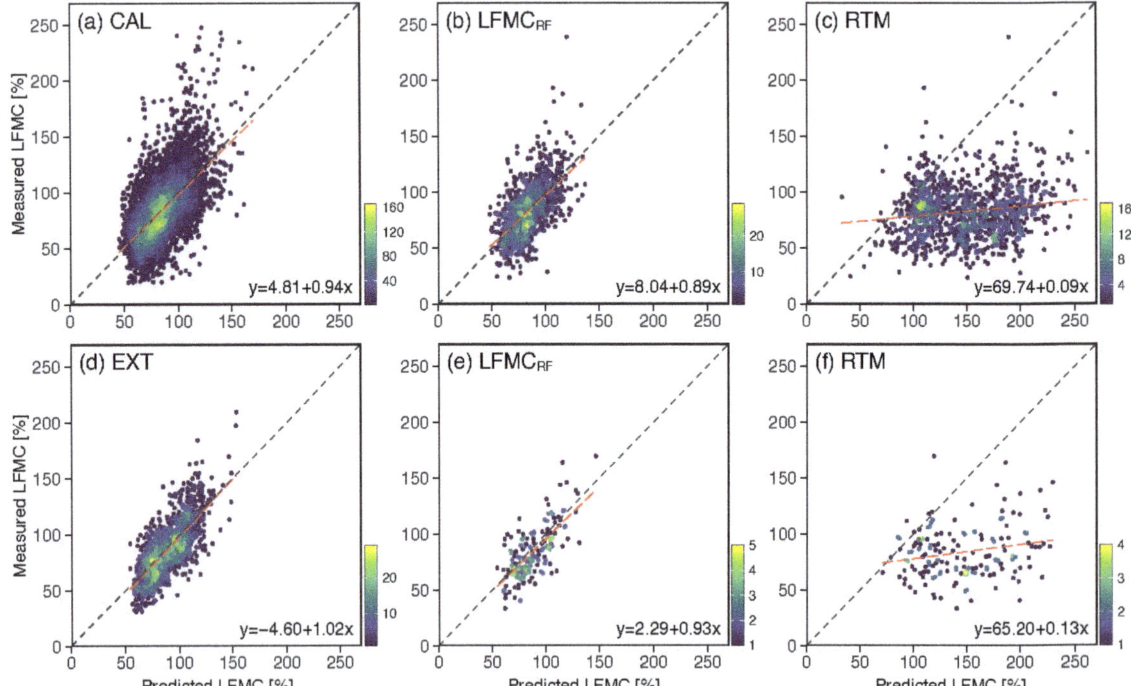

Figure 4. LFMC field measurements versus predictions from CAL (upper plots) and EXT (lower plots): all predictions (**a**,**d**), LFMC$_{RF}$ predictions made on the same data points available in RTM (**b**,**e**), and the corresponding RTM (**c**,**f**). Dashed black line and red line indicates the expected 1:1 relationship and the fitted linear model, respectively. Color scale indicates point density.

LFMC$_{RF}$ had better performance than RTM-based estimates when comparing against the same validation samples. In fact, poor correlation and large errors between observed and predicted values occurred in RTM simulations (Table 2). RTM systematically overpredicted LFMC when LFMC exceeded ~76% (Figure 4c,f). Negative values of VE$_{CV}$ (−10.15 and −8.98) indicated that these LFMC estimates were less accurate than using the mean of observations as predictions. Otherwise, the LFMC$_{RF}$ estimations used for this comparison showed the same level of accuracy as in the previous sections (Figure 4b,e), given that they were subsets of predictions from CAL and EXT.

3.4. Evaluation across Vegetation Types

Assessing the performance across vegetation types, LFMC$_{RF}$ reached better results in EXT (RMSE: 12–17%; CCC: 0.6–0.7) than in CAL (RMSE: 18–23%; CCC: 0.5–0.6) for all fuel types (Table 2). This coincides with previous results and may be because of the greater range in LFMC variation observed in the CAL dataset (Figure S2). Forests showed the smallest errors in CAL (MBE = 0.87%; RMSE = 18.32%), but the largest in EXT (MBE = 7.40%; RMSE = 16.87%). Grasslands obtained the best performance within the EXT validation (Table 2). However, they represented < 3% of the validation records and were mainly concentrated (~80% of the total) in Jul–Aug, where the model performed better (Figure 5c,d). In both cases, LFMC$_{RF}$ significantly underpredicted LFMC in shrublands (MBE −7.76 to −4.62). Temporally, the smallest errors (RMSE: 16–19%) were obtained during the hottest months (Jul–Aug), where field samples were primarily collected (Figure 5a,b), and also in winter months (Jan–Feb), matching with the lowest LFMC variability (Figure S2d). Forests showed larger stability during the entire year in both RMSE (Figure 5c) and LFMC measurements (Figure S2c). Contrarily, the performance of the model greatly fluctuated in grasslands. Grasslands reported the largest RMSE (36.7%) in May, one of the wettest months of the Mediterranean region, when fires are scarce, declining to an RMSE of 14.9% (Figure 5c) during the driest month.

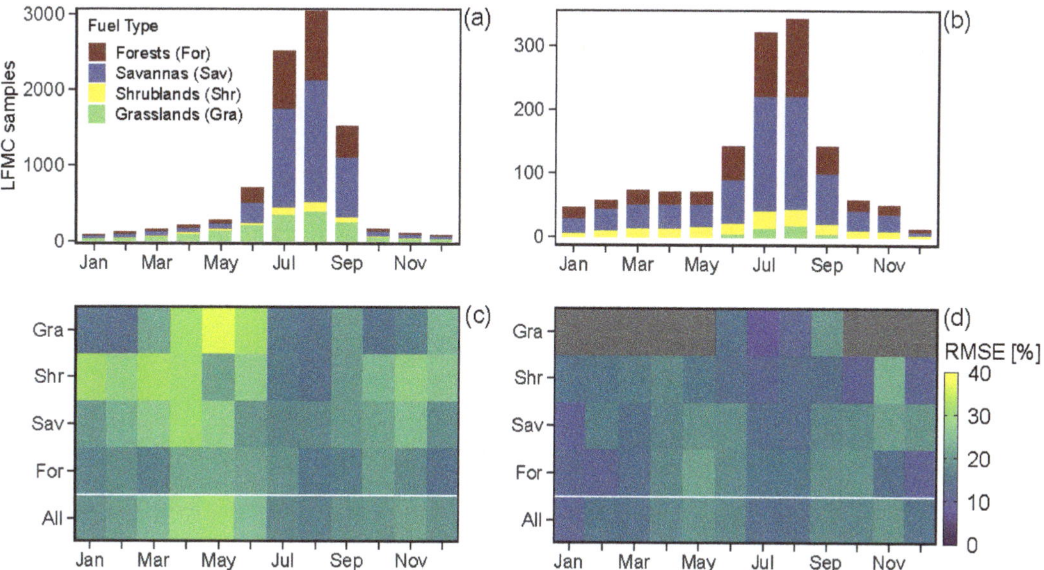

Figure 5. Number of testing samples and RMSE from CAL (**a,c**) and EXT (**b,d**) by vegetation type and month of the year. Gray cells in c and d indicate no data available. The IGBP classes from MCD12Q1 were aggregated by the vegetation functional type to which they belong.

3.5. Marginal Effects of the Predictors

Partial dependence plots exposed different patterns on the variation of LFMC estimates (Figure 6). VARI and DOY_SIN exerted the strongest effects on predictions. LFMC$_{RF}$ estimates monotonically increased as the VARI values increased. Conversely, LFMC generally monotonically decreased with increases of DOY_SIN, indicating that the highest LFMC values occurred in spring (−1) and the lowest in late summer (1). LST had non-significant effects on the LFMC estimations up to 20 °C but then presented a clear negative relationship. NR5 showed a concave shape, with marked increases at higher values of NR5 (>0.3; last decile). NDTI, NR3, and DOY_COS showed little effects on the predictions of LFMC, but they were still considered informative. The partial dependence of DOY_COS on the LFMC prediction may have been masked by the marginal effects of LST, as they were highly correlated (Figure S3).

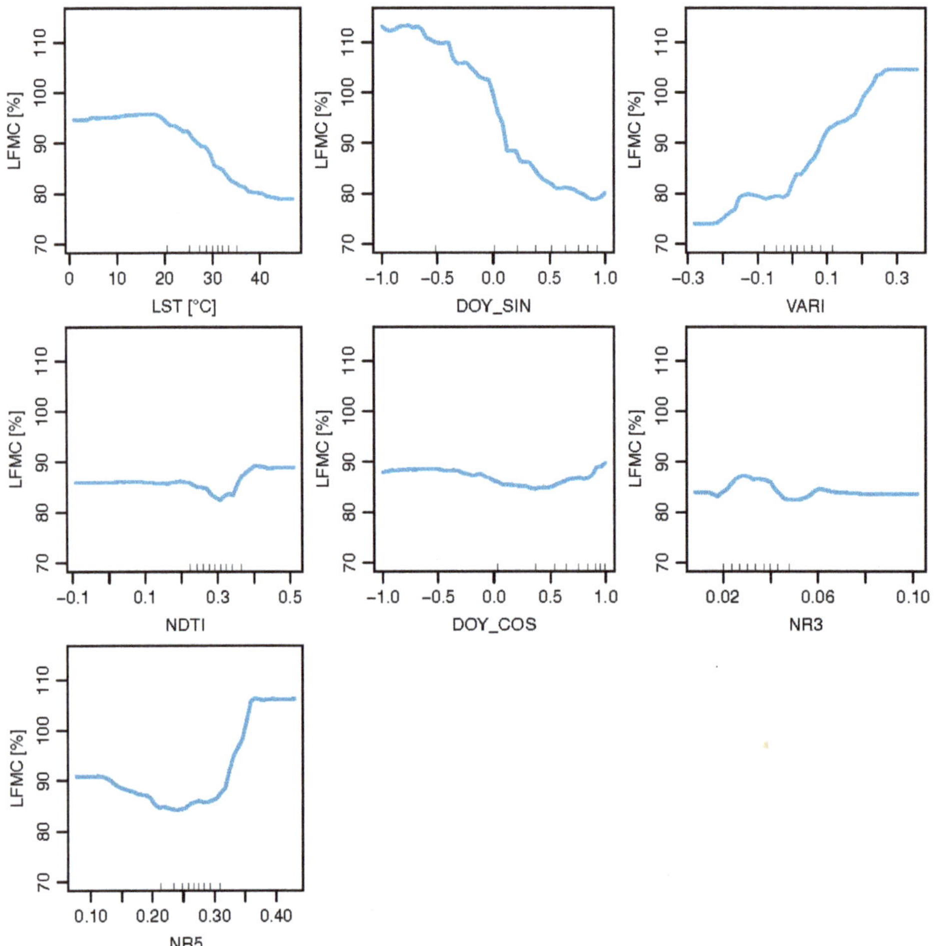

Figure 6. Partial dependence plots from the fitted model. Blue lines describe the average effect of a given predictor in the LFMC estimates. Small lines in the x axis indicate the deciles of the predictor values.

4. Discussion

We propose a novel method to estimate LFMC from remote sensing at the subcontinental scale by means of a selected set of remote sensing predictors and the RF algorithm. $LFMC_{RF}$ outperforms current approaches used in the Western Mediterranean basin in terms of validation errors and provides a solid alternative to predict LFMC over a wide range of environmental conditions using a simple but robust model with a unique formulation. In the next sections, we discuss the contribution of each selected predictor, the general and the spatiotemporal performance of the model, as well as their potential applicability and future improvements.

4.1. Selected Predictors

The key explanatory features derived from the FFS process were the variables derived from the day of the year (DOY_COS, DOY_SIN), LST, VARI, and NDTI, along with nadir reflectance bands 3 (blue) and 5 (NIR) to a minor degree.

DOY_SIN and DOY_COS had a significant influence on the LFMC estimates due to the seasonal variation in LFMC. In general, LFMC dynamics follow the distribution of the balance between evapotranspiration and rainfall in the Mediterranean region [21,26,34]. DOY_SIN partly reflects the average annual pattern in soil water availability and acts as a complement of the SI, maintaining the periodicity of LFMC within the year. Similarly, DOY_COS reflects changes in the temperature and is more related to vegetation surface temperature, which is measured by the LST [22,27].

As we expected, LST was a key factor explaining LFMC, and it showed a negative relationship with LFMC when temperatures were above ~20 °C [22,27,28]. LST is a key determinant of the energy balance of the vegetation, and its difference with air temperature is related to evapotranspiration and water losses [59]. Such differences between air temperature and LST depend on the density of vegetation cover, and previous works have shown strong relationships when combining LST and a vegetation index (e.g., [27]), as was done here. DOY_COS and LST are complementary because the former keeps the inter-annual variation of LFMC trends, while the latter provides better spatial information (that is, local deviations from the average trends) [27]. The partial dependence of LFMC on LST was similar to that reported in previous studies in that LST only affected LFMC after a certain temperature threshold [28]. LST is related to VPD [60], which is a variable that can also affect plant water content as a primary driver of evapotranspiration [61]. The importance of LST may thus be related to the fact that VPD significantly acts on leaf moisture content after a certain threshold is reached. Therefore, it is also possible that LST could be reflecting local differences in surface temperature and VPD. Further work is needed to fully understand the mechanisms by which LST affects LFMC.

VARI combines different visible wavelength bands (blue–green–red), and it has the ability to detect chlorophyll content and leaf structure variations, which are indirectly associated with changes in canopy moisture [19]. Several authors [15,19,23,26,30] have shown that VARI is one of the best indices for predicting LFMC on different vegetation types, and we also demonstrated a notable dependency of the $LFMC_{RF}$ estimations (Figure 6). Other authors found stronger correlations with indices that include SWIR [37] and NIR [21,62] bands predicting LFMC at local scales.

Reductions in chlorophyll content can result from water shortage but also from changes in leaf age, nutrient deficiency, health, and phenological stages [33,63]. Introducing NDTI from SWIR bands in the spectral region greatly sensitive to plant water content [33,64], was necessary to correct for VARI changes not driven by the moisture status of plants. Moreover, Wang et al. [63] described a connection (r = 0.45) between NDTI and dry matter content of vegetation. Dry matter weight is the denominator of the LFMC equation and could lead to variations in the spectral response and LFMC due to plant seasonal growth, independently of drought changes [30,65].

On the other hand, NIR (NR5, centered at 1240 nm) is partly influenced by water content but also by leaf internal structure and dry biomass [33,65]. This particularity may explain the concave effect that this variable had on predictions. Water loss produces an increment of NR5 as a result of lower water absorption [64]. However, at certain species and LFMC levels, water stress leads to leaf cell structure changes (reducing reflective areas by wilting) and leaf curling, which cause a decrease in NR5 [24,64].

We acknowledge that topography could have affected our results as it alters microclimatic variables influencing LFMC, such as solar radiation. However, a previous study that used reflectance bands as main explanatory variables [34] indicated a rather small effect on LFMC estimations with an RMSE improvement of ~1%.

4.2. Model Performance Assessment

Generalization errors of the $LFMC_{RF}$ (RMSE: 16–20%; MAE: 13–15%) were lower than in other studies attempting to model LFMC at large spatial scales. For instance, Zhu et al. [34] reported an overall RMSE of 27.9% using a similar spatial validation strategy but for the contiguous US. They also achieved an RMSE of 22.7% performing a standard cross-validation, which normally results in higher accuracy because the training and testing sets are not spatially independent. $LFMC_{RF}$ also showed smaller RMSE than did Rao et al. [31]

(25%), who used the same spatial approach as Zhu et al. [34] but ignored multi-species sites with high LFMC seasonal variation, where predictions tend to be more uncertain.

The proposed model tended to underestimate large values and overestimate small values of LFMC (Figure S6). Poor performance of an RF-based model towards the extremes is a well-known problem within RF models [35]. Nonetheless, similar problems were also reported in previous works based on machine learning [34,36], classical regression [23,26] and RTM simulation [20] methods. One reason for the systematic bias at high moisture levels can be the lower sensitivity of optical spectra to capture changes in water content while the vegetation gets wet [25,38]. Our strategy to address this problem was to assess LFMC over a very wide range, such that extreme values, those where LFMC estimation is problematic, are largely out of range. The lower level of LFMC in this study was 20%, but fuel moisture below 30% often corresponds to dead fuel (e.g., cured grass) and is thus beyond our scope, since we were interested in LFMC [24]. Similarly, the higher LFMC values (above 200%) may be related to harvested samples with the presence of primary tissues from a new vegetative period [21], plant parts other than leaves (e.g., fruits, flowers), or the inadequately inclusion of samples collected after rain or dew events [20].

The LFMC$_{RF}$ showed a better performance that RTM predictions from Quan et al. [38]. The RTM-based estimates were highly biased with a strong tendency to overpredict beyond 76% LFMC. This coincided with the results reported by Marino et al. [26], who found an identical pattern starting at the threshold of 65% using the RTM developed by Yebra et al. [20]. This demonstrates a better predictive power for our data-driven approach, even though physically-based approaches are expected to be more precise when applied to sites not used for calibration [24]. At any rate, we acknowledge that comparing a regional dataset like this one against a global dataset is not entirely fair, given the scale gap, but our results highlight that the RMSE of the global RTM hinders any local application for operational purposes.

The critical LFMC level associated with fire occurrence in the Mediterranean forests, and other parts of the world occurs around 100% [5–7]. Our model improves current products, but MAE around the critical threshold of 100% LFMC is still ~13%. Differences of 10% in LFMC estimation from field measurements are generally acceptable for fire management [26]. However, these results indicate that there is still room for further improvements, particularly towards the critical threshold, so as to avoid reporting of false fire alerts or omission of danger situations [19].

4.3. Evaluation across Vegetation Types

The predictive errors obtained by the LFMC$_{RF}$ within the training period for forests/savannas (RMSE 18–20%), shrublands (RMSE ~21%), and grasslands (RMSE ~23%) were similar between them and comparable to those reported by other studies for the same vegetation types (forests/savannas 22–32%, shrublands 14–29%, grasslands 29–49% [18,20,31,34,38].) Despite the methodological differences, this comparison demonstrates that a single model can be as accurate or even better than formalizing a model for each fuel class separately. This could be due to the RF architecture that allows using the spectral and thermal information itself to discriminate between vegetation functional types. Furthermore, misclassification problems of the land cover products used to differentiate between fuel classes can further increase the uncertainty of the LFMC estimates [20,34].

In general, we observed that the uncertainty of the LFMC predictions (Figure 5) depended on the range of LFMC values for testing and their local and temporal variability (Figure S2). For example, forests showed more stable behavior in both LFMC dynamics and prediction agreement. Deep root systems in trees reduce the seasonal LFMC variation [2]. On the contrary, grasslands reported the highest errors in spring (the wettest part of the year) and the lowest in the driest periods (summer, when fires are more likely). These patterns overlapped with the monthly maximum and minimum values of LFMC, that is, larger LFMC errors under higher LFMC values and smaller LFMC errors under lower values. Shrublands instead had a low temporal variability but presented a significant bias (MBE from -5 to -8%), likely because of the high proportion of large LFMC values (>120%) in their ground-truth sample distributions (16.4% of the total ground-truth samples). The error associated with predictions outside the training period (EXT) was similar to that from the CAL dataset (Figure 5). However, RMSE was slightly lower with the EXT dataset because of the lower LFMC variability in the EXT dataset relative to CAL. We thus conclude that the fitted model with historical data can be safely applied in future situations without the need for frequent readjustment, but with careful interpretation in the wettest months and for LFMC values below 30%.

4.4. Applicability and Potential Improvements

The relatively coarse resolution (~500 m) of the final product is appropriate for landscape-scale use and does not guarantee smaller-scale applications. Each individual pixel normally contains information from a mixture of vegetation canopy layers, species, surface litter, and soil elements with different properties that cannot be unambiguously separated [10,20]. We acknowledge that a limitation to this study is that we did not explicitly assess the representativeness of the samples within the site. We therefore took into account small-scale heterogeneity by implementing an $NDVI_{CV}$ filter, as in Quan et al. [38]. However, we did not observe any significant improvement after applying this filter, likely indicating that sample areas were relatively homogenous. In any case, sub-pixel variation and the scale mismatch between sample-plot size and pixel resolution hinder establishing relationships between field observations and satellite-derived variables, introducing uncertainties into the predictions. The latter could be solved using higher spatial resolution data (e.g., Sentinel-2 or Landsat) [26,36,37,62], but these satellites usually have lower revisit frequency disabling near-real-time usage and introducing a time lag between the images and the sampling date [26]. Future work should extend the use of our methods to these newer satellites because historical LFMC field data currently available is not yet sufficient to achieve this goal.

Further progress will come from joining our approach with microwave remote sensing data. Microwave observations (active and passive) can also detect changes in vegetation water content but are less sensitive to atmospheric conditions (e.g., clouds) than optical wavelengths [30] and have the ability to penetrate deeper into the canopies [31]. The recently available non-commercial radar data supplied by the Sentinel-1 A/B Synthetic Aperture Radar (SAR) may represent a great opportunity to infer the improvement of LFMC models at the operational level [31,32].

Sample representativeness is a general constrain in the empirical models [24]. In this study, field samples were not evenly distributed across the whole Mediterranean basin. They could be considered representative of the Western Mediterranean conditions since they were abundant in number (space and time) within their specific biome, as well as in species and environmental conditions. Thus, application of the $LFMC_{RF}$ should be limited to areas with similar characteristics, and LFMC estimates must be interpreted with caution in underrepresented areas (e.g., temperate zones). Despite that, the generated maps extend to the entire Mediterranean biome included in the Mediterranean basin, as well as some meridional areas of the temperate biomes of Europe (e.g., northern Spain) (Figure 2).

5. Conclusions

We successfully tested an RF algorithm as an approach to predict large-scale LFMC using the spectral and thermal information of MODIS and two static variables representing seasonal patterns. The $LFMC_{RF}$ is applicable to a wide variety of vegetation types, and the performance of the fitted model (MBE = 0.47%, RMSE = 19.9%, VE_{CV} = 0.37, CCC = 0.56) was comparable to that of other studies with similar purposes but with a higher degree of complexity than $LFMC_{RF}$, including the RTM-based methods with applications in the Mediterranean basin. The architecture of RF allows the introduction of new explanatory variables that would help to reduce the uncertainty in the predictions. LFMC maps were produced at 8-day intervals from 2001 to 2021. The final product provides a complete asset for studying the relationships between LFMC and the influencing factors that promote wildfire activity and fire regimes in the Mediterranean basin. Furthermore, after the imminent MODIS decommission, the new Visible Infrared Imaging Radiometer Suite (VIIRS) is expected to provide long-term continuity with better spatial resolution [24]. Continuous retrievals, either with MODIS or VIIRS, might be a valuable tool for quasi near-real-time fire risk assessment and for operational applications such as the mobilization of resources and people or the planning of preventive actions for fire mitigation (e.g., fuel reduction or prescribed burns).

Supplementary Materials: The following supporting information can be downloaded at: https://www.mdpi.com/article/10.3390/rs14133162/s1, Table S1. Performance metrics from the focal mean and simple pixel extraction comparison; Table S2. Spectral vegetation indices used to estimate LFMC; Table S3. Land cover classes from samples used in the study; Table S4. Boundaries of the RF hyperparameters grid-search space, adjusted parameters for the Forward Feature Selection (FFS) process and optimized hyperparameters for the final model; Figure S1. LFMC ground samples overall and by country distributions; Figure S2. Mean and standard deviation (SD) matrices from CAL and EXT of the LFMC field measurements shown by fuel type and month of the year, and the overall of each one; Figure S3. Correlation matrix between LFMC and predictive variables; Figure S4. Performance metrics profiles from the general model performance assessment (MP) alternative with the selected variables and the $NDVI_{CV}$ filter applied to the entire dataset and only to the training partition; Figure S5. LFMC field observations versus predictions from the CAL validation theoretically rejected by the 0.3 $NDVI_{CV}$ threshold; Figure S6. Residuals between predictions and observations against the LFMC observations and their marginal density distributions for CAL and EXT. This SM are distributed in 8 sections of additional methods and analyses: S1, Land surface temperature; S2, Data extraction method [66]; S3, Spectral vegetation indices; S4, Land cover definitions; S5, Model parametrization [67]; S6, Data description; S7, $NDVI_{CV}$ filter; S8, Additional prediction analysis.

Author Contributions: Conceptualization, À.C.C., P.G.-M. and V.R.d.D.; investigation, À.C.C.; formal analysis, À.C.C.; writing—original draft preparation, À.C.C.; writing—review and editing, P.G.-M. and V.R.d.D. All authors have read and agreed to the published version of the manuscript.

Funding: This study was funded by the MICINN (RTI2018-094691-B-C31), European Union's Horizon 2020-Research and Innovation Framework Programme under grant agreement no. 101003890 project FirEUrisk, the National Natural Science Foundation of China (U20A20179, 31850410483), and the talent proposals in Sichuan Province (2020JDRC0065) from Southwest University of Science and Technology (18ZX7131).

Data Availability Statement: All the scripts and data generated and analyzed during the current study are publicly available in the Github repository, https://github.com/fuegologos/lfmc_rf. The final product of the weekly LFMC maps is fully available at 10.5281/zenodo.6784663.

Acknowledgments: We remain indebted to Xingwen Quan for providing us the RTM-based LFMC data.

Conflicts of Interest: The authors declare no of interest.

References

1. Bradstock, R.A. A Biogeographic Model of Fire Regimes in Australia: Current and Future Implications. *Glob. Ecol. Biogeogr.* **2010**, *19*, 145–158. [CrossRef]
2. Resco de Dios, V. *Plant-Fire Interactions: Applying Ecophysiology to Wildfire Management, Managing Forest Ecosystems*; Managing Forest Ecosystems; Springer International Publishing: Cham, Switzerland, 2020; Volume 36, ISBN 978-3-030-41191-6.
3. Jolly, W.; Johnson, D. Pyro-Ecophysiology: Shifting the Paradigm of Live Wildland Fuel Research. *Fire* **2018**, *1*, 8. [CrossRef]
4. Nelson, R.M. Water Relations of Forest Fuels. In *Forest Fires: Behavior and Ecological Effects*; Johnson, E.A., Miyanishi, K., Eds.; Academic Press: San Diego, CA, USA, 2001; pp. 79–149.
5. Dennison, P.E.; Moritz, M.A. Critical Live Fuel Moisture in Chaparral Ecosystems: A Threshold for Fire Activity and Its Relationship to Antecedent Precipitation. *Int. J. Wildland Fire* **2009**, *18*, 1021. [CrossRef]
6. Nolan, R.H.; Boer, M.M.; de Dios, V.R.; Caccamo, G.; Bradstock, R.A. Large-Scale, Dynamic Transformations in Fuel Moisture Drive Wildfire Activity across Southeastern Australia. *Geophys. Res. Lett.* **2016**, *43*, 4229–4238. [CrossRef]

7. Luo, K.; Quan, X.; He, B.; Yebra, M. Effects of Live Fuel Moisture Content on Wildfire Occurrence in Fire-Prone Regions over Southwest China. *Forests* **2019**, *10*, 887. [CrossRef]
8. IPCC. *Climate Change 2022: Impacts, Adaptation, and Vulnerability. Contribution of Working Group II to the Sixth Assessment Report of the Intergovernmental Panel on Climate Change*; Pörtner, H.-O., Roberts, D.C., Tignor, M., Poloczanska, E.S., Mintenbeck, K., Alegría, A., Craig, M., Langsdorf, S., Löschke, S., Möller, V., et al., Eds.; IPCC: Geneva, Switzerland, 2022.
9. Dupuy, J.; Fargeon, H.; Martin-StPaul, N.; Pimont, F.; Ruffault, J.; Guijarro, M.; Hernando, C.; Madrigal, J.; Fernandes, P. Climate Change Impact on Future Wildfire Danger and Activity in Southern Europe: A Review. *Ann. For. Sci.* **2020**, *77*, 35. [CrossRef]
10. Chuvieco, E.; Aguado, I.; Salas, J.; García, M.; Yebra, M.; Oliva, P. Satellite Remote Sensing Contributions to Wildland Fire Science and Management. *Curr. For. Rep.* **2020**, *6*, 81–96. [CrossRef]
11. Boer, M.M.; Nolan, R.H.; De Dios, V.R.; Clarke, H.; Price, O.F.; Bradstock, R.A. Changing Weather Extremes Call for Early Warning of Potential for Catastrophic Fire. *Earth's Future* **2017**, *5*, 1196–1202. [CrossRef]
12. Gabriel, E.; Delgado-Dávila, R.; De Cáceres, M.; Casals, P.; Tudela, A.; Castro, X. Live Fuel Moisture Content Time Series in Catalonia since 1998. *Ann. For. Sci.* **2021**, *78*, 44. [CrossRef]
13. Martin-StPaul, N.; Pimont, F.; Dupuy, J.L.; Rigolot, E.; Ruffault, J.; Fargeon, H.; Cabane, E.; Duché, Y.; Savazzi, R.; Toutchkov, M. Live Fuel Moisture Content (LFMC) Time Series for Multiple Sites and Species in the French Mediterranean Area since 1996. *Ann. For. Sci.* **2018**, *75*, 70. [CrossRef]
14. Van Wagner, C.E. Development and Structure of the Canadian Forest Fire Weather Index System. Canadian Forestry Service. *For. Technol. Rep.* **1987**, *35*, 37.
15. Caccamo, G.; Chisholm, L.A.; Bradstock, R.A.; Puotinen, M.L.; Pippen, B.G. Monitoring Live Fuel Moisture Content of Heathland, Shrubland and Sclerophyll Forest in South-Eastern Australia Using MODIS Data. *Int. J. Wildland Fire* **2012**, *21*, 257. [CrossRef]
16. Ruffault, J.; Martin-StPaul, N.; Pimont, F.; Dupuy, J.-L.L. How Well Do Meteorological Drought Indices Predict Live Fuel Moisture Content (LFMC)? An Assessment for Wildfire Research and Operations in Mediterranean Ecosystems. *Agric. For. Meteorol.* **2018**, *262*, 391–401. [CrossRef]
17. Martin, M.S.; Bonet, J.A.; De Aragón, J.M.; Voltas, J.; Coll, L.; De Dios, V.R. Crown Bulk Density and Fuel Moisture Dynamics in Pinus Pinaster Stands Are Neither Modified by Thinning nor Captured by the Forest Fire Weather Index. *Ann. For. Sci.* **2017**, *74*, 51. [CrossRef]
18. Jurdao, S.; Yebra, M.; Guerschman, J.P.; Chuvieco, E. Regional Estimation of Woodland Moisture Content by Inverting Radiative Transfer Models. *Remote Sens. Environ.* **2013**, *132*, 59–70. [CrossRef]
19. Yebra, M.; Chuvieco, E.; Riaño, D. Estimation of Live Fuel Moisture Content from MODIS Images for Fire Risk Assessment. *Agric. For. Meteorol.* **2008**, *148*, 523–536. [CrossRef]
20. Yebra, M.; Quan, X.; Riaño, D.; Larraondo, P.R.; van Dijk, A.I.J.M.; Cary, G.J. A Fuel Moisture Content and Flammability Monitoring Methodology for Continental Australia Based on Optical Remote Sensing. *Remote Sens. Environ.* **2018**, *212*, 260–272. [CrossRef]
21. Argañaraz, J.P.; Landi, M.A.; Bravo, S.J.; Gavier-Pizarro, G.I.; Scavuzzo, C.M.; Bellis, L.M. Estimation of Live Fuel Moisture Content From MODIS Images for Fire Danger Assessment in Southern Gran Chaco. *IEEE J. Sel. Top. Appl. Earth Obs. Remote Sens.* **2016**, *9*, 5339–5349. [CrossRef]
22. Chuvieco, E.; Cocero, D.; Riaño, D.; Martin, P.; Martínez-Vega, J.; de la Riva, J.; Pérez, F. Combining NDVI and Surface Temperature for the Estimation of Live Fuel Moisture Content in Forest Fire Danger Rating. *Remote Sens. Environ.* **2004**, *92*, 322–331. [CrossRef]
23. Peterson, S.H.; Roberts, D.A.; Dennison, P.E. Mapping Live Fuel Moisture with MODIS Data: A Multiple Regression Approach. *Remote Sens. Environ.* **2008**, *112*, 4272–4284. [CrossRef]
24. Yebra, M.; Dennison, P.E.; Chuvieco, E.; Riaño, D.; Zylstra, P.; Hunt, E.R.; Danson, F.M.; Qi, Y.; Jurdao, S. A Global Review of Remote Sensing of Live Fuel Moisture Content for Fire Danger Assessment: Moving towards Operational Products. *Remote Sens. Environ.* **2013**, *136*, 455–468. [CrossRef]
25. Yebra, M.; Chuvieco, E. Linking Ecological Information and Radiative Transfer Models to Estimate Fuel Moisture Content in the Mediterranean Region of Spain: Solving the Ill-Posed Inverse Problem. *Remote Sens. Environ.* **2009**, *113*, 2403–2411. [CrossRef]
26. Marino, E.; Yebra, M.; Guillén-Climent, M.; Algeet, N.; Tomé, J.L.; Madrigal, J.; Guijarro, M.; Hernando, C. Investigating Live Fuel Moisture Content Estimation in Fire-Prone Shrubland from Remote Sensing Using Empirical Modelling and RTM Simulations. *Remote Sens.* **2020**, *12*, 2251. [CrossRef]
27. García, M.; Chuvieco, E.; Nieto, H.; Aguado, I. Combining AVHRR and Meteorological Data for Estimating Live Fuel Moisture Content. *Remote Sens. Environ.* **2008**, *112*, 3618–3627. [CrossRef]
28. McCandless, T.C.; Kosovic, B.; Petzke, W. Enhancing Wildfire Spread Modelling by Building a Gridded Fuel Moisture Content Product with Machine Learning. *Mach. Learn. Sci. Technol.* **2020**, *1*, 035010. [CrossRef]
29. Sow, M.; Mbow, C.; Hély, C.; Fensholt, R.; Sambou, B. Estimation of Herbaceous Fuel Moisture Content Using Vegetation Indices and Land Surface Temperature from MODIS Data. *Remote Sens.* **2013**, *5*, 2617–2638. [CrossRef]
30. Fan, L.; Wigneron, J.-P.; Xiao, Q.; Al-Yaari, A.; Wen, J.; Martin-StPaul, N.; Dupuy, J.-L.; Pimont, F.; Al Bitar, A.; Fernandez-Moran, R.; et al. Evaluation of Microwave Remote Sensing for Monitoring Live Fuel Moisture Content in the Mediterranean Region. *Remote Sens. Environ.* **2018**, *205*, 210–223. [CrossRef]
31. Rao, K.; Williams, A.P.; Flefil, J.F.; Konings, A.G. SAR-Enhanced Mapping of Live Fuel Moisture Content. *Remote Sens. Environ.* **2020**, *245*, 111797. [CrossRef]

32. Wang, L.; Quan, X.; He, B.; Yebra, M.; Xing, M.; Liu, X. Assessment of the Dual Polarimetric Sentinel-1A Data for Forest Fuel Moisture Content Estimation. *Remote Sens.* **2019**, *11*, 1568. [CrossRef]

33. Ceccato, P.; Flasse, S.; Tarantola, S.; Jacquemoud, S.; Grégoire, J.-M. Detecting Vegetation Leaf Water Content Using Reflectance in the Optical Domain. *Remote Sens. Environ.* **2001**, *77*, 22–33. [CrossRef]

34. Zhu, L.; Webb, G.I.; Yebra, M.; Scortechini, G.; Miller, L.; Petitjean, F. Live Fuel Moisture Content Estimation from MODIS: A Deep Learning Approach. *ISPRS J. Photogramm. Remote Sens.* **2021**, *179*, 81–91. [CrossRef]

35. Kuhn, M.; Johnson, K. *Applied Predictive Modeling*; Springer: New York, NY, USA, 2013; ISBN 978-1-4614-6848-6.

36. Adab, H.; Kanniah, K.D.; Beringer, J. Estimating and Up-Scaling Fuel Moisture and Leaf Dry Matter Content of a Temperate Humid Forest Using Multi Resolution Remote Sensing Data. *Remote Sens.* **2016**, *8*, 961. [CrossRef]

37. Costa-Saura, J.M.; Balaguer-Beser, Á.; Ruiz, L.A.; Pardo-Pascual, J.E.; Soriano-Sancho, J.L. Empirical Models for Spatio-Temporal Live Fuel Moisture Content Estimation in Mixed Mediterranean Vegetation Areas Using Sentinel-2 Indices and Meteorological Data. *Remote Sens.* **2021**, *13*, 3726. [CrossRef]

38. Quan, X.; Yebra, M.; Riaño, D.; He, B.; Lai, G.; Liu, X. Global Fuel Moisture Content Mapping from MODIS. *Int. J. Appl. Earth Obs. Geoinf.* **2021**, *101*, 102354. [CrossRef]

39. Yebra, M.; Scortechini, G.; Badi, A.; Beget, M.E.; Boer, M.M.; Bradstock, R.; Chuvieco, E.; Danson, F.M.; Dennison, P.; de Dios, V.R.; et al. Globe-LFMC, a Global Plant Water Status Database for Vegetation Ecophysiology and Wildfire Applications. *Sci. Data* **2019**, *6*, 155. [CrossRef]

40. Schaaf, C.B.; Wang, Z. *MCD43A4 MODIS/Terra+Aqua BRDF/Albedo Nadir BRDF Adjusted Ref Daily L3 Global–500m V006 [Data Set]*; NASA EOSDIS Land Processes DAAC: Sioux Falls, SD, USA, 2015. [CrossRef]

41. Wan, Z. New Refinements and Validation of the Collection-6 MODIS Land-Surface Temperature/Emissivity Product. *Remote Sens. Environ.* **2014**, *140*, 36–45. [CrossRef]

42. Sulla-Menashe, D.; Gray, J.M.; Abercrombie, S.P.; Friedl, M.A. Hierarchical Mapping of Annual Global Land Cover 2001 to Present: The MODIS Collection 6 Land Cover Product. *Remote Sens. Environ.* **2019**, *222*, 183–194. [CrossRef]

43. Gorelick, N.; Hancher, M.; Dixon, M.; Ilyushchenko, S.; Thau, D.; Moore, R. Google Earth Engine: Planetary-Scale Geospatial Analysis for Everyone. *Remote Sens. Environ.* **2017**, *202*, 18–27. [CrossRef]

44. Dinerstein, E.; Olson, D.; Joshi, A.; Vynne, C.; Burgess, N.D.; Wikramanayake, E.; Hahn, N.; Palminteri, S.; Hedao, P.; Noss, R.; et al. An Ecoregion-Based Approach to Protecting Half the Terrestrial Realm. *Bioscience* **2017**, *67*, 534–545. [CrossRef]

45. Breiman, L. Random Forests. *Mach. Learn.* **2001**, *45*, 5–32. [CrossRef]

46. Meyer, H.; Reudenbach, C.; Hengl, T.; Katurji, M.; Nauss, T. Improving Performance of Spatio-Temporal Machine Learning Models Using Forward Feature Selection and Target-Oriented Validation. *Environ. Model. Softw.* **2018**, *101*, 1–9. [CrossRef]

47. Krstajic, D.; Buturovic, L.J.; Leahy, D.E.; Thomas, S. Cross-Validation Pitfalls When Selecting and Assessing Regression and Classification Models. *J. Cheminform.* **2014**, *6*, 10. [CrossRef] [PubMed]

48. Pejović, M.; Nikolić, M.; Heuvelink, G.B.M.; Hengl, T.; Kilibarda, M.; Bajat, B. Sparse Regression Interaction Models for Spatial Prediction of Soil Properties in 3D. *Comput. Geosci.* **2018**, *118*, 1–13. [CrossRef]

49. Li, J. Assessing Spatial Predictive Models in the Environmental Sciences: Accuracy Measures, Data Variation and Variance Explained. *Environ. Model. Softw.* **2016**, *80*, 1–8. [CrossRef]

50. Lin, L.I.-K. A Concordance Correlation Coefficient to Evaluate Reproducibility. *Biometrics* **1989**, *45*, 255. [CrossRef] [PubMed]

51. Friedman, J.H. Greedy Function Approximation: A Gradient Boosting Machine. *Ann. Stat.* **2001**, *29*, 1189–1232. [CrossRef]

52. R Core Team. *R: A Language and Environment for Statistical Computing*; R Foundation for Statistical Computing: Vienna, Austria, 2022. Available online: https://www.R-project.org/ (accessed on 28 June 2022).

53. Wright, M.N.; Ziegler, A. Ranger: A Fast Implementation of Random Forests for High Dimensional Data in C++ and R. *J. Stat. Softw.* **2017**, *77*, 1–17. [CrossRef]

54. Liaw, A.; Wiener, M. Classification and Regression by RandomForest. *R News.* **2002**, *2*, 18–22.

55. Hijmans, R.J. *Raster: Geographic Data Analysis and Modeling*, R Package Version 3.5-15; 2022. Available online: https://CRAN.R-project.org/package=raster (accessed on 28 June 2022).

56. Pebesma, E. Simple Features for R: Standardized Support for Spatial Vector Data. *R J.* **2018**, *10*, 439–446. [CrossRef]

57. Microsoft-Corp; Weston, S. *DoParallel: Foreach Parallel Adaptor for the "Parallel" Package*, R package version 1.0.17; 2022. Available online: https://CRAN.R-project.org/package=doParallel (accessed on 28 June 2022).

58. Meyer, H. *CAST: "caret" Applications for Spatial-Temporal Models*, R Package Version 6.0-92; 2022. Available online: https://CRAN.R-project.org/package=caret (accessed on 28 June 2022).

59. Vidal, A.; Pinglo, F.; Durand, H.; Devaux-Ros, C.; Maillet, A. Evaluation of a Temporal Fire Risk Index in Mediterranean Forests from NOAA Thermal IR. *Remote Sens. Environ.* **1994**, *49*, 296–303. [CrossRef]

60. Hashimoto, H.; Dungan, J.; White, M.A.; Yang, F.; Michaelis, A.; Running, S.W.; Nemani, R. Satellite-Based Estimation of Surface Vapor Pressure Deficits Using MODIS Land Surface Temperature Data. *Remote Sens. Environ.* **2008**, *112*, 142–155. [CrossRef]

61. Balaguer-Romano, R.; Díaz-Sierra, R.; De Cáceres, M.; Cunill-Camprubí, À.; Nolan, R.H.; Boer, M.M.; Voltas, J.; de Dios, V.R. A Semi-Mechanistic Model for Predicting Daily Variations in Species-Level Live Fuel Moisture Content. *Agric. For. Meteorol.* **2022**, *323*, 109022. [CrossRef]

62. García, M.; Riaño, D.; Yebra, M.; Salas, J.; Cardil, A.; Monedero, S.; Ramirez, J.; Martín, M.P.; Vilar, L.; Gajardo, J.; et al. A Live Fuel Moisture Content Product from Landsat TM Satellite Time Series for Implementation in Fire Behavior Models. *Remote Sens.* **2020**, *12*, 1714. [CrossRef]

63. Wang, L.; Hunt, E.R.; Qu, J.J.; Hao, X.; Daughtry, C.S.T. Remote Sensing of Fuel Moisture Content from Ratios of Narrow-Band Vegetation Water and Dry-Matter Indices. *Remote Sens. Environ.* **2013**, *129*, 103–110. [CrossRef]

64. Chuvieco, E.; Riaño, D.; Aguado, I.; Cocero, D. Estimation of Fuel Moisture Content from Multitemporal Analysis of Landsat Thematic Mapper Reflectance Data: Applications in Fire Danger Assessment. *Int. J. Remote Sens.* **2002**, *23*, 2145–2162. [CrossRef]

65. Bowyer, P.; Danson, F.M. Sensitivity of Spectral Reflectance to Variation in Live Fuel Moisture Content at Leaf and Canopy Level. *Remote Sens. Environ.* **2004**, *92*, 297–308. [CrossRef]

66. Congalton, R.G.; Green, K. *Assessing the Accuracy of Remotely Sensed Data*; CRC Press: Boca Raton, FL, USA, 2019. [CrossRef]

67. Meyer, H.; Reudenbach, C.; Wöllauer, S.; Nauss, T. Importance of spatial 100 predictor variable selection in machine learning applications—Moving from data 101 reproduction to spatial prediction. *Ecol. Model.* **2019**, *411*, 108815. [CrossRef]

remote sensing

Article

Quantifying Lidar Elevation Accuracy: Parameterization and Wavelength Selection for Optimal Ground Classifications Based on Time since Fire/Disturbance

Kailyn Nelson *, Laura Chasmer and Chris Hopkinson

Department of Geography and Environment, University of Lethbridge, 401 University Drive, Lethbridge, AB T1K 3M4, Canada
* Correspondence: kailyn.nelson@uleth.ca

Abstract: Pre- and post-fire airborne lidar data provide an opportunity to determine peat combustion/loss across broad spatial extents. However, lidar measurements of ground surface elevation are prone to uncertainties. Errors may be introduced in several ways, particularly associated with the timing of data collection and the classification of ground points. Ground elevation data must be accurate and precise when estimating relatively small elevation changes due to combustion and subsequent carbon losses. This study identifies the impact of post-fire vegetation regeneration on ground classification parameterizations for optimal accuracy using TerraScan and LAStools with airborne lidar data collected in three wavelengths: 532 nm, 1064 nm, and 1550 nm in low relief boreal peatland environments. While the focus of the study is on elevation accuracy and losses from fire, the research is also highly pertinent to hydrological modelling, forestry, geomorphological change, etc. The study area includes burned and unburned boreal peatlands south of Fort McMurray, Alberta. Lidar and field validation data were collected in July 2018, following the 2016 Horse River Wildfire. An iterative ground classification analysis was conducted whereby validation points were compared with lidar ground-classified data in five environments: road, unburned, burned with shorter vegetative regeneration (SR), burned with taller vegetative regeneration (TR), and cumulative burned (both SR and TR areas) in each of the three laser emission wavelengths individually, as well as combinations of 1550 nm and 1064 nm and 1550 nm, 1064 nm, and 532 nm. We find an optimal average elevational offset of ~0.00 m in SR areas with a range (RMSE) of ~0.09 m using 532 nm data. Average accuracy remains the same in cumulative burned and TR areas, but RMSE increased to ~0.13 m and ~0.16 m, respectively, using 1550 nm and 1064 nm combined data. Finally, data averages ~0.01 m above the field-measured ground surface in unburned boreal peatland and transition areas (RMSE of ~0.19 m) using all wavelengths combined. We conclude that the 'best' offset for depth of burn within boreal peatlands is expected to be ~0.01 m, with single point measurement uncertainties upwards of ~0.25 m (RMSE) in areas of tall, dense vegetation regeneration. The importance of classification parameterization identified in this study also highlights the need for more intelligent adaptative classification routines, which can be used in other environments.

Keywords: elevation; airborne laser scanning; peatland; carbon; accuracy; change detection; disturbance

Citation: Nelson, K.; Chasmer, L.; Hopkinson, C. Quantifying Lidar Elevation Accuracy: Parameterization and Wavelength Selection for Optimal Ground Classifications Based on Time since Fire/Disturbance. *Remote Sens.* **2022**, *14*, 5080. https://doi.org/10.3390/rs14205080

Academic Editors: Leonor Calvo, Elena Marcos, Susana Suarez-Seoane and Víctor Fernández-García

Received: 7 September 2022
Accepted: 8 October 2022
Published: 11 October 2022

Publisher's Note: MDPI stays neutral with regard to jurisdictional claims in published maps and institutional affiliations.

1. Introduction

Boreal peatlands contain considerable carbon (C) stores and have acted as a long-term sink for atmospheric C since the Holocene [1,2]. However, with climate change, many of these peatland regions are drying and becoming more vulnerable to wildland fire [3–5], which are increasing in both frequency and severity [4,6]. There is interest in quantifying the contribution of peat combustion to atmospheric C [7–9]. Improving estimations of C loss during wildland fire is especially critical in boreal environments, where soil combustion can account for up to ~90% of C loss [7].

In recent years, several studies [6,10–12] have described the loss of C from wildland fire in peatlands; however, there are methodological limitations for estimating C loss across a broad range of peatland and boreal ecosystems. Fieldwork is labor-intensive and time-consuming and cannot survey the full range of environmental variations that influence the loss of C from fire in peatland landscapes [8,13,14]. Optical remote sensing is often utilized to estimate burn severity and is particularly useful in its ability to cover broad spatial extents (e.g., [13,15]); however, optical remote sensing of the understory and ground surface is occluded by the pre-fire vegetation canopy and any remaining post-forest canopy—a limitation in assessing burn severity as well as pre-fire conditions [8,13,16]. Therefore, these sensing techniques cannot easily measure ground surface elevation and cannot measure depth of burn, an essential component of biomass loss [9].

Airborne and Unpiloted Aerial Vehicle (UAV) lidar provide an opportunity to resolve both the lack of spatial coverage of field data and reduced ability to determine ground elevation from high spatial resolution optical remote sensing due to occlusion. Lidar is useful for measuring ground surface elevations and vegetation structural characteristics across a range of land cover types, including boreal peatlands (e.g., [17–19]). This capability allows for not only the quantification of pre-fire fuels and post-fire ecosystem regeneration in the study of wildland fire (e.g., [9,20,21]), but also in forestry (e.g., [19,22]), urban planning and road design (e.g., [13]), hydrological modelling (e.g., [23]), mapping and modelling of land cover distribution (i.e., wetlands) (e.g., [24,25]), monitoring of permafrost thaw (e.g., [9]), soil erosion (e.g., [26]) and flooding (e.g., [27]). A benefit of the use of lidar is the ability to measure both canopy structure, understory, and ground surface elevation [13]. Multi-temporal, pre- and post-fire lidar data also enable quantification of biomass losses from fire and post-fire vegetation regeneration (e.g., [28]). As laser pulse returns can measure ground surface elevation, the technology is particularly useful for determining surface elevation changes, such as depth of burn during wildland fire, quantification of erosion, and impacts of permafrost slaw if pre- and post-disturbance lidar data are available. However, despite its utility, questions arise on the accuracy of lidar data for determining elevation (and therefore depth of burn, C losses, etc.) associated with different ground classification routines and also, the efficacy of lidar-based observations as time since fire increases. Ref. [13] suggest that most error is introduced during the classification stage; however, custom, environment-specific ground classification parameterization can improve DEM accuracy [19,29]. Due to the need for accurate ground surface data when quantifying relatively small changes in elevation from combustion, erosion, slumping, permafrost thaw, and anthropogenic disturbance, the quantification of ground classification routines specific to land cover and vegetation growth that result in the least error is required. There is also an urgent need for more accurate measurements of soil combustion and overall C losses from boreal peatlands and their potential influence on the global climate system [11,30].

Based on the necessity for accurate ground elevation data for estimating depth of peatland burn in pre- and post-fire lidar data, this study aims to: (a) identify how post-fire vegetation regeneration impacts optimal ground classification configurations using industry-standard software: TerraScan (Terrasolid, Helsinki, Finland) and LAStools (Rapid-Lasso, Gilching, Germany, GmbH); and (b) compare multispectral lidar emission wavelength(s) (532 nm (green); 1064 nm (near infrared); 1550 nm (shortwave infrared); 1064 and 1550 nm combined; and, 532, 1064, and 1550 nm combined) in burned and unburned boreal peatlands and transition zones in western Canada. The overall goal is to provide recommendations for ground classification of lidar data across a range of vegetation regeneration required for quantifying subtle changes in elevation, including depth of burn from fire (in bi-temporal, pre- and post-fire lidar data). While this research focuses on wildland fire, recommendations will also be useful for hydrological modelling, forestry applications, and land surface engineering/mining/cut-fill operations.

2. Materials and Methods

2.1. Study Area

The study area is located about 30 km south of Fort McMurray, Alberta (centre: 12N 482464E 6260554N) in the Boreal Plains ecozone of Canada (Figure 1) [31]. The region is characterized by flat to slightly undulating terrain with some hummocky zones. It is dominated by bog and fen peatlands (dominant wetland classes in Alberta [32]), aspen (*Populus tremuloides*) uplands, and black spruce (*Picea mariana*) lowlands/transition zones [31]. Extensive forestry and oil exploration and extraction occur within the region, as do subsistence and commercial hunting and fishing and minor agricultural practices [31].

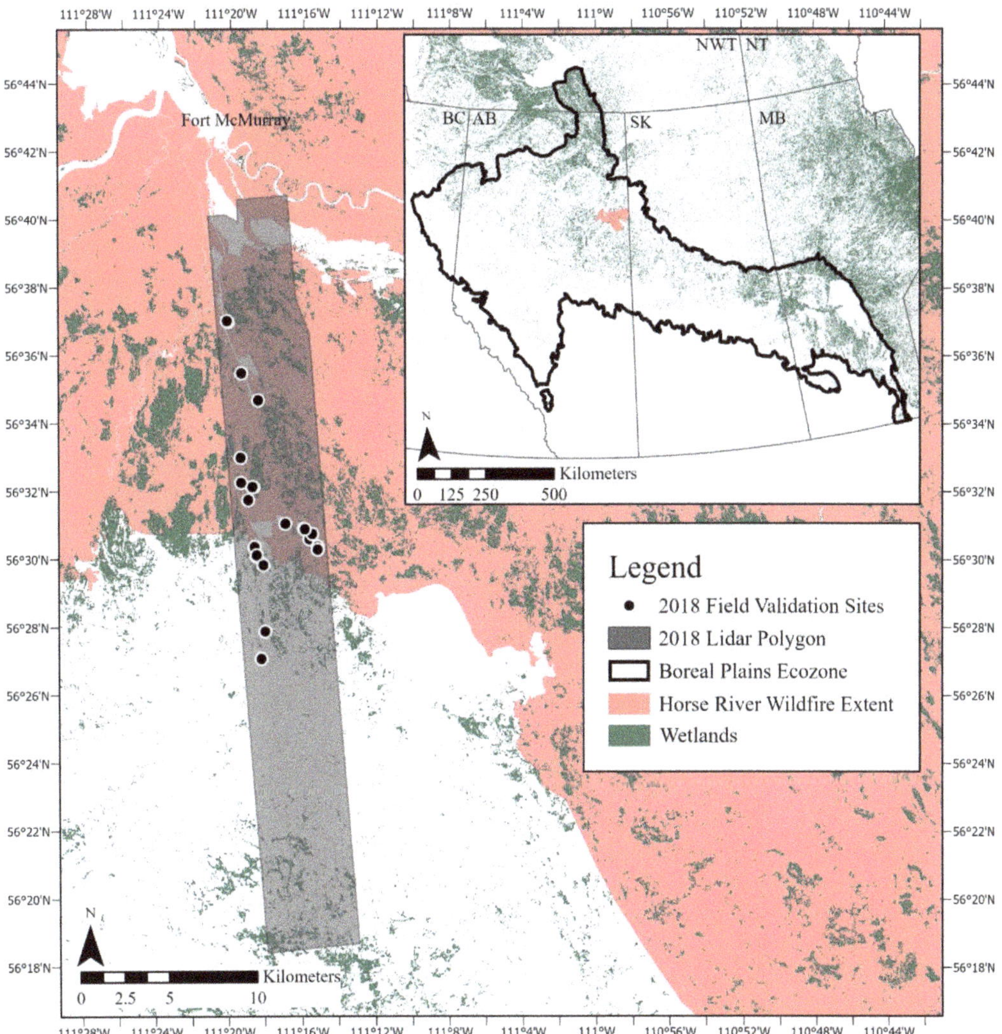

Figure 1. Map illustrating the extent of the Horse River Wildfire within the Boreal Plains Ecozone (inset), which extends across Canada from northern British Columbia (BC) and into Alberta (AB), Saskatchewan (SK), and Manitoba (MB) and the study area, including lidar survey polygon and field validation transects/plots.

The study covers a 20,441-ha area south of Fort McMurray, extending beyond the area burned by the Horse River Wildfire in 2016 (Figure 1). The Horse River Wildfire, covering approximately 600,000 ha, ignited 7 km outside Fort McMurray on 1 May 2016, under hazardous conditions—uncharacteristically hot (~35 °C), dry, and windy (~43 km hr^{-1}) weather. The fire was declared under control on 4 July 2016; however, smoldering peat burn continued for approximately 15 months before being extinguished [33,34]. The burned region includes a variety of burn severities and levels of vegetative regeneration since the fire, from little to no regeneration to significant vegetation growth (Figure 2). This allows for the opportunity to compare laser pulse interactions and ground elevation accuracies across a range of conditions akin to timing of lidar data collection following fire.

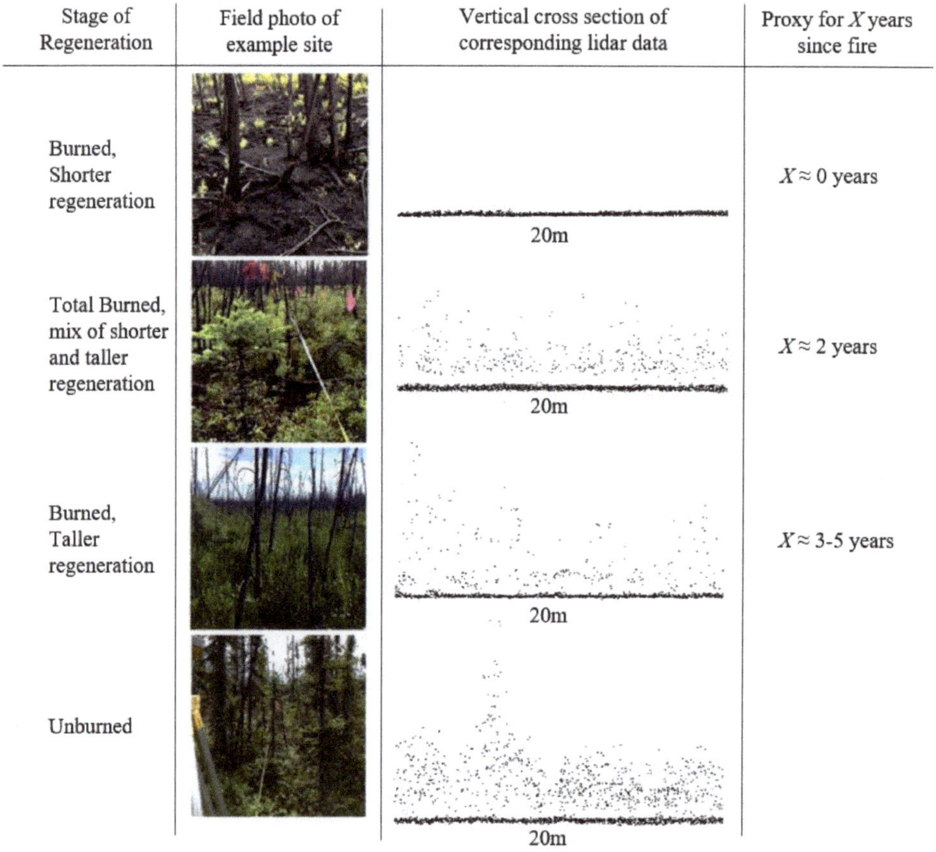

Stage of Regeneration	Field photo of example site	Vertical cross section of corresponding lidar data	Proxy for X years since fire
Burned, Shorter regeneration		20m	$X \approx 0$ years
Total Burned, mix of shorter and taller regeneration		20m	$X \approx 2$ years
Burned, Taller regeneration		20m	$X \approx 3\text{-}5$ years
Unburned		20m	

Figure 2. Four vegetation categories used to represent time since fire with field photos and lidar point clouds.

In sites with shorter regeneration (SR), post-fire vegetation heights averaged 0.20–0.35 m (Figure 2). Dominant vegetation was primarily *Sphagnum* spp. And feathermoss (*Pleurozium* spp.), with subdominant vegetation consisting of mosses, herbs, and low-lying herb species such as Labrador tea (*Rhododendron groenlandicum*), reindeer lichen (*Cladonia rangiferina*), bog cranberry (*Vaccinium oxycoccos*), cloudberry (*Rubus chamaemorus*), and horsetail (*Equisetum fluviatile*). In sites with taller post-fire vegetative regeneration (TR), above-surface vegetation heights averaged 0.40–1.00 m (Figure 2). While dominant vegetation included some similar species as the SR sites such as feathermoss and *Sphagnum* spp., sites were also dominated by more woody vegetation and tall shrubs, such as willow (*Salix* spp.), bog, shrub, and paper birch (*Betula pumila, glandulosa,* and *papyrifera*), black spruce (*Picea mariana*), fireweed (*Chamaenerion angustifolium*), horsetail (*Equisetum* spp.), rose (*Rosa acicularis*), trembling aspen (*Populus tremuloides*), and raspberry (*Rubus idaeus*).

2.2. Data Acquisition

Airborne lidar data were collected in July 2018, two years following the Horse River Wildfire, using a Titan multispectral lidar (Teledyne Optech, Inc., Vaughan, ON, Canada). The Titan collects data using three laser emission wavelengths (channels): 1550 nm (shortwave infrared (SWIR); channel 1), which is 3.5° forward of nadir; 1064 nm (near-infrared (NIR); channel 2), which is emitted at nadir; and 532 nm (green; channel 3) which is 7° forward of nadir (Figure 3a) [18]. The survey was flown at ~1000 m above ground, with scan angles of ±25 degrees, a pulse repetition frequency of 100 kHz per channel (300 kHz total), and a 50% flightline overlap. This survey configuration resulted in average point densities of 4.8 pts m^{-2}, 4.2 pts m^{-2}, and 2.1 pts m^{-2} for channels 1, 2, and 3, respectively. As laser scan lines are not spatially coincident, the 50% overlap reduces gaps, especially prevalent along scan line edges, such that validation points do not exist within one or two channels, introducing bias (Figure 3b).

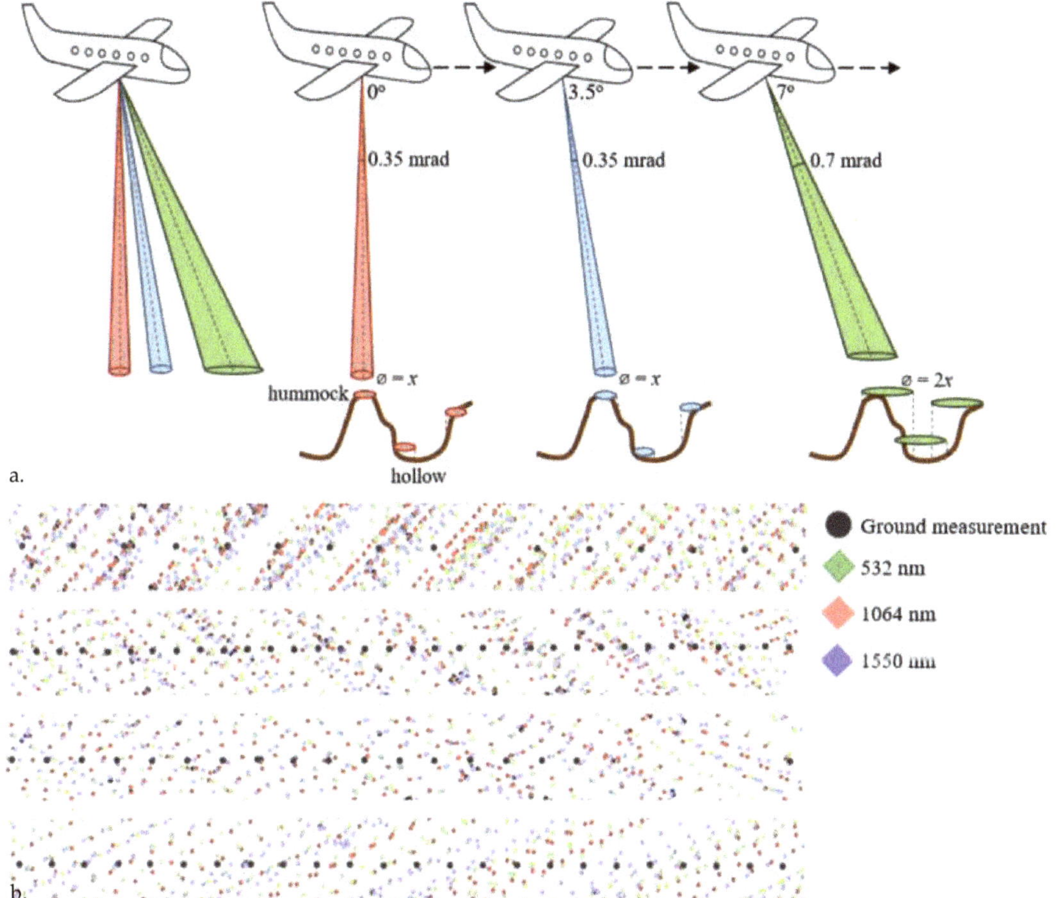

Figure 3. (**a**) Illustration of lidar laser beam angles, beam divergence, and impact on footprint diameter (Ø) in peatlands with variable microtopography (hollows and hummocks); (**b**) Samples of validation transects and lidar data demonstrating spatial distribution of validation points throughout the three channels. Note: microtopography in (**a**) has been exaggerated for demonstration purposes.

Field data were collected coincident with the 2018 lidar data collection for the calibration and validation of lidar data. To select sample sites, the study area was first stratified into scales of influence: (a) burned versus non-burned areas within and proximal to the Horse River wildfire; (b) within burned areas, different classes of burn severity (minimum, medium, and severe) determined visually from optical remote sensing imagery as well as through on-the-ground assessments at the time of field data collection; and, (c) peatland type (treed and open bogs; rich and poor fens) determined from optical remote sensing imagery and on-the-ground assessments. Data were collected along ~30 m transects in 18 burned and 6 unburned peatland sites. Transects intersected upland-peatland transition zones and peatlands. Global Navigation Satellite System (GNSS) ground elevation validation points were collected in burned and unburned landscapes. To validate post-fire ground surface elevations [9], GNSS stations were placed at the beginning and end of each transect and were left to run for the duration of sampling (>1 h for centimeter accuracy). Precise Point Positioning (PPP) was used to process these end points. A level was used at one- (burned sites) or two- (unburned sites) meter intervals to determine ground elevation relative to the GNSS base stations. A total of 708 ground elevations were measured: 130 in unburned and 578 in burned peatlands with variable rates of vegetation regeneration. Post Processed Kinematic (PPK) GNSS elevation locations were also collected along two road surfaces (n = 2655) to ensure the elevational accuracy of airborne lidar data in areas of flat terrain without any overstory canopy influences on ground surface elevation [17,35].

2.3. Data Processing

Lidar returns from road surfaces were compared with PPK GNSS survey points and vertically batch-shifted to ensure the average offset between lidar data and calibration points was zero [35] using Bentley Microstation TerraSolid Terrascan software version 021.011 (Terrasolid, Helsinki, Finland) [36]. Isolated or outlier points were removed, and an iterative ground classification analysis was then conducted through which ground vs. non-ground returns were classified. The five ground cover types (road, unburned, short vegetation regeneration (SR), and tall vegetation regeneration (TR), and cumulative burned (all regeneration stages)) were analyzed separately to identify optimal ground return classifications for each type, optimized for relatively flat to slightly undulating boreal peatland and transitional environments with micro-topographic hummocks and hollows.

2.3.1. TerraScan

Within TerraScan, six ground classification parameters can be readily modified: (a) 'max building size', which sets the grid size for seed ground point selection; (b) 'terrain angle', which is the maximum slope between a seed point and a candidate point; (c) 'iteration angle', which is the maximum angle that a point can be added to the ground classification; (d) 'iteration distance', which is the maximum distance that a point can be added to the ground classification; (e) 'reduce iteration angle', a binary choice which reduces the number of unnecessary points added to the surface in areas of high point density by reducing the number of points that are added to the surface if edge length is longer than all triangle edges; and (f) 'stop triangulation', another binary choice which reduces the number of unnecessary points added to the surface by not processing points within a triangle if edge length is longer than all triangle edges [37,38] (Table 1). Adjusting each ground classification parameter results in morphological differences in the resultant ground-classified data.

Table 1. TerraScan ground classification parameter modifications used in ground classification iteration analysis.

Ground Classification	Max Building Size	Terrain Angle	Iteration Angle	Iteration Distance	Reduce Iteration Angle When Edge Length<	Stop Triangulation When Edge Length<
1	60	50	6	1.4	5	-
2	60	55	6	1.4	5	-
3	60	60	6	1.4	5	-
4	60	65	6	1.4	5	-
5	60	70	6	1.4	5	-
6	60	75	6	1.4	5	-
7	60	80	6	1.4	5	-
8	60	88	6	1.4	5	-
9	60	88	2	1.4	5	-
10	60	88	2	0.5	5	-
11	60	88	2	1	5	-
12	60	88	2	1.5	5	-
13	60	88	2	2	5	-
14	60	88	5	1.4	5	-
15	60	88	5	0.5	5	-
16	60	88	5	1	5	-
17	60	88	5	1.5	5	-
18	60	88	5	2	5	-
19	60	88	10	1.4	5	-
20	60	88	10	0.5	5	-
21	60	88	10	1	5	-
22	60	88	10	1.5	5	-
23	60	88	10	2	5	-
24	60	88	15	1.4	5	-
25	60	88	15	0.5	5	-
26	60	88	15	1	5	-
27	60	88	15	1.5	5	-
28	60	88	15	2	5	-
29	60	88	15	1.5	-	-
30	60	88	15	1.5	1	-
31	60	88	15	1.5	2	-
32	60	88	15	1.5	10	-
33	60	88	15	1.5	5	0.5
34	60	88	15	1.5	5	2
35	60	88	15	1.5	5	5
36	60	88	15	1.5	5	0.25

Thirty-six different ground classification parameterizations were developed by making adjustments to TerraScan's classification parameters (Table 1). Classifications 1–28 were produced by iterating through adjustments to the four primary parameters: 'Max Building Size', 'Terrain Angle', 'Iteration Angle', and 'Iteration Distance'. Classifications 29–36 were developed by refining an optimal classification (27) using 'Reduce Iteration Angle When Edge Length<' and 'Stop Triangulation When Edge Length<' (Table 1). Each of the classification parameterizations was used to produce ground surfaces in the channels and channel combinations tested, resulting in 180 distinct ground surfaces. Each surface output was compared to ground elevation field validation data (elevation collected along the road, unburned peatlands, TR peatlands, SR peatlands, and total burned peatlands) using TerraScan's control report function (see Section 2.4) for a total of 740 ground classification tests.

2.3.2. LAStools

LASground, from LAStools (RapidLasso GmbH, Gilching, Germany), offers five defined ground classifications that, like TerraScan, use an adaptive TIN algorithm to classify ground points. Two were tested (excluding urban settings): 'nature' and 'wilderness'. These settings differ in their step size (cell size within which lowest point becomes initial ground point) and 'bulge' (height allowance for TIN to "bulge" above planar surface). These can be refined using the options 'default', 'fine', 'extra', 'ultra', and 'hyper', resulting in ten different readily accessible ground classification parameterizations (lettered A–J; Table 2). Using refining options intensifies the search for seed ground points—this is often most useful for ground surfaces with steep hills [39]. Each of the ten different classification parameterizations was used to produce ground surfaces in the channels and channel combinations tested, resulting in 50 distinct ground surfaces. These were brought into TerraScan for the quantification of control point statistics. Like the TerraScan analysis, each ground surface was compared with field elevation measurements from road, unburned peatlands, TR peatlands, SR peatlands, and burned peatlands cumulatively, for a total of 250 ground classification tests.

Table 2. LASTools ground classification parameter modifications used in ground classification iteration analysis.

	Ground Classification	Refinement	Bulge (m)	Step Size (m)	Subgrid for Initial Ground Points
A	Nature	Default	0.5	5	=step size
B	Nature	Fine	0.5	5	Default granularity × 4
C	Nature	Extra Fine	0.5	5	Fine granularity × 4
D	Nature	Ultra-Fine	0.5	5	Extra Fine granularity × 4
E	Nature	Hyper Fine	0.5	5	Ultra-Fine granularity × 4
F	Wilderness	Default	0.3	3	=step size
G	Wilderness	Fine	0.3	3	Default granularity × 4
H	Wilderness	Extra Fine	0.3	3	Fine granularity × 4
I	Wilderness	Ultra-Fine	0.3	3	Extra Fine granularity × 4
J	Wilderness	Hyper Fine	0.3	3	Ultra-Fine granularity × 4

2.4. Vertical Accuracy Assessment

Ground classification outputs performed using TerraScan ($n = 36$) and LAStools software ($n = 10$) in each of the three available laser emission wavelengths and wavelength combinations were compared with field-measured elevations from road ($n = 2655$), unburned peatlands ($n = 130$), burned peatlands ($n = 578$), TR peatlands ($n = 267$), and SR peatlands ($n = 269$). Validation data were segregated into TR vs. SR vegetative regeneration based on average measured vegetation height and dominant species per plot (1 m × 1 m with three elevation measures through the center of each plot, perpendicular to the transect; Figure 2). The separation of peatlands based on vegetation provides an opportunity to quantify elevation accuracy from lidar across a range of environmental characteristics, including unburned with a full understory, burned with no or shorter regeneration (SR; a proxy for lidar data collected immediately post-fire), and burned with tall regeneration (TR; a proxy for data collection several years post-fire). Validation data were distributed throughout the study area, and the number of validation elevations measured in the field exceeded the minimum ($n = 20$) and the recommended ($n = 30$) suggested for each vegetation cover type by the American Society for Photogrammetry and Remote Sensing [13,40,41]. All ground-classified data were examined in TerraScan, where validation point elevations were compared with lidar point elevations, a standard methodology for lidar vertical accuracy assessments [41,42]. Through TerraScan's control report function, lidar points were used to interpolate a surface using a Triangulated Irregular Network (TIN). As it is unlikely that a lidar point exists at the same x, y location as a validation point, validation points were compared to their x, y location on the TIN surface [43,44]. Control point statistics, including the difference in elevation between control points and lidar ground returns (dz; average, maximum, and minimum), standard deviation, and root mean square error (RMSE), were quantified via TerraScan's control report function [44].

To identify the optimal ground classification for each vegetation cover type, classified outputs were assessed based on RMSE (commonly used to determine accuracy) [41,45,46] and by absolute average dz ($|\overline{dz}|$), while also being mindful of point density. Optimal ground classification statistics ($\overline{dz}$ and standard deviation (SD)) were then used to determine total error ($\overline{dz} \pm$ SD) when using multitemporal lidar data to assess ground surface elevation changes (pre- and post-fire). The uncertainties associated with multi-temporal surface elevation measurements are independent of one another, so the propagated error (SD) was calculated through quadrature, (Equation (1)), where Q is the average over- or under-estimation of surface elevation change, $\in a$ is cumulative SD, $\overline{dz}(b)$ and $\overline{dz}(c)$ are the average deviations of lidar classified ground surface from the measured ground surface at times b and c, and $\in b$ and $\in c$ are the SDs of ground surface measurements at two points in time (i.e., pre- and post-fire). Note that the average deviation ($\overline{dz}$) is used, and not absolute average deviation ($|\overline{dz}|$).

$$Q \pm \in a = \left(\overline{dz}(b) + \overline{dz}(c)\right) \pm \sqrt{(\in b)^2 + (\in c)^2} \tag{1}$$

3. Results

The results demonstrate wavelength dependencies and optimal ground classification parameterizations for each vegetation cover type tested within TerraScan and LAStools.

3.1. Differences between Ground-Surveyed Road Elevations and Lidar-Measured Road Ground Classifications

The optimal ground return classification aims to observe the lowest differences in ground surface elevations between field validation and nearby laser returns in each wavelength. Based on the flat, non-vegetated road surface GNSS measurements, we found that neither classification parameter choice (Tables 1 and 2), nor wavelength, significantly impacted the quality of the ground classification along road surfaces (Figures 4 and 5; Table S1). Using both TerraScan and LAStools, the $|\overline{dz}|$ from the measured elevations ranged from 0.00 to 0.02 m, and RMSE from 0.04 m to 0.05 m.

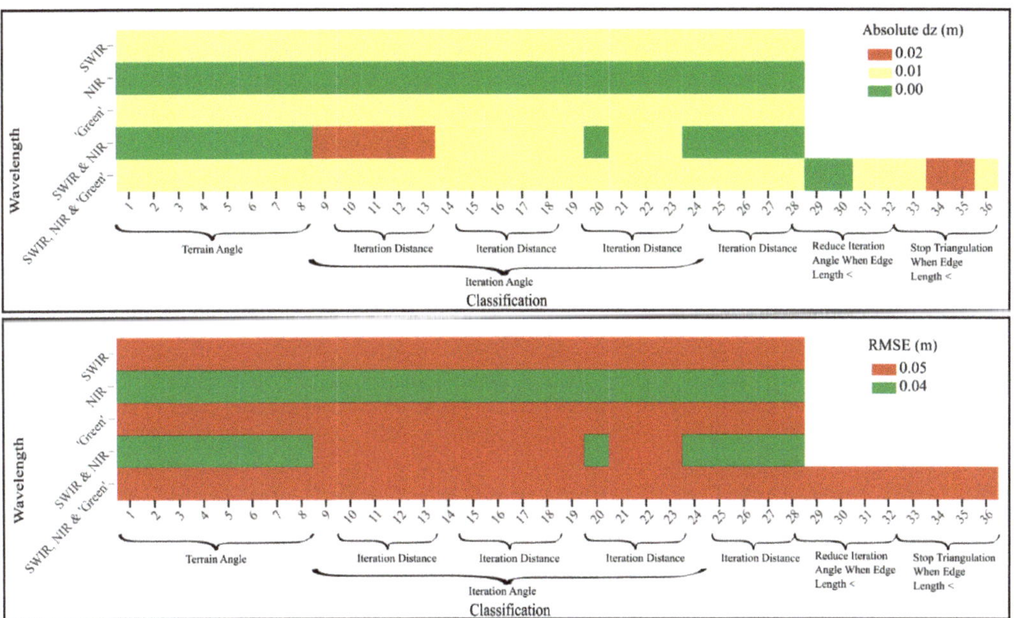

Figure 4. Ground classification results ($|\overline{dz}|$ and RMSE) along a flat road surface for baseline comparisons using parameterization methods in Table 1. Classifications were conducted in TerraScan.

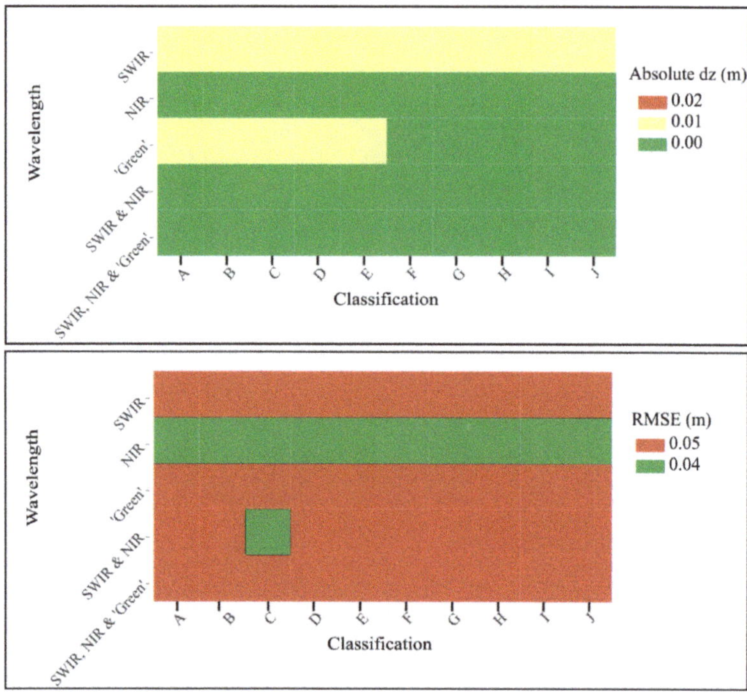

Figure 5. Ground classification results ($|\overline{dz}|$ and RMSE) along a flat road surface for baseline comparisons using parameterization methods in Table 2. Classifications were conducted in LAStools.

This provides a baseline for comparisons to ground classifications in varying vegetation regeneration stages and demonstrates that any changes observed in ground classification accuracies result from different parameterizations responding differently to variable vegetation regeneration.

3.2. Differences between Field-Measured Elevation and Lidar Return Ground Classification in Shorter Vegetative Regeneration Peatlands

In SR peatlands and transitional areas (representing characteristics that are similar to peatlands that have been surveyed soon after a fire), we found that the ground classifications that produced the most accurate results in TerraScan were classifications 14 through 18 (a slight reduction in iteration angle from six to five as compared to default) with laser pulse emission at 532 nm (Figure 6 and Table 1). These all produced a lidar-derived ground-classified elevation with an $|\overline{dz}|$ = 0.00 m (RMSE = 0.09 m). However, the point density was below 1 point m^{-2}; (0.86 points m^{-2}; Table S1). If a higher point density were required, using all three laser pulse emissions (IR, NIR and Visible) and increasing iteration angle from 6 to 15, as well as reducing iteration angle when edge length < 1.0, 2.0, or turned off (classifications 29–31) produce nearly-as-accurate ground surfaces with $|\overline{dz}|$ = 0.01 m (RMSE = 0.09 m) and 4.06, 3.94, and 3.32 points m^{-2}, respectively (Figure 6; Tables 1 and S1). In more typically used lidar systems that collect data at 1064 nm (NIR), the optimal classifications were 24–28 (iteration angle increased from 6 to 15°; Table 1), which resulted in $|\overline{dz}|$s slightly elevated above the true ground surface ($|\overline{dz}|$ = 0.03 m; RMSE = 0.10 m; 1.03 points m^{-2}; Figure 6 and Table S1).The poorest ground classifications for SR areas were those within which iteration angle was narrowed to 2°.

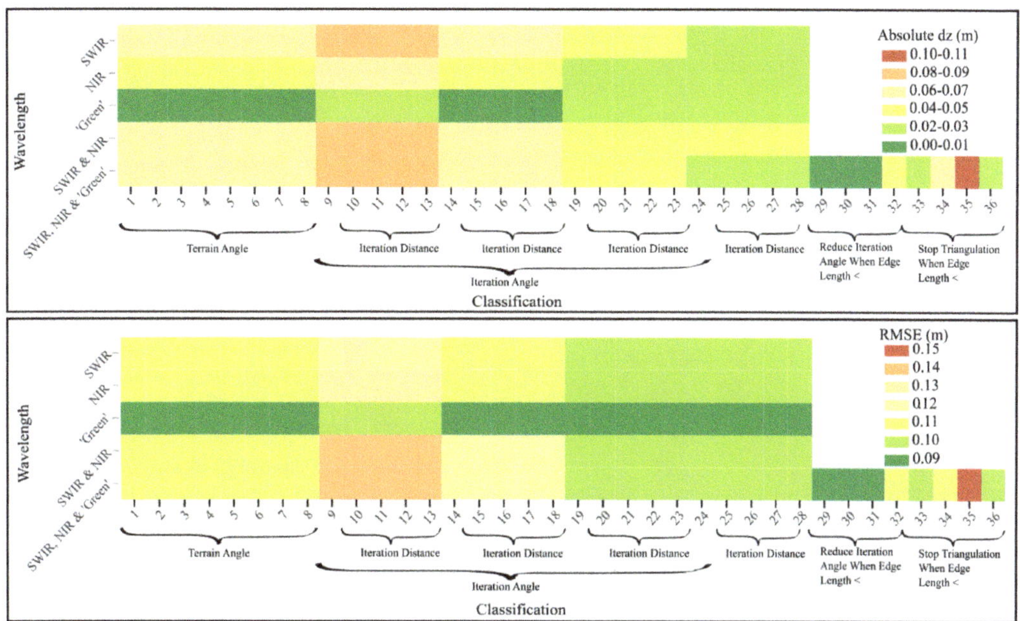

Figure 6. Ground classification results ($|\overline{dz}|$ and RMSE) in burned peatlands with low vegetation regeneration two years post-fire (as a proxy for immediately post-fire). Classifications were conducted in TerraScan.

Using LAStools, ground classifications F–J using lidar data collected at all three wavelengths were optimal (Figure 7 and Table 2). These classifications produced lidar-measured ground elevations that did not, on average, deviate from the true ground surface ($|\overline{dz}|$ = 0.00 m; RMSE = 0.09 m; 3.70–3.72 points m^{-2}; Figure 7 and Table S1). In this case, where all channels were used, refinement did not impact the ground-classified surface's accuracy or point density (m^{-2}). In a lidar system that collects data in the 1064 nm wavelength, optimal classifications were G–J, which produced lidar-measured ground classifications with $|\overline{dz}|$ = 0.01 m (RMSE = 0.09 m; 1.3–1.31 points m^{-2}; Figure 7 and Table 2). However, in this landscape, when using LAStools, the most significant difference in classification accuracy was due to the channel with which the data were collected; changes within a channel were negligible (i.e., within the 1064 nm data: $|\overline{dz}|$ remained at 0.01 m regardless of classification, RMSE only varied by 0.01 m (0.09–0.10 m), and point density varied from 1.24–1.31 points m^{-2} (Figure 7 and Table S1). While the optimal classifications from TerraScan and LAStools were comparable, the poorest classifications from each were notably different. The classifications produced in LAStools had an $|\overline{dz}|$ ranging from 0.00–0.03 m and an RMSE ranging from 0.09–0.10 m, whereas TerraScan classifications had an $|\overline{dz}|$ ranging from 0.00–0.09 m and an RMSE ranging from 0.09–0.15 m (Figures 5 and 6).

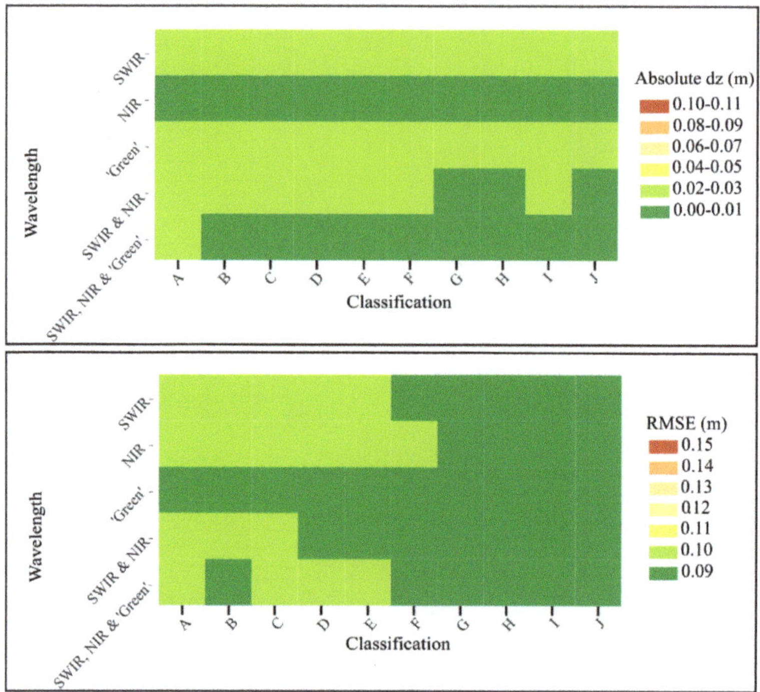

Figure 7. Ground classification results ($|\overline{dz}|$ and RMSE) in burned peatlands with low vegetation regeneration two years post-fire (as a proxy for immediately post-fire). Classifications were conducted in LAStools.

In summary, for a well-calibrated and locally controlled (e.g., over a nearby highway surface) airborne lidar survey we can expect a spatially averaged difference in elevation of <0.01 m with a range of ~0.09 m in areas of burned ground surface with no to low vegetation regeneration using optimal classifications in both TerraScan and LAStools. However, appropriate parameterization in TerraScan is dependent on the channel(s) available and required point density (Tables 1 and S1).

3.3. Differences between Field-Measured Elevation and Lidar Ground Classification across All Burned Peatlands (Cumulative Shorter Vegetative Regeneration and Taller Vegetative Regeneration Sites)

As vegetation growth increases in the two years following wildland fire, and vegetation regeneration varies from low (as in SR sites) to high (as in TR sites), optimal ground classification parameters change. In TerraScan, the greatest similarity (and lowest error) when comparing lidar ground-classified returns with the surveyed ground elevation in all burned sites combined (a proxy for ~2 years post-fire) was found when the iteration angle was increased from 6 to 15, but iteration distance was reduced to 0.5 or 1.0 from 1.4 default (parametrization methods 25 or 26) using both 1550 nm and 1064 nm data (Figure 8; Tables 1 and S1). These classifications produced lidar-measured ground elevations with an $|\overline{dz}|$ = 0.00 m (RMSE = 0.13 m; 1.88 points m^{-2}). For a typical 1064 nm laser emission wavelength system, classifications 20 through 23 produced optimal results (iteration angle increased from six to ten), also producing ground elevations with an $|\overline{dz}|$ = 0.00 m but with a slightly higher RMSE and lower point density (RMSE = 0.14 m; 1.03 points m^{-2}; Figure 8 and Table 1). As with SR areas, the ground classifications with the lowest accuracy in cumulative burned areas were those whose iteration angle was narrowed to 2.

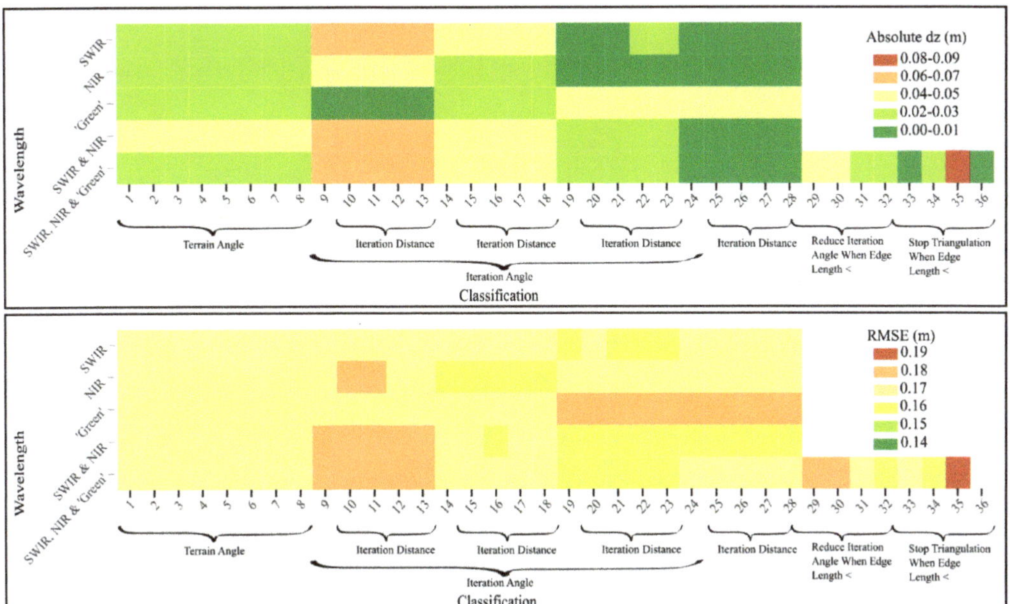

Figure 8. Ground classification results ($|\overline{dz}|$ and RMSE) in burned peatlands unsegregated based on vegetation regeneration two years post-fire (true representation of two years post-fire). Classifications were conducted in TerraScan.

When compared with optimal ground classification for laser returns in SR landscapes, the optimal ground classification used for total burned areas provides results with comparable $|\overline{dz}|$, but slightly higher uncertainty (SR RMSE = 0.09 m), as would be expected with an increase in vegetation height and height variability.

As in SR areas, classification of returns to ground and elevation accuracy in areas representative of vegetation two years post-fire (cumulative burn areas) depended more on laser pulse emission wavelength than on classification parameters applied in LAStools; however, channel optimization differed. The ground classifications that produced the most accurate results for burned surfaces compared with measured elevations were A through C, E, G, and J (Table 2), using 1550 nm data ($|\overline{dz}|$ = 0.00; RMSE = 0.14; point density = 1.27–1.34 points m^{-2}; Figure 9; Table 2); however, within a given wavelength, all parameterizations produced similar results (for example, using 1550 nm data, $|\overline{dz}|$ = 0.00–0.01 m; RMSE = 0.14–0.15 m; points m^{-2} = 1.27–1.34). By using combined 1550 nm and 1064 nm data, similar results are produced ($|\overline{dz}|$ = 0.01–0.02 m; RMSE = 0.14) but point density increases to 2.4–2.61 points m^{-2}. Similarly to SR landscapes, optimal classifications from TerraScan and LAStools were negligibly different; however, the poorest classification from LAStools was more accurate than that of TerraScan ($|\overline{dz}|$ = 0.06 m; RMSE = 0.15 m and $|\overline{dz}|$ = 0.09 m; RMSE = 0.16 m, respectively; Figures 7 and 8).

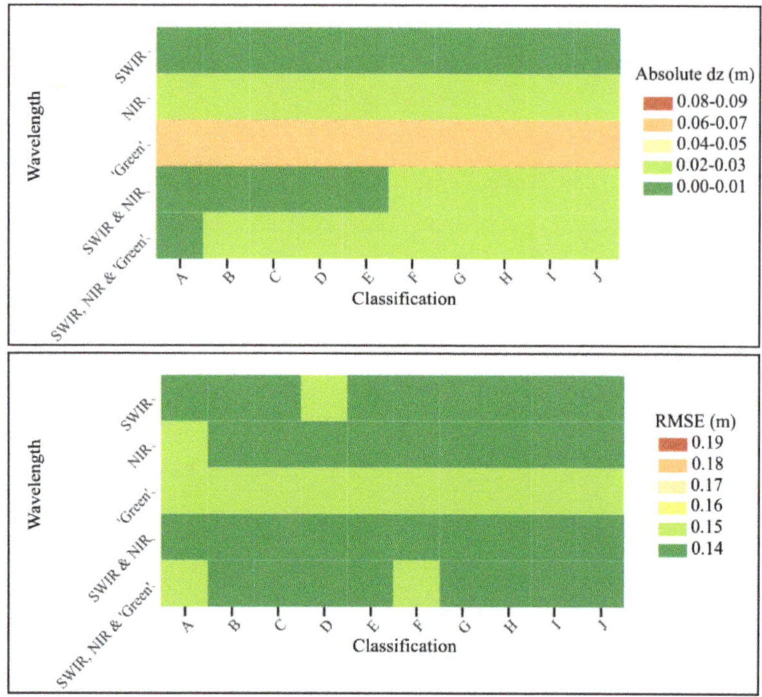

Figure 9. Ground classification results ($|\overline{dz}|$ and RMSE) in burned peatlands unsegregated based on vegetation regeneration two years post-fire (true representation of two years post-fire). Classifications were conducted in LASTools.

In the case of more typical 1064 nm lidar systems, classifications B through J produced ground elevations with an $|\overline{dz}|$ of 0.02 m, an RMSE of 0.14 m, and 1.24–1.31 points m^{-2} compared with field-measured; however, even the "poorest" classification (greatest difference from measured) produced from 1064 nm data showed nearly identical results ($|\overline{dz}|$ = 0.02; RMSE = 0.15; Figure 9 and Table 2), emphasizing the importance of channel selection during data collection over classification parameterization choice when using LAStools.

In summary, two years post-fire, we can expect an average elevational accuracy of ~0.00 m with a range of ~0.13 m using combined 1550 nm and 1064 nm data with TerraScan. Using LAStools, spatially averaged elevational accuracy is comparable at ~0.00 m but with a slightly higher range of ~0.14 m when measured at 1550 nm.

3.4. Differences between Field-Measured Elevation and Lidar Return Ground Classification in Taller Vegetative Regeneration Peatlands

In areas with the greatest vegetation growth since fire (proxies for >2 years post-burn), the most accurate ground classification in TerraScan was classification 20, using combined 1550 nm and 1640 nm data (Figure 10; Tables 1 and S1). For this classification, the iteration angle was increased from six to ten, and the iteration distance was reduced from 1.4 m to 0.5 m when compared with default parameters. This resulted in a lidar ground classification output with a ground classification accuracy of $|\overline{dz}|$ = 0.00 m (RMSE = 0.16 m; 1.54 points m^{-2}). In the case of more typical airborne lidar emitting laser pulses at 1064 nm, this classification still resulted in the most optimal ground surface (with a point density of >1 point m^2); however, the lidar-measured ground surface sat ~0.03 m above measured ($|\overline{dz}|$ = 0.03 m; RMSE = 0.17 m; 1.03 points m^{-2}). More accurate classification schemes were identified, with fewer (in this case, 0.75) points m^{-2} (classifications 14 through 18;

$|\overline{dz}|$ = 0.00 m; RMSE = 0.17 m; Figure 10; Table 1). Generally, the least accurate results were similar to those for cumulative burned areas—those within which the iteration angle was reduced to two; however, in TR areas, emission wavelength made the most difference to accuracy, with the lowest accuracy classifications produced by 532 nm data.

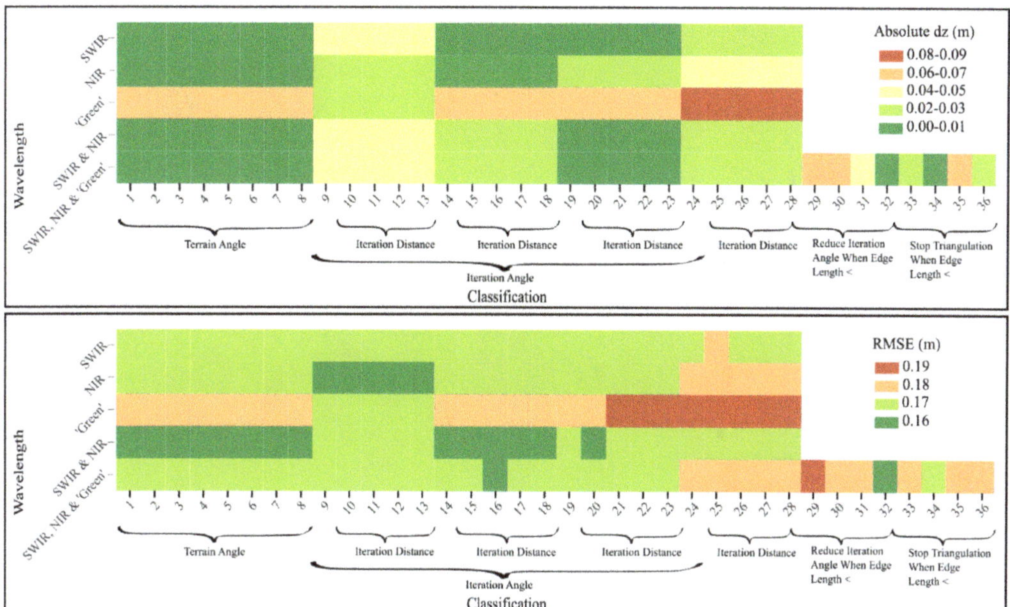

Figure 10. Ground classification results ($|\overline{dz}|$ and RMSE) in burned peatlands with tall vegetation regeneration two years post-fire (as a proxy for 3+ years post-fire). Classifications were conducted in TerraScan.

Using LAStools, the optimal ground classifications were A through E ('Nature' classifications with any level of refinement) using combined 1550 nm and 1064 nm data (Figure 11; Tables 2 and S1). The lidar-measured ground classified returns had an $|\overline{dz}|$ = 0.03 m from field-measured (RMSE = 0.18 m; 2.40–2.42 points m^{-2}). Comparable results were achieved using only 1550 nm data, but point density was reduced to 1.27–1.34 points m^{-2}. As with earlier vegetation regeneration stages, classification accuracy depended more on the wavelength than on classification parameters with LAStools. For systems using only 1064 nm emission wavelengths, the lidar-measured ground surface sits slightly above the measured ground ($|\overline{dz}|$ = 0.04–0.05 m; RMSE = 0.18 m; 1.24–1.31 points m^{-2}). Similar to TerraScan results, the poorest classifications were those produced with 532 nm data.

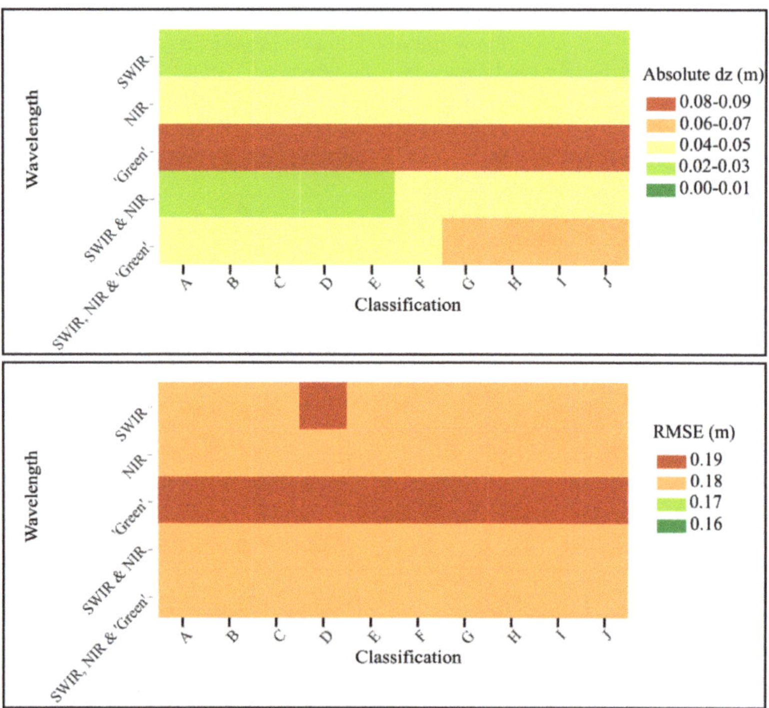

Figure 11. Ground classification results ($|\overline{dz}|$ and RMSE) in burned peatlands with tall vegetation regeneration two years post-fire (as a proxy for 3+ years post-fire). Classifications were conducted in LAStools.

In summary, unlike areas with lower vegetation regeneration, in TR areas, the optimal classifications from TerraScan and LAStools differed, with TerraScan classifications generally performing better (Figures 10 and 11). Using TerraScan we can still expect an average elevation accuracy of ~0.00 m; however, the accuracy at a given point is expected to have a range of ~0.16 m using combined 1550 nm and 1640 nm data with TerraScan. Using the same lidar data but with LAStools, the average elevational accuracy is slightly lower at ~0.03 m, with a slightly increased RMSE of ~0.18 m.

3.5. Differences between Field-Measured Elevation and Lidar Return Ground Classification in Unburned Peatlands

Finally, in unburned areas (or areas that have fully regenerated post-fire), the optimal ground classifications were 9 through 13 using all wavelengths combined (Figure 12; Tables 1 and S1). By reducing the iteration angle to two, these classifications resulted in an $|\overline{dz}|$ = 0.01 m and an RMSE = 0.19 m; however, point density was relatively low for this particular survey (0.64 points m^{-2}). When a threshold of 1.0 points m^{-2} is set, the optimal ground classifications shifted to 14 through 18, with their iteration angles set to 5° (Table S1). These classifications have a point density of 1.16 points m^{-2}, but an $|\overline{dz}|$ = 0.05 m (Figure 12; Table 1). In the case of a lidar emitting at 1064 nm, the optimal classifications are the same as those when all wavelengths are combined (9 through 13); however, classification accuracies notably decline with a threshold of 1.0 points m^{-2}. Classifications 19 through 23 provide the best outputs, in this case, with an $|\overline{dz}|$ = 0.09 m (RMSE = 0.22 m; 1.03 points m^{-2}). The least accurate classifications are those produced using lidar data collected at 532 nm, where the iteration angle is 15 (widest angle tested).

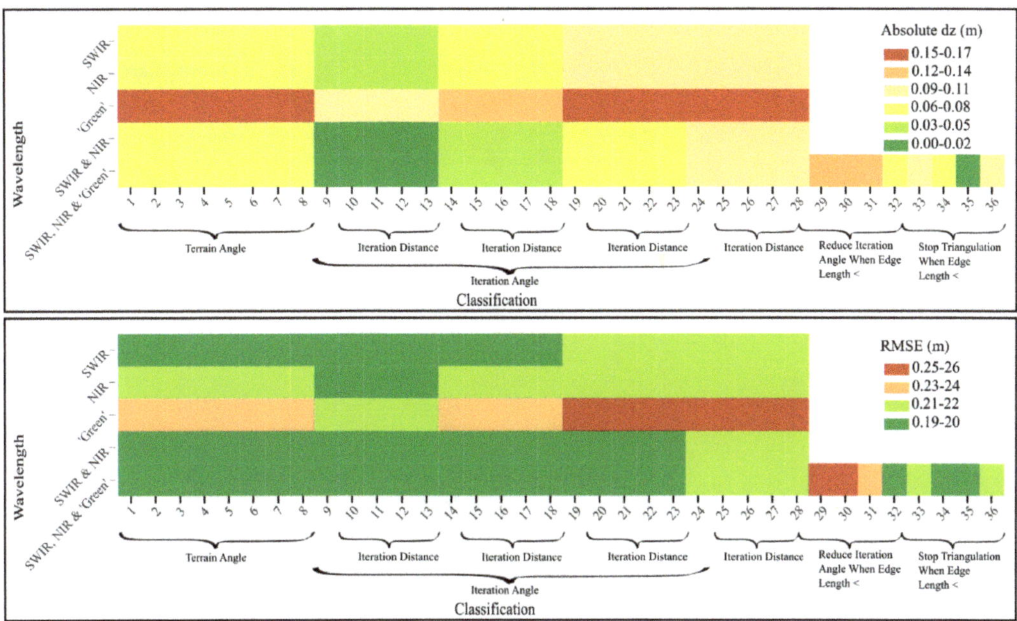

Figure 12. Ground classification results ($|\overline{dz}|$ and RMSE) in unburned peatlands. Classifications were conducted in TerraScan.

Unlike in burned landscapes, where classification was not the dominant control on accuracy compared with emission wavelength in LAStools, there was a distinction between classifications in unburned areas. In these zones, classification B using data collected at 1064 nm was optimal (Figure 13 and Table 2), resulting in the lowest difference between field-measured and ground classified returns. This classification resulted in an $|\overline{dz}|$ = 0.09 m (RMSE = 0.21 m; 1.24 points m^{-2}). As with other vegetation regeneration stages, data collected at 532 nm resulted in the least accurate ground classifications, regardless of parameterization, followed by use of all channels combined.

In summary, as with TR areas, the optimal classifications from TerraScan and LAStools differed in unburned boreal peatlands. We can expect a spatially averaged elevation accuracy of ~0.01 m (or ~0.05 m with a threshold of point density > 1 point m^{-2}) with an accuracy at a given point within an RMSE of ~0.19 m when using all channels combined, processed in TerraScan (Figure 12). Using LAStools, accuracy was poorer, with an average elevation accuracy of ~0.09 m with an RMSE of ~0.21 m at 1064 nm (Figure 13).

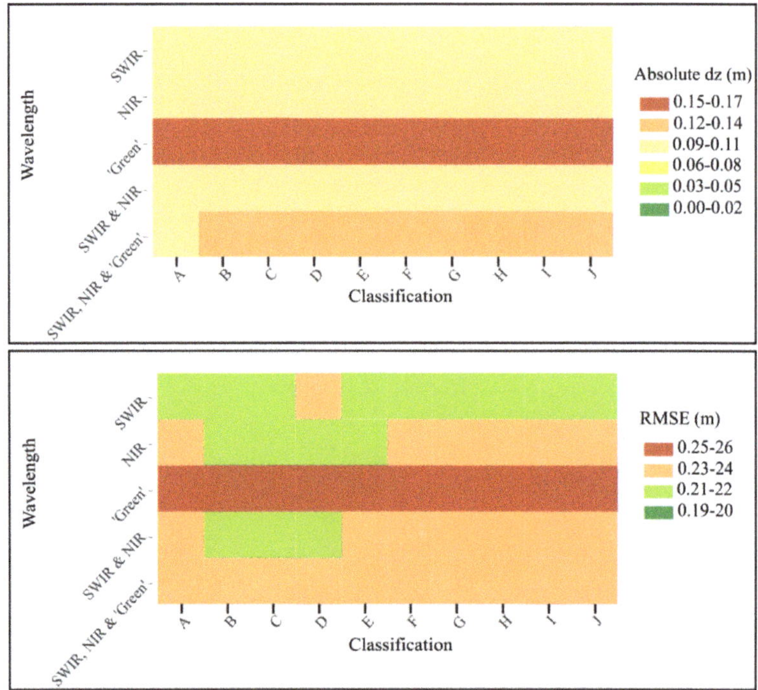

Figure 13. Ground classification results ($|\overline{dz}|$ and RMSE) in unburned peatlands. Classifications were conducted in LAStools.

3.6. Expected Ground Surface Elevation Accuracies of Lidar Data in the Years following Wildland Fire

By isolating taller versus shorter vegetative regeneration regions and using these as proxies for time since fire, we can estimate the elevation accuracy of lidar data collection in the years following wildland fire. Within the first year since fire (which we assume is the 'SR' category), lidar-measured ground elevation accuracies would be expected to be approximately 0.00 ± 0.09 m ($\overline{dz} \pm$ standard deviation (SD); Figure 14a and Table S1). For this classification, as well as the others used, below, SD was equal to RMSE. Approximately two years post-fire (at the time of this lidar data collection, assumed to include burned area surveyed–SR and TR), we would expect to see ground elevation accuracy of approximately 0.00 ± 0.13 m (Figure 14a and Table S1). As vegetation growth continues beyond the third year post-fire ('TR' category), the elevation accuracy of the lidar-measured ground classified points would be reduced to approximately 0.00 ± 0.16 m (Figure 14a and Table S1). In unburned areas, elevation accuracy would be approximately 0.01 ± 0.19 m (Figure 14a and Table S1).

As the highest uncertainty (SD) was associated with unburned (pre-fire) areas, the RMSEs associated with surface elevation changes were minimally different depending on the stage of post-fire vegetation regeneration (Figure 14b). By using optimal ground classification schemes, the post-fire $\overline{dz}$s were consistently 0.00 m, and pre-fire was ~0.01 m (Figures 4–13; Table S1). As such, the standard offset for elevation change was ~0.01 m, on average, regardless of vegetation regeneration. We found that if lidar data were collected immediately post-fire (i.e., SR areas), SD of the elevation change (depth of burn) was 0.21 m (Figure 14b). If data were collected ~ two years post-fire (i.e., cumulative burned areas), SD was 0.23 m (Figure 14b). Furthermore, if lidar data were collected three or more years post-fire (i.e., TR areas), SD was 0.25 m (Figure 14b).

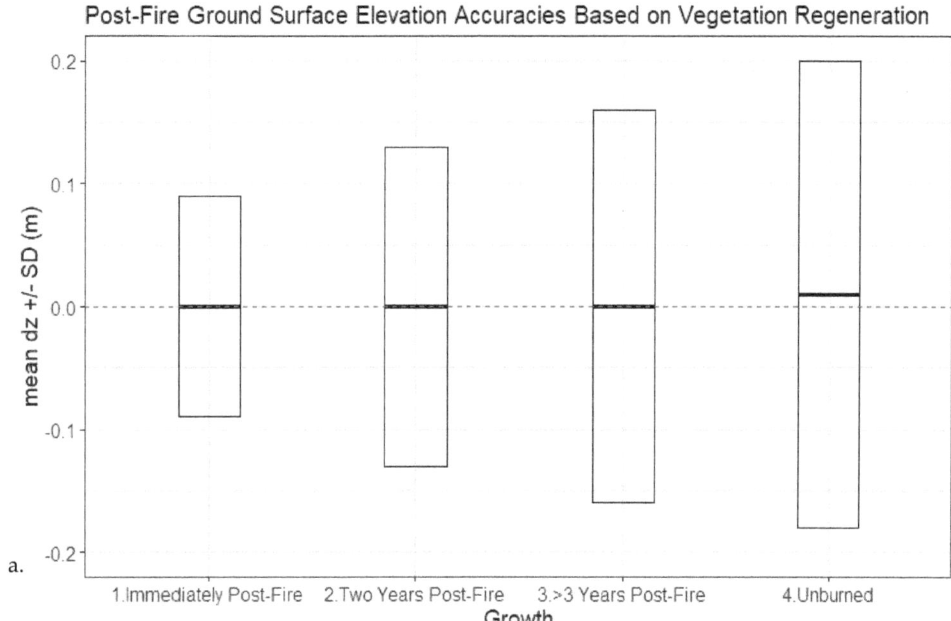

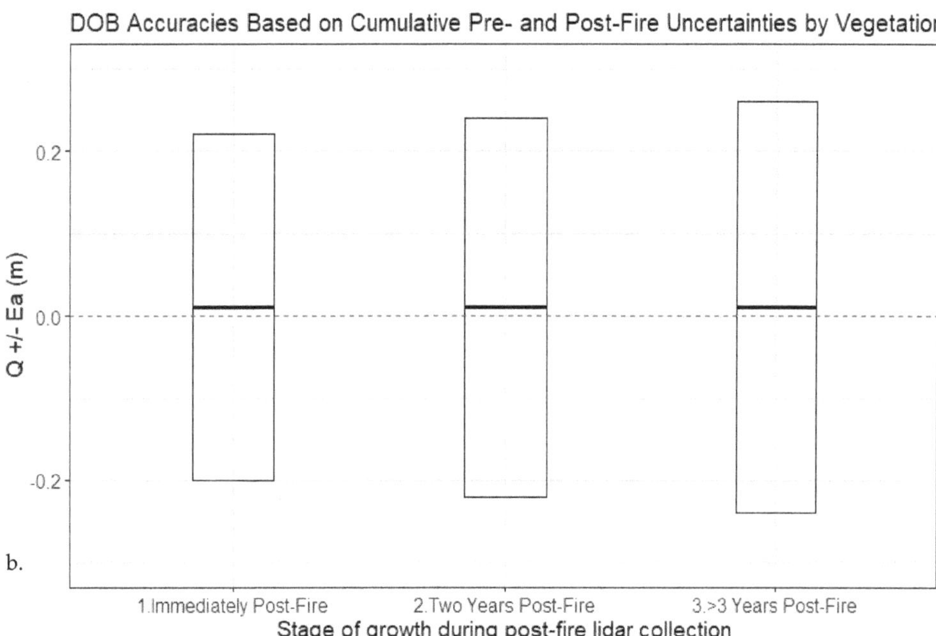

Figure 14. (**a**) Expected ground elevation accuracies of lidar data in the years following wildland fire in boreal peatlands; (**b**) Expected depth of burn (DOB) accuracies of lidar data in the years following wildland fire in boreal peatlands, assuming pre-fire lidar data were collected in "unburned conditions", where Q = average over- or under-estimation of surface elevation change, and Ea is cumulative error (SD). Note: for all measurements used, SD was equal to RMSE.

3.7. Wavelength Dependency of Ground Classification Accuracy as Varies by Vegetation Regeneration

By optimizing parameterizations, highly accurate ground classifications in low relief environments are possible for any wavelength or combination used. However, regardless of processing software, local vegetation characteristics still influence which wavelength produces optimal ground classification results. When applying the TerraScan ground classification parameterizations to road surfaces, optimal ground classification parameterizations derived from 1064 nm data had the lowest error compared with ground control measurements. However, wavelength-associated differences in ground surface elevation were negligible: $|\overline{dz}|$ s ranged from 0.00 to 0.02 m (on average for all combination of wavelengths) and RMSEs from 0.04 to 0.05 m (Figure 15a and Table S1). In unburned and TR areas, the combination of 1550 nm and 1064 nm wavelengths resulted in the most accurate ground classifications, with the least variability based on parameterization optimizations (Figure 15e,c). The least wavelength-dependent vegetation regeneration stage was that of cumulative burned areas with variable regeneration heights (a proxy for ~2 years post-fire; Figure 15d). While 1064 nm data resulted in the most accurate ground classifications with the least variability based on the parameterization scheme, the use of any wavelength individually resulted in similar levels of accuracy. However, the accuracy of ground classifications became more variable when wavelengths were combined (either 1550 nm and 1064 nm; or, 1550 nm, 1064 nm, and 532 nm). In SR zones, lidar-based ground elevations measured at 532 nm were the most accurate, with the lowest variability based on parameterizations (Figure 15b).

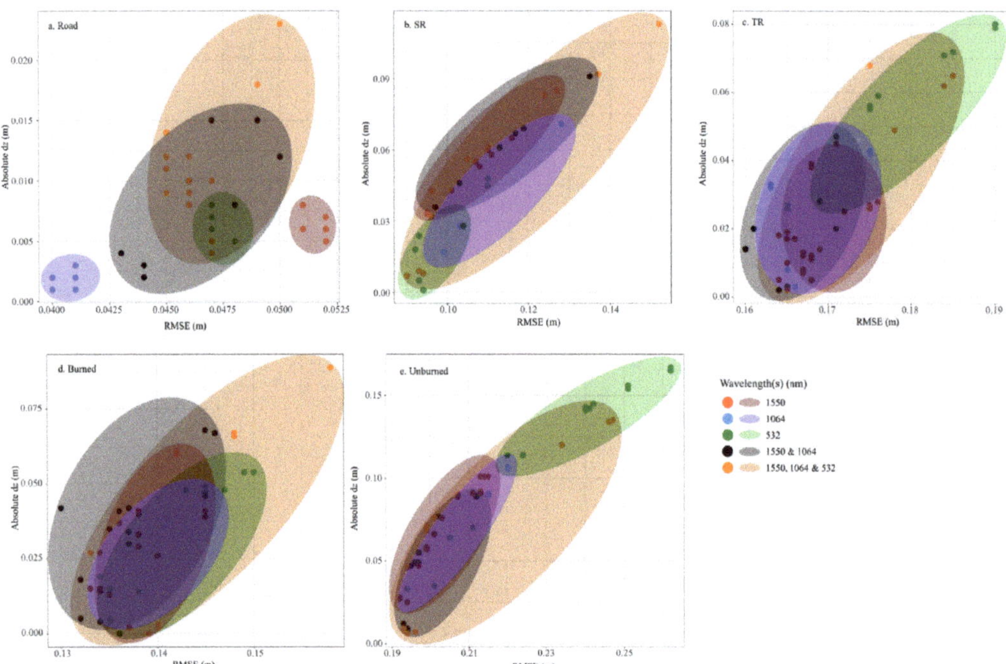

Figure 15. Ground classification results by wavelength along/in: (**a**) roads, (**b**) burned peatlands with short regeneration, (**c**) burned peatlands with tall regeneration, (**d**) all burned peatlands, and (**e**) unburned peatlands two years post-fire. Results were identified using TerraScan as determined by lidar channel. Each point represents an iterative parameter set. Note: axis range varies by plot.

In LAStools, wavelength selection impacted ground classification accuracy more than in TerraScan (Figures 15 and 16). Interestingly, processing software affects wavelength dependency for a given regeneration stage, particularly in SR areas (Figures 15b and 16b). Along road surfaces, the use of data collected at 1064 nm, or both 1550 nm and 1064 nm, resulted in slightly lower RMSEs; however, $|\overline{dz}|$ differences were negligible, and RMSEs varied by only ~0.01 m (Figure 16a). In unburned areas, 1064 nm data resulted in ground classifications with the lowest $|\overline{dz}|$s and RMSEs (Figure 16e). Data collected using the 1550 nm wavelength provided the most accurate classifications in both cumulative burned and TR areas (Figure 16d,c). In SR areas, all wavelengths combined provided the most accurate ground classifications (Figure 16b).

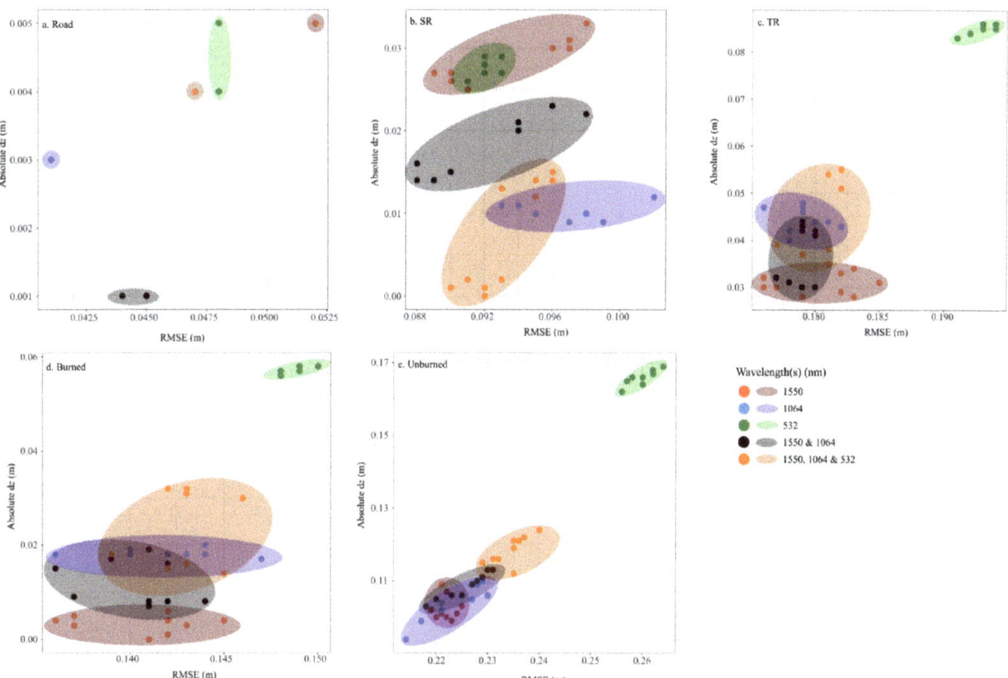

Figure 16. Ground classification results by wavelength along/in: (**a**) roads, (**b**) burned peatlands with short regeneration, (**c**) burned peatlands with tall regeneration, (**d**) all burned peatlands, and (**e**) unburned peatlands two years post-fire. Results were identified using LAStools as determined by lidar channel. Each point represents an iterative parameter set. Note: axis range varies by plot

4. Discussion

Despite the utility of lidar data for measuring ground surface elevation, time-series lidar data pre- and immediately post-fire is relatively rare but is increasing in availability with the application of lidar for understanding the impacts of wildland fire on ecosystems. This study provides an opportunity to assess the potential for error during the classification of lidar returns as "ground" and the accuracy of ground elevations for use in DEMs. It also identifies the ground classification errors associated with scorched patches/new vegetation regeneration (SR) and early post-fire regeneration stages, indicating time since disturbance.

Previous studies have found that errors in ground classification most often occur as a result of steep terrain [19,41,47] or dense vegetation [19,29,48]. Here, the study area is relatively flat, and so the potential for error that results from sloped terrain is greatly reduced (e.g., [19]). However, where vegetation has regenerated post-fire, stems and foliage can be relatively dense. Further, burned peatland surfaces can have significant microtopographical variability (hummocks and hollows) that may be difficult for the lidar ground classification to differentiate from short, dense vegetation (e.g., [49]). Our results show that within a given vegetation height range, the accuracies of lidar ground classifications (provided the parameters are set logically) do not deviate greatly from the most to least accurate ground classification of laser returns compared with ground control (Table S1). For example, within SR zones, using the least accurate classification scheme/wavelength combination results in classified returns that were, on average, 0.11 m below measured ground elevation, with a RMSE of 0.15 m. For many applicationns, such as canopy height measurements in forested environments, this may be sufficient (e.g., [50]); however, it is important to achieve the best accuracy possible when determining ground surface elevations for applications including combustion from wildland fire (e.g., using pre- and post-fire datasets) and hydrological modelling (e.g., [19]). This is due to the need for accurate quantification of slight differences in the elevation surface from pre-post fire or spatial changes in local surface topography.

We found that as vegetation regenerates post-fire, both optimal parameterizations and the wavelength used for lidar data collection differ at varying growth stages. In SR areas, or immediately post-fire, laser pulse emissions at 532 nm or using all wavelengths combined (532 nm, 1064 nm, and 1550 nm) provide the most accurate ground classifications compared with measured (Figures 4, 5, 15 and 16; Table S1). In these areas where vegetation regeneration was minimal, laser pulses emitted at 532 nm better characterized the ground surface due to the dominance of moss cover in measurement plots with little overlying vegetation in these peatland environments. It should be noted that channel dependencies are likely a result of both wavelength as well as pulse geometry (beam angle and footprint). Pulses emitted at 532 nm have lower energy receipt, a wider footprint, and a tendency to reflect from green vegetation above the ground surface (Figure 3a) [51,52]. As such, ground classified returns at 532 nm were the least accurate in unburned and TR areas (Figures 15 and 16).

As vegetation heights increased, the data from 1064 nm, 1550 nm, or both wavelengths combined, produced the most accurate elevations, noting that the addition of the 532 nm wavelength reduced accuracy even when combined with the other two (Figures 15 and 16). In cumulative burned areas, the use of 1550 nm provided the most optimal ground classifications, while in TR areas, either using 1550 nm data or combined 1064 nm and 1550 nm data had the least error and variability based on parameterizations (Figures 15 and 16). In cumulative burn areas where ground classification was conducted over regions of highly variable vegetation heights and densities, classifications that used 1064 nm data resulted in vegetation misclassified as ground [17,53]. Typical lidar systems with 1064 nm laser wavelength emission have greater reflectance from short vegetation and mosses [54]. Therefore, in areas with low, dense vegetation, there may be energy transmission losses as the laser pulse intercepts and reflects from vegetation instead of the ground surface [9,17,51,53]. This was also observed in [19] who found that low, dense vegetation was more commonly misclassified as ground, such that the classification was less accurate than in areas with tall overstory vegetation and reduced understory, e.g., in some forests. This phenomenon can be seen in the classifications of unburned peatlands, where 1064 nm data were optimal for ground classifications as this resulted in the lowest error (Figures 15 and 16). Here, it is likely that lidar was penetrating through canopies to low-lying understory vegetation (ground dominated by mosses) and the ground surface.

As with channel selection, the parameterization settings that optimized ground classifications depended on the dominant and sub-dominant vegetation heights found within plots. In TerraScan, we found that changes to three parameters impacted ground classifications: iteration angle, iteration distance, and the ability to reduce iteration angle when edge length exceeds a set distance (Tables 1 and S1). In burned landscapes with little vegetation regeneration (SR), adjustments made to the iteration angle improved the accuracy of the lidar-measured ground elevations compared with field-measured (Tables 1 and S1). Using 532 nm wavelength data (most accurate but low point density), the iteration angle was slightly reduced from six to five degrees in the optimal classification. Using all wavelengths combined, or only the 1064 nm wavelength data, the iteration angle was increased to 15 and 10 degrees, respectively (Table S1). The classifications using 532 nm data were more optimal using a smaller iteration angle (best for flat landscapes), likely due to the large pulse footprint. However, by increasing the iteration angle when using all wavelengths or 1064 nm data, the classification was better able to retain the steep transitions between hummocks and hollows, which would be most significant in areas with low regeneration, resulting in higher accuracy ground classification. A larger iteration angle allows the ground classification routine to adhere to surface elevation variability [38] by including returns that follow the microtopographic morphology of the hummocks and hollows. Despite the general topography of peatlands and transitions into forests being relatively flat, we are able to optimize the classification to account for highly localized topographic variability by setting the iteration angle to a greater angle within a confined area [10,38].

As vegetation heights increased (cumulative burned and TR areas), maintaining a high iteration angle but reducing iteration distance improved results. Iteration distance, which is the maximum height at which a point will be added to the ground classification [38], may be optimized at a longer distance in landscapes with little vegetation (e.g., for lidar surveys in the months following a fire). This will better account for the discrepancies in dz from returns from hummocks vs. hollows. However, in areas with greater vegetation regeneration and fewer scorched gaps between vegetation patches, a larger iteration distance of 1.4 m (vs. 0.5 m) may result in some returns from short vegetation being included in the ground classification. This increases the lidar ground classified elevation surface above the measured ground surface elevation, thereby increasing the inaccuracy of the DEM (Tables 1 and S1). Optimal parameterizations change notably in unburned areas or in areas with complete vegetation regeneration, post-fire. Interestingly, the optimal classifications in these regions had a reduced iteration angle of two degrees, which was the iteration angle that resulted in the poorest classifications in all post-burn analyses, even in TR areas. As peat combustion can enhance elevation differences between hummocks and hollows in peatlands [55], unburned peatlands often have less undulating moss ground surfaces. Further, reducing the iteration angle in unburned areas reduces the likelihood of including low herbaceous or shrubby vegetation in the ground classification. However, by reducing the iteration angle, the tendency to add more points into the ground class is also reduced, thereby reducing the point density. We found that by increasing the iteration angle to $5°$, point density was increased to >1 point m^{-2}; however, this increased the $\overline{dz}$ as the routine included low-lying vegetation as ground.

When using LAStools, we found that classification parameters were less important than the channel (laser pulse emission/reception wavelength as well as pulse geometry) with which data were collected. In SR and TR areas, ground elevation classifications were slightly improved when step size was set to 5 ('Nature' setting) instead of 3 ('Wilderness' setting) but were not impacted by the level of refinement. The ground classification was slightly better in unburned landscapes when step size was five and refinement was set to 'Fine', where the initial ground point search grid is four times more refined than the step size (Tables 2 and S1).

The results of this study demonstrate that not only do optimal classification parameters differ based on vegetation structures and environmental conditions within these low-relief ecosystems, but also that the same ground classification parameters can be optimal for specific environmental conditions (e.g., burned, low, moderate vegetation regeneration), but far less accurate for a different set of environmental conditions. For example, in TerraScan, classifications 8 through 13 provided the most accurate classifications in unburned landscapes but were the least accurate in cumulative burned and SR areas. There may be more variable ground topography in a burned landscape, which is not optimally parameterized using the same classification. With reduced iteration angle, areas with highly variable microtopography result in underestimation of ground surface elevation because returns from hummocks are excluded from the classification because the angular differences between returns in hollows vs. hummocks is too great. Steep angular classification of returns from the tops of the *Sphagnum* hummocks resulted in points being added to short vegetation instead of the ground class. This emphasizes the need to classify returns not only by land cover type but also by vegetation characteristics within those land covers for the most optimal/accurate classification parameterization. For example, classification to wetland class, burned and unburned, and within the burned class, high versus low regeneration (which can be determined based on the lidar data derivatives).

5. Conclusions

This study demonstrates the importance of optimizing classification parameters from default settings and optimizing parameters for different land cover types and vegetation structural characteristics. While it is important to iterate/optimize methods with any new data set, the results here provide parameters that can be used in burned and unburned boreal peatland environments and as a starting point for parameterization in similar environments. We provide optimal parameterization for boreal peatlands along a post-fire regeneration trajectory, with "SR" representing peatlands soon after burn, unsegmented burned land cover representing two years post-fire, "TR" representing several years post-fire, and unburned vegetation covers representing land cover later in the regeneration trajectory. However, we suggest that there is a need for the development of accessible adaptative classification procedures, which can be used to (a) filter the landscape by attributes and (b) identify optimal parameterizations.

From this study, we conclude an expected 'best' average accuracy for depth of burn (pre-fire elevation minus post-fire elevation) within peatlands will have a spatially averaged error (dz) of ~0.01 m, indicating that soil organic matter loss (determined most simply as a change in elevation) would be over-estimated by an average of 0.01 m. Using the adventitious roots method, the average uncertainty is 0.004 m to 0.04 m (e.g., [54,56,57]) at the tree base. However, airborne lidar methods provide an opportunity to quantify spatially continuous elevational variability between trees and across a broad range of peatlands and environmental characteristics. In addition, we demonstrate that lidar surveys completed in the years following combustion do not become significantly less accurate for quantifying depth of burn overall (offset at ~0.01 m and RMSEs ranging from ~0.21 m to ~0.25 m in SR to TR areas; Figure 14). We suggest that lidar data collected up to three years post-fire can be utilized for depth of burn analyses without significant differences in cumulative errors associated with laser pulse interactions in the understory.

Supplementary Materials: The following supporting information can be downloaded at: https: //www.mdpi.com/article/10.3390/rs14205080/s1, Table S1: Ground Classification Results.

Author Contributions: Conceptualization, K.N., L.C. and C.H.; methodology, K.N., L.C. and C.H.; formal analysis, K.N.; investigation, K.N.; resources, L.C. and C.H.; data curation, K.N.; writing— original draft preparation, K.N.; writing—review and editing, K.N., L.C. and C.H.; visualization, K.N.; supervision, L.C.; project administration, L.C.; funding acquisition, L.C., C.H. and K.N. All authors have read and agreed to the published version of the manuscript.

Funding: This research was funded by the Natural Sciences and Engineering Research Council (NSERC) Discovery Grant and University of Lethbridge Start-Up Funding to L. Chasmer. The Teledyne Optech Inc. Titan multispectral lidar was purchased using Western Economic Diversification Canada funding to C. Hopkinson. GNSS equipment was purchased using Canadian Foundation for Innovation funds provided to C. Hopkinson. Research was also funded by post-graduate support of K. Nelson, including the NSERC Canada Graduate Scholarship–Doctoral (CGS D), the Alberta Innovates Graduate Student Scholarship, the Nexen Fellowship in Water Research, and the University Of Lethbridge School Of Graduate Studies, as well as funding from Canada Wildfire (NSERC SPG-N).

Data Availability Statement: Supporting data can be found in Supplementary Information Table S1.

Acknowledgments: The authors would like to thank Linda Flade, Emily Jones, Craig Mahoney, and Maxim Okhrimenko for their field and lidar data collection/processing assistance.

Conflicts of Interest: The authors declare no conflict of interest.

References

1. Frolking, S.; Roulet, N.T. Holocene radiative forcing impact of northern peatland carbon accumulation and methane emissions. *Glob. Change Biol.* **2007**, *13*, 1079–1088. [CrossRef]
2. Yu, Z.C. Northern peatland carbon stocks and dynamics: A review. *Biogeosciences* **2012**, *9*, 4071–4085. [CrossRef]
3. Flannigan, M.; Cantin, A.S.; De Groot, W.J.; Wotton, M.; Newbery, A.; Gowman, L.M. Global wildland fire season severity in the 21st century. *For. Ecol. Manag.* **2013**, *294*, 54–61. [CrossRef]
4. Miller, C.A.; Benscoter, B.W.; Turetsky, M.R. The effect of long-term drying associated with experimental drainage and road construction on vegetation composition and productivity in boreal fens. *Wetl. Ecol. Manag.* **2015**, *23*, 845–854. [CrossRef]
5. Walker, X.J.; Mack, M.C.; Johnstone, J.F. Stable carbon isotope analysis reveals widespread drought stress in boreal black spruce forests. *Glob. Change Biol.* **2015**, *21*, 3102–3113. [CrossRef]
6. Kohlenberg, A.J.; Turetsky, M.R.; Thompson, D.K.; Branfireun, B.A.; Mitchell, C.P. Controls on boreal peat combustion and resulting emissions of carbon and mercury. *Environ. Res. Lett.* **2018**, *13*, 035005. [CrossRef]
7. Van der Werf, G.R.; Randerson, J.T.; Giglio, L.; Collatz, G.J.; Mu, M.; Kasibhatla, P.S.; Morton, D.C.; DeFries, R.S.; Jin, Y.; van Leeuwen, T.T. Global fire emissions and the contribution of deforestation, savanna, forest, agricultural, and peat fires (1997–2009). *Atmos. Chem. Phys.* **2010**, *10*, 11707–11735. [CrossRef]
8. Thompson, D.K.; Waddington, J.M. A Markov chain method for simulating bulk density profiles in boreal peatlands. *Geoderma* **2014**, *232*, 123–129. [CrossRef]
9. Chasmer, L.E.; Hopkinson, C.D.; Petrone, R.M.; Sitar, M. Using multitemporal and multispectral airborne lidar to assess depth of peat loss and correspondence with a new active normalized burn ratio for wildfires. *Geophys. Res. Lett.* **2017**, *44*, 11851–11859. [CrossRef]
10. Mickler, R.A.; Welch, D.P.; Bailey, A.D. Carbon emissions during wildland fire on a North American temperate peatland. *Fire Ecol.* **2017**, *13*, 34–57. [CrossRef]
11. Lin, S.; Liu, Y.; Huang, X. Climate-induced Arctic-boreal peatland fire and carbon loss in the 21st century. *Sci. Total Environ.* **2021**, *796*, 148924. [CrossRef] [PubMed]
12. Morison, M.; van Beest, C.; Macrae, M.; Nwaishi, F.; Petrone, R. Deeper burning in a boreal fen peatland 1-year post-wildfire accelerates recovery trajectory of carbon dioxide uptake. *Ecohydrology* **2021**, *14*, e2277. [CrossRef]
13. Aguilar, F.J.; Mills, J.P. Accuracy assessment of LiDAR-derived digital elevation models. *Photogramm. Rec.* **2008**, *23*, 148–169. [CrossRef]
14. Hokanson, K.J.; Lukenbach, M.C.; Devito, K.J.; Kettridge, N.; Petrone, R.M.; Waddington, J.M. Groundwater connectivity controls peat burn severity in the boreal plains: Groundwater controls peat burn severity. *Ecohydrology* **2016**, *9*, 574–584. [CrossRef]
15. Whitman, E.; Parisien, M.-A.; Holsinger, L.M.; Park, J.; Parks, S.A. A method for creating a burn severity atlas: An example from Alberta, Canada. *Int. J. Wildland Fire* **2020**, *29*, 995–1008. [CrossRef]
16. Hudak, A.T.; Morgan, P.; Bobbitt, M.J.; Smith AM, S.; Lewis, S.A.; Lentile, L.B.; Robichaud, P.R.; Clark, J.T.; McKinley, R.A. The relationship of multispectral satellite imagery to immediate fire effects. *Fire Ecol.* **2007**, *3*, 64–90. [CrossRef]
17. Hopkinson, C.; Chasmer, L.; Sass, G.; Creed, I.; Sitar, M.; Kalbfleisch, W.; Treitz, P. Vegetation class dependent errors in lidar ground elevation and canopy height estimates in a boreal wetland environment. *Can. J. Remote Sens.* **2005**, *31*, 191–206. [CrossRef]
18. Ekhtari, N.; Glennie, C.; Fernandez-Diaz, J.C. Classification of airborne multispectral lidar point clouds for land cover mapping. *IEEE J. Sel. Top. Appl. Earth Obs. Remote Sens.* **2018**, *11*, 2068–2078. [CrossRef]
19. Moudrý, V.; Klápště, P.; Fogl, M.; Gdulová, K.; Barták, V.; Urban, R. Assessment of LiDAR ground filtering algorithms for determining ground surface of non-natural terrain overgrown with forest and steppe vegetation. *Measurement* **2020**, *150*, 107047. [CrossRef]
20. Andersen, H.E.; McGaughey, R.J.; Reutebuch, S.E. Estimating forest canopy fuel parameters using LIDAR data. *Remote Sens. Environ.* **2005**, *94*, 441–449. [CrossRef]
21. McCarley, T.R.; Hudak, A.T.; Sparks, A.M.; Vaillant, N.M.; Meddens, A.J.; Trader, L.; Francisco, M.; Kreitler, J.; Boschetti, L. Estimating wildfire fuel consumption with multitemporal airborne laser scanning data and demonstrating linkage with MODIS-derived fire radiative energy. *Remote Sens. Environ.* **2020**, *251*, 112114. [CrossRef]
22. Chen, Y.; Zhu, X.; Yebra, S.; Harris, N. Development of a predictive model for estimating forest surface fuel load in Australian eucalypt forests with LiDAR data. *Environ. Model. Softw.* **2017**, *97*, 61–71. [CrossRef]
23. O'Neil, G.L.; Saby, L.; Band, L.E.; Goodall, J.L. Effects of LiDAR DEM smoothing and conditioning techniques on a topography-based wetland identification model. *Water Resour. Res.* **2019**, *55*, 4343–4363. [CrossRef]
24. Millard, K.; Richardson, M. Wetland mapping with LiDAR derivatives, SAR polarimetric decompositions, and LiDAR–SAR fusion using a random forest classifier. *Can. J. Remote Sens.* **2013**, *39*, 290–307. [CrossRef]
25. O'Neil, G.L.; Goodall, J.L.; Watson, L.T. Evaluating the potential for site-specific modification of LiDAR DEM derivatives to improve environmental planning-scale wetland identification using Random Forest classification. *J. Hydrol.* **2018**, *559*, 192–208. [CrossRef]
26. Neugirg, F.; Kaiser, A.; Schmidt, J.; Becht, M.; Haas, F. Quantification, analysis and modelling of soil erosion on steep slopes using LiDAR and UAV photographs. *Proc. Int. Assoc. Hydrol. Sci.* **2015**, *367*, 51–58. [CrossRef]
27. Escobar Villanueva, J.R.; Iglesias Martínez, L.; Pérez Montiel, J.I. DEM generation from fixed-wing UAV imaging and LiDAR-derived ground control points for flood estimations. *Sensors* **2019**, *19*, 3205. [CrossRef] [PubMed]

28. Campbell, M.J.; Dennison, P.E.; Kerr, K.L.; Brewer, S.C.; Anderegg, W.R. Scaled biomass estimation in woodland ecosystems: Testing the individual and combined capacities of satellite multispectral and lidar data. *Remote Sens. Environ.* **2021**, *262*, 112511. [CrossRef]

29. Schmid, K.A.; Hadley, B.C.; Wijekoon, N. Vertical accuracy and use of topographic LIDAR data in coastal marshes. *J. Coast. Res.* **2011**, *27*, 116–132. [CrossRef]

30. Nelson, K.; Thompson, D.; Hopkinson, C.; Petrone, R.; Chasmer, L. Peatland-fire interactions: A review of wildland fire feedbacks and interactions in Canadian boreal peatlands. *Sci. Total Environ.* **2021**, *769*, 145212. [CrossRef]

31. Downing, D.J.; Pettapiece, W.W. *Natural Regions and Subregions of Alberta*; Pub. No. T/852; Government of Alberta Publish: Edmonton, AB, Canada, 2006; 264p.

32. Alberta Environment and Sustainable Resource Development (ESRD). *Alberta Wetland Classification System*; Water Policy Branch, Policy and Planning Division: Edmonton, AB, Canada, 2015.

33. MNP LLP. *A Review of the 2016 Horse River Wildfire*; Forestry Division, Alberta Agriculture and Forestry: Edmonton, AB, Canada, 2017. Available online: https://www.alberta.ca/assets/documents/Wildfire-MNP-Report.pdf (accessed on 20 July 2021).

34. Institute for Catastrophic Loss Reduction. *Fort McMurray Wildfire: Learning from Canada's Costliest Disaster*; Institute for Catastrophic Loss Reduction: Toronto, ON, Canada, 2019; p. 9. Available online: https://www.zurichcanada.com/-/media/project/zwp/canada/docs/english/weather/fort-mcmurray-report_canada.pdf (accessed on 16 May 2021).

35. Csanyi, N.; Toth, C.K. Improvement of lidar data accuracy using lidar-specific ground targets. *Photogramm. Eng. Remote Sens.* **2007**, *73*, 385–396. [CrossRef]

36. Axelsson, P. DEM generation from laser scanner data using adaptive TIN models. *Int. Arch. Photogramm. Remote Sens.* **2000**, *33*, 110–117.

37. GeoCue Group. TerraScan Ground Parameters. 2020. Available online: https://support.geocue.com/terrascan-ground-filter-parameters/ (accessed on 10 February 2021).

38. Terrasolid Ltd. TerraScan User Guide. Terrasolid Ltd., Batch Processing Reference > Classification Routines > Points > Ground. 2021. Available online: https://terrasolid.com/guides/tscan/crground.html (accessed on 10 November 2021).

39. Rapidlasso GmbH. Lasground_New README. 2021. Available online: https://lastools.github.io/download/lasground_new_README.txt (accessed on 5 December 2021).

40. ASPRS. *ASPRS Guidelines: Vertical Accuracy Reporting for Lidar Data Version 1.0*; American Society for Photogrammetry and Remote Sensing Lidar Committee (PAD): Baton Rouge, LA, USA, 2004.

41. Pourali, S.; Arrowsmith, C.; Chrisman, N.; Matkan, A. Vertical accuracy assessment of LiDAR ground points using minimum distance approach. In Proceedings of the Research Locate 14, Canberra, Australia, 7–9 April 2014.

42. Maune, D.; Black, T.; Constance, E. DEM user requirements. In *Digital Elevation Model Technologies and Applications: The DEM Users Manual*, 2nd ed.; American Society for Photogrammatry and Remote Sensing: Bethesda, MD, USA, 2007; pp. 449–473.

43. Goulden, T.; Hopkinson, C.; Jamieson, R.; Sterling, S. Sensitivity of DEM, slope, aspect and watershed attributes to LiDAR measurement uncertainty. *Remote Sens. Environ.* **2016**, *179*, 23–35. [CrossRef]

44. GeoCue Group. Control Point Statistics in TerraScan: TerraScan, Versions 002.001 and Above. 2017. Available online: https://support.geocue.com/wp-content/uploads/2017/03/Control-Point-Statistics-in-TerraScan.pdf (accessed on 10 February 2021).

45. Carlisle, B.H. Modelling the spatial distribution of DEM error. *Trans. GIS* **2005**, *9*, 521–540. [CrossRef]

46. Goulden, T.; Hopkinson, C. The forward propagation of integrated system component errors within airborne lidar data. *Photogramm. Eng. Remote Sens.* **2010**, *76*, 589–601. [CrossRef]

47. Goulden, T.; Hopkinson, C. Mapping simulated error due to terrain slope in airborne lidar observations. *Int. J. Remote Sens.* **2014**, *35*, 7099–7117. [CrossRef]

48. Hopkinson, C.; Chasmer, L.E.; Zsigovics, G.; Creed, I.F.; Sitar, M.; Treitz, P.; Maher, R.V. Errors in LIDAR ground elevations and wetland vegetation height estimates. *Int. Arch. Photogramm. Remote Sens. Spat. Inf. Sci.* **2004**, *36*, 108–113.

49. Brubaker, K.M.; Myers, W.L.; Drohan, P.J.; Miller, D.A.; Boyer, E.W. The use of LiDAR terrain data in characterizing surface roughness and microtopography. *Appl. Environ. Soil Sci.* **2013**, *2013*, 891534. [CrossRef]

50. Tinkham, W.T.; Smith, A.M.; Hoffman, C.; Hudak, A.T.; Falkowski, M.J.; Swanson, M.E.; Gessler, P.E. Investigating the influence of LiDAR ground surface errors on the utility of derived forest inventories. *Can. J. For. Res.* **2012**, *42*, 413–422. [CrossRef]

51. Hopkinson, C.; Chasmer, L.; Gynan, C.; Mahoney, C.; Sitar, M. Multisensor and multispectral lidar characterization and classification of a forest environment. *Can. J. Remote Sens.* **2016**, *42*, 501–520. [CrossRef]

52. Okhrimenko, M.; Coburn, C.; Hopkinson, C. Multispectral lidar: Radiometric calibration, canopy spectral reflectance, and vegetation vertical SVI profiles. *Remote Sens.* **2019**, *11*, 1556. [CrossRef]

53. Reddy, A.D.; Hawbaker, T.J.; Wurster, F.; Zhu, Z.; Ward, S.; Newcomb, D.; Murray, R. Quantifying soil carbon loss and uncertainty from peatland wildfire using multi-temporal LiDAR. *Remote Sens. Environ.* **2015**, *170*, 306–316. [CrossRef]

54. Gerrand, S.; Aspinall, J.; Jensen, T.; Hopkinson, C.; Collingwood, A.; Chasmer, L. Partitioning carbon losses from fire combustion in a montane valley, Alberta Canada. *For. Ecol. Manag.* **2021**, *496*, 119435. [CrossRef]

55. Benscoter, B.W.; Greenacre, D.; Turetsky, M.R. Wildland fire as a key determinant of peatland microtopography. *Can. J. For. Res.* **2015**, *45*, 1132–1136. [CrossRef]

56. Kasischke, E.S.; Johnstone, J.F. Variation in postfire organic layer thickness in a black spruce forest complex in interior Alaska and its effects on soil temperature and moisture. *Can. J. For. Res.* **2005**, *35*, 2164–2177. [CrossRef]
57. Boby, L.A.; Schuur, E.A.; Mack, M.C.; Verbyla, D.; Johnstone, J.F. Quantifying fire severity, carbon, and nitrogen emissions in Alaska's boreal forest. *Ecol. Appl.* **2010**, *20*, 1633–1647. [CrossRef]

 *remote sensing*

Article

Evaluating a New Relative Phenological Correction and the Effect of Sentinel-Based Earth Engine Compositing Approaches to Map Fire Severity and Burned Area

Adrián Israel Silva-Cardoza [1,†], Daniel José Vega-Nieva [2,†], Jaime Briseño-Reyes [2,*], Carlos Ivan Briones-Herrera [2], Pablito Marcelo López-Serrano [2], José Javier Corral-Rivas [2], Sean A. Parks [2,3] and Lisa M. Holsinger [2,3]

[1] Maestría en Geomática Aplicada a Recursos Forestales y Ambientales, Facultad de Ciencias Forestales, Universidad Juárez del Estado de Durango, Río Papaloapan y Blvd, Durango S/N Col. Valle del Sur, Durango C.P. 34120, Mexico; adkrdoza@hotmail.com

[2] Facultad de Ciencias Forestales, Universidad Juárez del Estado de Durango, Río Papaloapan y Blvd, Durango S/N Col. Valle del Sur, Durango C.P. 34120, Mexico; daniel.vega@ujed.mx (D.J.V.-N.); 51580@alumnos.ujed.mx (C.I.B.-H.); p_lopez@ujed.mx (P.M.L.-S.); jcorral@ujed.mx (J.J.C.-R.); sean.parks@usda.gov (S.A.P.); lisa.holsinger@usda.gov (L.M.H.)

[3] USDA Forest Service, Aldo Leopold Wilderness Research Institute, Rocky Mountain Research Station, Missoula, MT 59801, USA

* Correspondence: jaime.briseno@ujed.mx; Tel.: +52-618-158-6627

† These authors contributed equally to this work.

Citation: Silva-Cardoza, A.I.; Vega-Nieva, D.J.; Briseño-Reyes, J.; Briones-Herrera, C.I.; López-Serrano, P.M.; Corral-Rivas, J.J.; Parks, S.A.; Holsinger, L.M. Evaluating a New Relative Phenological Correction and the Effect of Sentinel-Based Earth Engine Compositing Approaches to Map Fire Severity and Burned Area. *Remote Sens.* **2022**, *14*, 3122. https://doi.org/10.3390/rs14133122

Academic Editors: Leonor Calvo, Elena Marcos, Susana Suarez-Seoane and Víctor Fernández-García

Received: 18 May 2022
Accepted: 21 June 2022
Published: 29 June 2022

Abstract: The remote sensing of fire severity and burned area is fundamental in the evaluation of fire impacts. The current study aimed to: (i) compare Sentinel-2 (*S2*) spectral indices to predict field-observed fire severity in Durango, Mexico; (ii) evaluate the effect of the compositing period (1 or 3 months), techniques (average or minimum), and phenological correction (constant offset, *c*, against a novel relative phenological correction, *rc*) on fire severity mapping, and (iii) determine fire perimeter accuracy. The Relative Burn Ratio (*RBR*), using *S2* bands 8a and 12, provided the best correspondence with field-based fire severity (*FBS*). One-month *rc* minimum composites showed the highest correspondence with *FBS* ($R^2 = 0.83$). The decrease in R^2 using 3 months rather than 1 month was $\geq$0.05 (0.05–0.15) for *c* composites and <0.05 (0.02–0.03) for *rc* composites. Furthermore, using *rc* increased the R^2 by 0.05–0.09 and 0.10–0.15 for the 3-month *RBR* and *dNBR* compared to the corresponding *c* composites. *Rc* composites also showed increases of up to 0.16–0.22 and 0.08–0.11 in kappa values and overall accuracy, respectively, in mapping fire perimeters against *c* composites. These results suggest a promising potential of the novel relative phenological correction to be systematically applied with automated algorithms to improve the accuracy and robustness of fire severity and perimeter evaluations.

Keywords: Sentinel-2; post-fire severity; initial fire assessment; soil burn severity; fire perimeter; image compositing; vegetation phenology

1. Introduction

The evaluation of burned area and fire severity is fundamental for understanding a variety of ecological processes (e.g., [1–4]), including fire emissions and carbon cycling (e.g., [5,6]), post-fire erosion (e.g., [7–9]), tree mortality (e.g., [10–12]), post-fire recovery trajectories (e.g., [13–16]), and ecosystem resilience (e.g., [17,18]). In spite of the great progress in fire severity and burned area evaluations using medium-resolution sensors, several challenges remain for the development of more standardized, automated methods of post-fire assessment. Among others, these challenges include: (i) the assessment of spectral indices, particularly for Sentinel-2, to map burned area perimeters and measure fire severity and fire effects by strata, namely vegetation and soil; (ii) the evaluation of the effect of compositing periods and techniques, particularly on automated platforms such as Google Earth Engine (e.g., [19]), on the accuracy for mapping initial assessments of fire severity and burned area perimeters; (iii) the quantification of the effect of non-fire-related temporal and spatial variations in spectral indices [20–22] and the development of simple, readily automated methods of phenological correction that account for the spatial and temporal heterogeneity of such effects [23].

1.1. Evaluation of Spectral Indices to Map Fire Severity and Perimeter, Particularly for Sentinel-2

Regarding this challenge, several spectral indices using medium-resolution sensors have been developed [3]. These include several formulations of *NBR* indices based on NIR/SWIR bands (e.g., [24–27]), and more recently, spectral indices including Sentinel-2 red-edge bands such as *CIre* [28,29] or *BAIS2* [30]. Nonetheless, the selection of the most appropriate indices from Sentinel-2 to map fire severity and perimeters is still an open research question, with a large variety of results depending on the ecosystems analyzed (e.g., [8,9,31–35]). Furthermore, while the majority of fire severity research has been developed in temperate to boreal ecosystems, there are comparatively fewer studies in areas with different climatic and vegetation characteristics. This is the case in Mexico, a very ecologically diverse country with high fire activity (e.g., [36–39]), where previous studies evaluating spectral indices, validated with field data, to map fire severity and burned area perimeters, are lacking.

1.2. Evaluation of Compositing Techniques and Period for Fire Severity and Perimeter Mapping

1.2.1. Comparison of the Effects of Compositing Techniques for Fire Severity and Burned area Mapping

Pixel-based compositing methods can provide a new analysis paradigm instead of depending on a single scene [40], resulting in an unprecedented opportunity for standardizing and automating fire severity and burned area evaluations [27,41–43]. Nevertheless, the image composite techniques analyzed in the literature have largely varied, depending on the study goal, location, and period analyzed. For example, mean composites have been used for extended assessments of fire severity [19,41,44,45] and mean, median, or medoid [46,47] composites have been also used to map burned area and recovery [47–50]. Likewise, minimum, or maximum composites [26,27,42,51–53] have been used for mapping fire perimeters and trends of burned area [54] or fire severity [55]. Because of this variety of approaches, there is a need for a formal evaluation of the effect of compositing techniques (i.e., average or minimum) and compositing periods on their accuracy in mapping fire perimeters and fire severity. Such an evaluation is particularly lacking for immediate post-fire assessments of fire impacts (hereafter termed "fire severity" following [1], [2] or [54], which propose the use of this term for initial severity assessments), which are fundamental for post-fire erosion emergency planning (e.g., [8,9,56]) but have received less attention from compositing analysis studies.

1.2.2. Assessment of the Influence of Compositing Period for Mapping Initial Fire Severity and Burned Area Perimeter

The post-fire image assessment period can largely influence what is being observed (e.g., [1,57]). Non-fire variations in spectral indices can be caused by natural oscillations in vegetation and soil moisture (e.g., [58–60]) or vegetation processes such as senescence, regrowth, and phenological changes in grasses, shrubs, and trees (e.g., [61,62]), both for broadleaves (e.g., [63]) and conifer species (e.g., [64]). Therefore, the values of a spectral index can vary considerably from slight changes in image dates, resulting in substantially different burn severity maps [21,22,54]. This can affect image accuracy in mapping fire perimeters [62] and fire severity [19,58]. Moreover, it could also influence the widely reported lack of transferability of fire severity and burned area thresholds between seasons and ecosystems [20,65–67].

However, although evidence of variability in the assessment of burn severity driven by image selection has been shown in many studies, paradoxically, little attention has been paid to quantifying the temporal and spatial variations in spectral fire severity indices [20,21,54] and how this impacts their suitability for mapping fire perimeters [23] and severity [22]. In addition, in contrast to the more widely analyzed Landsat satellite, there are relatively few studies examining the temporal variations in Sentinel spectral indices' accuracy in mapping burned area [68] and fire severity [22]. Furthermore, most of those relatively scarce studies have used paired-image techniques, contrasting with a scarcity of studies addressing temporal and phenological variations in fire severity spectral indices' accuracy using composite images [19,22]).

1.3. Development of a Spatially Variable Automated Phenological Correction

In spite of the frequent documentation of the above-mentioned phenological biases, most of the literature has generally failed to include any measure to quantify or correct for such non-fire phenological changes to standardize fire severity assessments [20,69]. The most notable exception is the constant offset method [41,57,67,70,71]. This approach consists in subtracting from the uncalibrated *dNBR* image a phenological constant offset, generally calculated as the average *dNBR* from the unburnt area surrounding the fire perimeter [70,72,73]. Because of the relative simplicity of this method, some steps have been taken towards its automatization for Landsat [19,41,74] and Sentinel images [75]. Nevertheless, in order to be representative, the average offset should be calculated in similar vegetation types to those where the fire occurred [41,74,76]. This strong assumption presents a challenge in heterogeneous fuels, making the constant offset method not advisable in those cases, as it would be compromised by the spatial heterogeneity of phenological effects [41,45]. Still, fuel heterogeneity is very common (e.g., [23]), resulting in frequent spatial and temporal variations in the phenological effects (e.g., [54,77]).

Although some attempts to propose spatially variable techniques of phenological correction have been made (e.g., [22,54,77]), in their current states, the majority of those approaches require supervision to either manually select control pixels (e.g., [54,77]) or to manually create phenologically matched composites [22]. In this sense, to the best of our knowledge, the use of the pre-fire Normalized Burn Ratio [78] (*NBRpre*), a widely used proxy of biomass [70], has not been yet tested to automatically stratify and sample for spatial phenological variations in unburned vegetation. Such a systematic stratification of phenological effects could lead to a standardized, automated, novel relative phenological correction procedure for analyzing and minimizing the spatial heterogeneity of phenological variations. This approach could overcome some of the limitations of the offset method, while benefiting from its relative simplicity and ease of automation.

Consequently, the present work aimed to evaluate and map fire severity and burned area perimeter through paired and composite spectral indices of S2-MSI images, the latter calculated with different periods, techniques, and phenological corrections (constant offset, c, against a novel spatially variable relative phenological correction, rc). Specifically, the goals of the study were: (i) to evaluate the correspondence of paired Sentinel-2 (S2) images with field data from wildfires in the temperate to semi-arid forests of Durango, Mexico, to select the best S2 spectral index to map the initial assessment of fire severity in the analyzed ecosystems; (ii) to analyze the effect of Google Earth Engine S2 composite period length (1 or 3 months), technique (average or minimum), and phenological correction (c and rc) for estimating field-observed fire severity; and finally, (iii) to quantify the effect of composite period length, technique, and phenological correction on the accuracy in mapping the burned area perimeter for each wildfire.

2. Materials and Methods

2.1. Study Area

This study was conducted in the Mexican physiographic province of *Sierra Madre Occidental* in Durango State (SMO), between 26°59′50″N, 22°20′26″S latitude North and 103°42′26″E, 107°14′12″W longitude West. The study area is dominated by conifers of the genera *Pinus, Juniperus, and Cupressus*, as well as grasslands and shrublands dominated by

Acacia and *Prosopis* species [79,80]. The climate ranges from semi-dry temperate to semi-cold subhumid and subhumid temperate according to the Köppen climate classification adapted for Mexico [81]. The average annual temperature ranges from 5 to 18 °C, with average annual precipitation ranging from 300 to 1500 mm [82]. The Durango SMO region is characterized by surfaces of large mesas that extend over the central part of the state, as well as mountain areas with slopes ranging from 20 to 80%. Altitude ranges from 800 to 3200 m [83]. We selected three fires that occurred between April and May 2019, covering a gradient in fuel types and climates ranging from semi-arid (Site 1) to humid (Site 2) and semi-humid (Site 3) (Table 1, Figure 1). Fire regimes in the study area include both high-frequency, low-severity fires in the denser, wetter pine and oak forests and lower-frequency, higher-severity fires in the more open, semi-arid forests [84]. Severe fires can affect species of timber value, such as pine stands, as well as tree and shrubby species used for charcoal production, such as *Quercus* or *Prosopis* species.

Table 1. Summary of wildfires selected in the Sierra Madre Occidental, Durango.

Fire	Start–End Date	Fire Size (ha) [1]	Fuelbed Types	Latitude N	Longitude W
Site 1	2019/04/26–2019/05/27	23,809	TU, TL, GR, GS	23°36′36″	104°33′28.8″
Site 2	2019/05/14–2019/06/28	2602	TL, TU	23°49′26″	105°40′44.4″
Site 3	2019/04/12–2019/05/09	7029	TL, TU, GR	25°36′25″	105°57′18″

[1] [85], fuelbed types [84] (Figure 1): TU = grass or shrub understory mixed with timber litter, TL = timber litter, GS = grass and shrublands, and GR = grasslands.

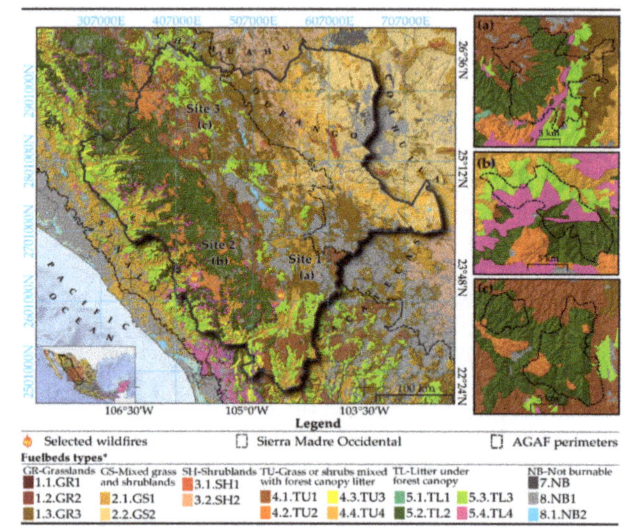

Figure 1. Location of the three 2019 wildfires analyzed in Durango State, Mexico, where (**a**) = Site 1, (**b**) = Site 2, (**c**) = Site 3, AGAF = aggregated active fire perimeter [75], fuelbed types (scale: 1:250,000) [84]: 1.1.GR1 = warm–humid tall grass, 1.2.GR2 = temperate–humid short grass and shrublands, 1.3.GR3 = semi-arid short grass and shrublands, 2.1.GS1 = semi-arid grass and shrublands, 2.2.GS2 = arid grass and shrublands, 3.1.SH1 = arid tall shrublands, 3.2.SH2 = arid short shrublands, 4.1.TU1 = subhumid temperate timber understory with dense shrubs, 4.2.TU2 = humid temperate–cold timber understory with dense pasture, 4.3.TU3 = subhumid timber understory with sparse grasslands, 4.4.TU4 = subhumid warm to semi-arid timber understory with sparse grass and shrublands, 5.1.TL1 = humid temperate short-needle coniferous timber litter, 5.2.TL2 = humid long-needle coniferous timber litter, 5.3.TL3 = subhumid to humid sclerophyll broadleaf timber litter, 5.4.TL4 = seasonally dry humid broadleaf timber litter, 7.NB = not burnable (bare ground), 8.NB1 = not forestry areas (e.g., urban development or agricultural), and 8.1.NB2 = water bodies.

The selected wildfires included a heterogeneous-severity fire (low–medium severity to crown fire patches) (Site 1), with an estimated burned area of 23,809 ha (the largest wildfire in Durango state for 2019) [85]. Fuels in Site 1 ranged from semi-arid grass and shrublands in the eastern part of the fire to a more subhumid to temperate conifer and broadleaf understory and litter in the west (Figure 1a). Site 2 was dominated by pine and oak timber litter (Figure 1b), affected by a low-to-medium-severity fire in a humid climate in the west of the SMO. Finally, Site 3 was a medium-to-high-severity fire (with frequent crown fire patches) in a subhumid and temperate–cold pine timber understory mixture with shrublands in NW Durango (Figure 1c) [84].

2.2. Field Sampling of Fire Severity

Fire severity field sampling was performed during the months of November and December 2019 after fires at the three locations. Field and satellite imagery dates were selected to represent the early conditions of fire severity immediately after the fire (i.e., the initial assessment) in order to minimize the fire obscuration effects caused by rapid regrowth that have been documented for similar semi-arid ecosystems (e.g., [61,71]). We measured a total of 115 circular plots (30 m in diameter), distributed in a stratified sampling scheme, measuring homogeneously burned patches—stratified by *dNBR* [72]—within the fire perimeter (85 plots), as well as unburned patches (30 control plots). At each fire location, we also stratified sampled plots by vegetation types (Figure 1) and pre-fire *NBR* (e.g., [86,87]), both for burned and unburned plots. The center of each plot was georeferenced using a high-precision GPS (Spectra Precision *MobileMapper*© 60, with an average precision of <2 m) [88].

The field protocol for fire severity was based on the methodology for temperate forest described by Silva-Cardoza et al. [86,87]. This field protocol is an adaptation of different published guides for fire severity and mapping evaluation, i.e., Key and Benson [57], Parson et al. [89], Vega et al. [90,91], and Rodriguez et al. [92]. Both vegetation and soil strata were measured (Table 2). For every tree with a diameter at breast height (DBH) > 7.5 cm, we measured the mean scorch height, total height, crown base height, crown diameter, and DBH. The scorched crown volume (%), defined as the sum of the estimated percent assessed for the scorched and consumed crown of each tree, was visually estimated [93]. At the time of field data collection, the overstory stratum was defined as every individual with DBH > 7.5 cm [94]. Once we processed field data, the overstory was separated by 75th percentile heights values on each plot sampled into canopy (>75th percentile) and subcanopy (<75th percentile) [95]. Understory strata were defined as small trees and shrubs with DBH < 7.5 cm (Table 2), given the availability of allometric equations which can be used to estimate biomass for the main species in the study area [96]. For these strata, we deliberately did not include severity measurements in herbaceous plants, given their ephemeral nature and fast recovery in the analyzed ecosystems, among other challenges documented in the literature [97], and focused our evaluation on the effects on small trees and shrubs. For those, we quantified frequencies by basal diameter classes 0–2.5, 2.5–5, and 5–7.5 cm and measured the average total height, crown base height, crown diameter, and estimated scorched crown volume for a representative sample for each basal diameter class.

Table 2. Strata levels used in the field inventory performed in this study.

Level	Stratum	Description
Vegetation	5. Overstory	Woody individuals with DBH > 7.5 cm.
	4. Canopy	Woody individuals with DBH > 7.5 cm and h > P75th on each plot.
	3. Subcanopy	Woody individuals with DBH > 7.5 cm and h < P75th on each plot.
	2. Understory	Small trees and shrubs with DBH < 7.5 cm.
Soil	1. Soil	Inert surface materials of litter, duff, and mineral soil.

DBH = diameter at breast height, h = total tree height (m) and P75 = 75th percentile height per individual for each plot.

The assessment of fire impacts on soil—hereafter named soil burn severity (SBS) following Parson et al. [89], Keely [98], Vega et al. [91], and Sobrino et al. [8]—was performed on 9 square quadrats (30 × 30 cm) per plot, at every 5 m in 3 transects of 15 m (at 0, 120, and 240° azimuth) from the plot center. In each quadrat, the cover, depth, and intact/charred state of the remaining litter and duff were measured. The color and depth of ash layer, the degree of surface fine root consumption, and the soil structure alteration were also evaluated at each quadrat. To classify SBS, the following levels were considered, following Vega et al. [91] and Sobrino et al. [8]: 0: unburned; 1: burned litter but limited duff consumption; 2: litter and duff layers totally charred covering unaltered mineral soil; 3: litter and duff completely consumed (bare soil), but mineral soil organic matter not consumed and structure unaffected; 4: bare soil, surface mineral soil organic matter consumed and soil structure affected; 5: bare soil and mineral soil structure and color largely altered. For each quadrat, the percentage of cover by SBS level (0–5) was registered. The weighted SBS average by cover was calculated for every quadrat, and the plot-level SBS was finally obtained as the average of all the measured quadrats per plot. The SBS levels (1–5) were scaled from 0 to 100, named *soil burn severity index (SSI)*, by multiplying by 20 to homogenize the levels with the vegetation strata severity scale.

Based on the analyzed strata, we calculated the following *field-based severity indices* (FBSs):

$$FSI = \frac{\sum(CCS + SCS + UCS + SSI)}{4} \tag{1}$$

where: *FSI* = Fire Severity Index, *CCS* = canopy crown scorch volume (%), *SCS* = subcanopy crown scorch volume (%), *UCS* = understory crown scorch volume (%), and *SSI* = soil burn severity index.

The structure of the *FSI* index (0–100) is relatively similar to the first proposed Composite Burn Index (*CBI* [57]), which averages the severity observed in all evaluated strata. In order to account for possible correlations in the crown damage between canopy and subcanopy strata, as well as potential correlations between understory severity and SBS (e.g., [90]), we tested a simplified FSI_2 index, using overstory strata (all trees with DBH > 7.5) and grouping the understory and soil fire severity as a surface stratum, following Equation (2):

$$FSI_2 = \frac{OCS + \frac{UCS+SSI}{2}}{2} \tag{2}$$

where: FSI_2 = Fire Severity Index 2, *OSC* = overstory crown scorch volume (%), *UCS* = understory crown scorch volume (%), and *SSI* = soil burn severity index.

We additionally evaluated a field severity index weighted by the cover of each stratum, named *weighted FSI*, resembling the *Geometrically structured CBI (GeoCBI)* proposed by De Santis and Chuvieco [99], by multiplying each *FSI* stratum by its corresponding cover:

$$wFSI = CCS * CC + SCS * SC + UCS * UC + SSI * SOC \tag{3}$$

where: *wFSI* = weighted Fire Severity Index, *CSC* = canopy crown scorch volume (%), *CC* = canopy cover (%), *SCS* = subcanopy crown scorch volume (%), *SC* = subcanopy cover (%), *UCS* = understory crown scorch volume (%), *UC* = understory cover (%), *SSI* = soil severity index, and *SOC* = soil cover (%), calculated as 100 − (*CCC* + *SCC* + *UCC*).

Finally, we also tested a simplified version of *wFSI*, named $wFSI_2$, with a similar structure to FSI_2, considering overstory and surface strata (soil and understory), weighted by their corresponding cover:

$$wFSI_2 = OSC * OC + \left(\frac{USC + SSI}{2}\right) * (100 - OC) \tag{4}$$

where: $wFSI_2$ = weighted Fire Severity Index 2, *OSC* = overstory crown scorch volume (%), *OC* = overstory cover (%), *UCS* = understory crown scorch volume (%), and *SSI* = soil severity index.

2.3. Remotely Sensed Data

2.3.1. Aggregated Active Fire Perimeters

In order to locate the areas of fire occurrence to retrieve Sentinel images, we used monthly aggregated active fire (AGAF) perimeters [75], downloaded from the Fire Danger Forecast System of Mexico (Sistema de Predicción de Peligro de Incendios Forestales de Mexico, SPPIF) (http://forestales.ujed.mx/incendios2/, accessed on 16 May 2022), [100,101]. AGAF perimeters are calculated in the SPPIF by applying a convex hull algorithm to active fire data from the Moderate Resolution Imaging Radiometer Spectrum (MODIS, 1 km) [102] and the Visible Infrared Imaging Radiometer Suite (VIIRS, V1375 m) [103], as described in Briones-Herrera et al. [75,104]. Historical monthly and 10-day AGAF perimeters for Mexico are available for download from the SPPIF for the period 2012–current day [101]. They are updated in the SPPIF at every MODIS and VIIRS satellite pass [104], provided in near real time by the *Comisión Nacional para el Conocimiento y Uso de la Biodiversidad* (CONABIO) [105].

2.3.2. Sentinel-2 Data

Data download and analysis for fire severity and burned area mapping involved three stages, as summarized in Figure 2: (i) *Section I* aimed to analyze the correspondence of the field-based severity indices (*FBSs*) and field severity data by stratum, with single-date paired images. The goal of this first analysis was to select the spectral index (*SI*) with the highest correspondence against *FBS* data for further evaluation in the subsequent stages. (ii) *Section II* aimed to evaluate the effect of the image compositing period (1 or 3 months), technique (average or minimum), and phenological correction (constant or relative, *c or rc*) for predicting field fire severity. The evaluated composites were downloaded from *Google Earth Engine* (*GEE*) using the selected *SI* from section I. (iii) *Section III* aimed to evaluate the *GEE* composites' accuracy in burned area perimeter mapping, by means of a kappa and overall accuracy analysis of burned/unburned points for each individual fire (Figure 2).

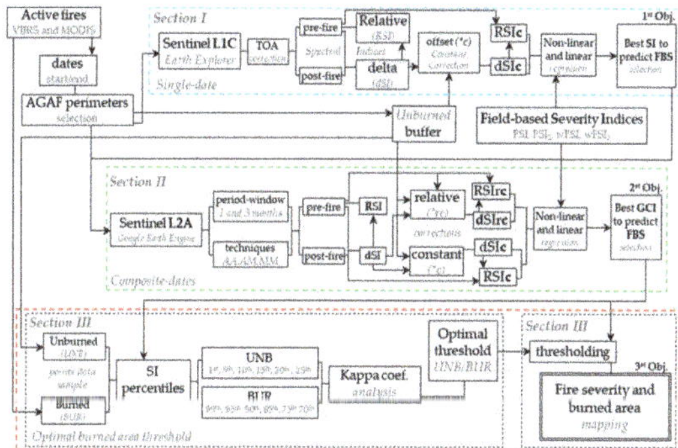

Figure 2. Summary of the study methodology for analyzing Sentinel-2 imagery performance for: (i) evaluating spectral indices from paired dates to predict field-observed fire severity (*Section I*); (ii) comparing image composites of varying length (1 or 3 months), composite technique (average or minimum), and phenological correction (*c or rc*) for predicting field fire severity (*Section II*) and for (iii) mapping burned area perimeter (*Section III*). Where: AGAF: aggregated active fire perimeters (2.3.1), *TOA*: top-of-atmosphere, *dSI*: differenced spectral index (Equation (29)), *RSI*: relative spectral index (Equation (31)), *c*: constant phenological correction (Equation (30)), *SI*: spectral index, *FBSs*: field-based fire severity indices, *FSI*: field severity index (Equation (1)), *FSI₂*: field severity index 2 (Equation (2)), *wFSI*: weighted field severity index (Equation (3)), *wFSI₂*: weighted field severity index 2 (Equation (4)), *AA, AM, MM*: compositing technique (Table 4), *rc*: relative phenological correction (Equation (32)), and *GCI*: Google Earth Engine composite index.

2.3.3. Section I: Comparison of Spectral Indices from Paired Images against Fire Severity

For the analysis of paired images, we downloaded single-date pre- and post-fire L1C images from the United States Geological Survey (USGS) Earth Explorer data portal (https://earthexplorer.usgs.gov/, accessed on 12 February 2022). We selected the images with the lowest cloud cover, closest to the dates before and after fire occurrence based on *MODIS* and *VIIRS* active fires dates for each fire (Table 3). The digital values of L1C-level images were transformed into apparent reflectance using top of atmosphere (TOA) correction, based on the Dark Object Subtraction (DOS) method, proposed by Chávez [106]. Earth Explorer images were processed in the Semi-Automatic classification plug-in proposed by Congedo [107] on QGIS software v3.16.16-Hannover [108].

Spectral Indices Analyzed from Paired Sentinel-2 Imagery

The Sentinel-2 spectral indices analyzed for fire severity discrimination are listed in Table 3. For all of the tested spectral indices *SI* (Table 3), we calculated the difference in spectral values between pre- and post-fire dates, *dSI*, following Equation (5):

$$dSI = \left(SI_{pre} - SI_{post} \right) \times 10^3 \tag{5}$$

where dSI = differenced spectral index [109] from paired pre- and post-*SI* images (SI_{pre} and SI_{post}, respectively).

In order to account for phenological and weather-related influences between the pre- and post-fire images (e.g., [41,70,72,73]), we calculated the phenologically corrected spectral indices, *dSIc*, for all the considered indices, following Equation (6). For the constant "offset" c [73] calculation, we used a buffer from 3 to 5 km from the aggregated active fire perimeter, following Briones-Herrera et al. [75].

$$dSI_c = dSI - c \tag{6}$$

where dSI_c = constant phenology − corrected dSI and c = constant phenological correction, calculated as the average *dSI* value in the unburned vegetation.

Table 3. Spectral indices derived from Sentinel L1C and L2A imagery calculated in this study.

Spectral Index	Equation [1]		Reference [2]
Burn Area Index for Sentinel-2	$BAIS21 = \left(1 - \sqrt{(B6 * B7 * B8)/B4}\right) * \left((B11 - B8/\sqrt{B11 + B8}) + 1\right)$	(7)	-
	$BAIS22 = \left(1 - \sqrt{(B6 * B7 * B8)/B4}\right) * \left((B12 - B8/\sqrt{B12 + B8}) + 1\right)$	(8)	-
	$BAIS23n = \left(1 - \sqrt{(B6 * B7 * B8a)/B4}\right) * \left(\left(B11 - B8a/\sqrt{B11 + B8A}\right) + 1\right)$	(9)	-
	$BAIS24n = \left(1 - \sqrt{(B6 * B7 * B8a)/B4}\right) * \left(\left(B12 - B8a/\sqrt{B12 + B8A}\right) + 1\right)$	(10)	[30]
Chlorophyll Index red-edge	$CIre = (B7/B5) - 1$	(11)	[28]
Normalized Burn Ratio	$NBR1 = (B11 - B12)/(B11 + B12)$	(12)	[26]
	$NBR2 = (B8 - B11)/(B8 + B11)$	(13)	[78]
	$NBR3 = (B8 - B12)/(B8 + B12)$	(14)	[31]
	$NBR4n = (B8a - B11)/(B8a + B11)$	(15)	[24]
	$NBR5n = (B8a - B12)/(B8a + B12)$	(16)	[31]
Normalized Difference Index	$NDI1re = (B6 - B5)/(B6 + B5 - 2 * B1)$	(17)	[110]
	$NDI2re = (B7 - B5)/(B7 + B5 - 2 * B1)$	(18)	[110]
Normalized Difference vegetation Index	$NDVI1re = (B6 - B5)/(B6 + B5)$	(19)	[111]
	$NDVI2re = (B7 - B5)/B7 + B5)$	(20)	[112]
	$NDVI3 = (B8 - B4)/(B8 + B4)$	(21)	[113]
	$NDVI4re = (B8 - B5)/(B8 + B5)$	(22)	[111]
	$NDVI5re = (B8 - B7)/(B8 + B7)$	(23)	[29]
	$NDVI6n = (B8a - B4)/(B8a + B4)$	(24)	[114]
	$NDVI7ren = (B8a - B5)/(B8a + B5)$	(25)	[29]
	$NDVI8ren = (B8a - B7)/(B8a + B7)$	(26)	[29]
Modified Simple Ratio	$MSR1re = ((B8/B5) - 1)/(\sqrt{B8 + B5} + 1)$	(27)	[115]
	$MSR2re = ((B8/B7) - 1)/(\sqrt{B8 + B7} + 1)$	(28)	-
	$MSR3ren = ((B8a/B5) - 1)/(\sqrt{B8a + B5} + 1)$	(29)	[29]
	$MSR4ren = ((B8a/B7) - 1)/(\sqrt{B8a + B7} + 1)$	(30)	-

[1] Sentinel-2 MSI spectral bands: B1 = ~443 nm, B4= ~665 nm, *red-edge bands (*re):* B5=~704 nm, B6 = ~740 nm, B7 = ~783 nm, *NIR:* B8 = ~833 nm, *NIR narrow (*n):* B8A = ~865 nm, *SWIR:* B11= ~1613 nm, B12= ~2202 nm. Spatial resolution bands = 20 m (B1, B4, and B8 were resampled) [116]. [2] References with "-" are index modifications tested in this study.

In addition, for the NBR_i indices considered (Equations (12)–(16), Table 3), we calculated the *Relativized Burn Ratio* (RBR) following Equation (31) [70]:

$$RSI_{ic} = dNBR_{ic} / \left(NBR_{ipre} + 1.001\right) \tag{31}$$

where RSI_{ic} = RBR with constant phenological correction, $dNBR_{ic}$: differenced NBR_i (Table 3) with constant phenological correction (Equation (6)), NBR_{ipre}: pre-fire NBR_i (Table 3) and i = 1, 2, 3, 4n, and 5n (i.e., NBR1, NBR2, NBR3, NBR4n, and NBR5n, Table 3).

2.3.4. Section II: Comparison of GEE Composites for Mapping Fire Severity

Sentinel-2 composites were generated and downloaded from Google Earth Engine (GEE), a multi-petabyte catalog of satellite imagery and geospatial datasets which is a cloud-based platform [117]. GEE composite indices were derived from Sentinel-2 processing level 2A: bottom-of-atmosphere [GEE dataset ID: ee.ImageCollection ("COPERNICUS/S2_SR")] imagery. SCL and Q60 bands were used to select the "optima" pixels (i.e., free of clouds, shadows, water, and/or snow) within a given period in each pre- and post-fire window.

GEE Composite Periods and Techniques Analyzed

Composite indices were generated with varying composite lengths of 1 or 3 months for the best *dSI* and *RSI* selected in *Section I*. Dates immediately before and after each fire, based on MODIS and VIIRS active fire data, were used to define each composite search period (Table S2). We tested average and minimum compositing since they are the most widely used approaches in the literature (e.g., [19,27,28,41,42,44,45,48]). For each 1- and 3- month composite, three combinations of *average* (A) and *minimum* (M) for the pre- and post- fire images (AA, AM, and MM) were tested (see Table 4). GEE code for calculating and downloading the composites was based on the modification of Parks et al. [41] for Sentinel-2 by Briones-Herrera et al. [75], initialized with active fire perimeters.

Table 4. Composite techniques analyzed for GEE Sentinel-L2A imagery.

Composite Technique	Pre-Fire	Post-Fire
AA	Average	Average
AM	Average	Minimum
MM	Minimum	Minimum

Constant and Relative Phenological Correction

For each of the 6 combinations of composite length (1 or 3 months) and technique (*AA, AM,* or *MM,* Table 4), two phenological corrections were compared, resulting in a total of 12 composites calculated for each *dSI* and *RSI* analyzed. The first correction method tested was the constant phenological correction c [73]. The phenological corrections were calculated for each of the composites analyzed by subtracting the c offset, following Equations (30) and (31). The constant offset c was calculated in *GEE* following Parks et al. [41] and using the 3 to 5 km buffer of the aggregated active fire perimeter, as proposed by Briones Herrera et al. [75], to calculate the offset as the average *dSI* from the unburned buffer.

The constant phenological correction method applied the same correction c to every pixel, therefore assuming that phenological changes in *dSI* for the unburned vegetation were homogeneous. Additionally, in order to consider the potential spatial heterogeneity in unburned *dSI* changes, we calculated the average *dSI* (e.g., dNBR) for every pre-*SI* (e.g., *NBRpre*) value in the unburned 3–5 km buffer. This allowed us to assess variations in vegetation phenological changes (i.e., unburned *dNBR*) between vegetation with varying characteristics, since *NBRpre* values are frequently used as proxies of total biomass or tree cover [70]. To calculate a spatially variable phenological correction, we assigned the average unburned *dSI* (e.g., dNBR) to each pre-*SI* (e.g., *NBRpre*) pixel for the entire image extent, resulting in a map of relative phenological correction (*rc*). Such *rc* maps represented the expected phenological change (i.e., unburned *dNBR*) in the absence of fire for each type of vegetation, as stratified by *NBRpre* values. Finally, we calculated the relative corrected *rc* indices, *dSIrc*, using Equation (32):

$$dSI_{rc} = dSI - rc \qquad (32)$$

where dSI_{rc} = relative corrected *dSI*, *dSI* = uncorrected differenced spectral index [109], and *rc* = relative phenological correction map.

Based on dSI_{rc}, we also calculated its correspondent in the relative version RSI_{rc} (Equation (33)) for all 6 of the composites analyzed.

$$RSI_{irc} = dNBR_{irc} / (NBR_{ipre} + 1.001) \qquad (33)$$

where RSI_{irc} = RBR with relative phenological correction, *i* = 1, 2, 3, 4*n*, and 5*n* (i.e., *RBR1rc*, *RBR2rc*, *RBR3rc*, *RBR4nrc*, and *RBR5nrc*, $dNBR_{iRc}$: differenced NBR_i (Table 3), with relative phenological correction (Equation (32)), NBR_{ipre} : pre-fire NBR_i (Table 3) and *i* = 1, 2, 3, 4*n*, and 5*n* (i.e., *NBR1*, *NBR2*, *NBR3*, *NBR4n*, and *NBR5n*, Table 3).

This resulted in a total of 24 composites analyzed for the two phenological corrections tested: *c* and *rc*.

2.4. Predicting Fire Severity from Spectral Indices and Composites

We evaluated linear and non-linear (exponential and power) regression models to predict paired-image spectral indices (*section I*) from field-measured severity variables by stratum (*OSC, CSC, SSC, USC,* and *SSI*) and from field severity indices, i.e., *FSI, FSI$_2$, wFSI,* and *wFSI$_2$* (see Equations (1)–(4)). For example, for a linear model, we tested the following regression (Equation (34)):

$$y = \beta_0 + \beta_1 \times \text{FBS} \qquad (34)$$

where *y* = Sentinel-2 spectral index (*SI*), *FBS* = field-based severity per strata, i.e., *OSC, CSC, SSC, USC,* and *SSI,* and field severity indices, i.e., *FSI, FSI$_2$, wFSI,* and *wFSI$_2$* (Equations (1)–(4)).

Models were fitted using the "*lm*" and "*nls*" packages from the R software [118]. Model goodness of fit was evaluated based on the squared correlation coefficient between the predicted and observed values (R^2) and their *Root Mean Square Error* (θ, RMSE) [119]. The best-performing spectral index was selected for further analysis in *Section II*, again using linear and non-linear (exponential and power) regression models, in this case to predict GEE composite indices of the selected absolute and relative spectral index (*dSI* and *RSI*) from the analyzed field severity indices. Composites' correspondence with field fire severity was also evaluated using model R^2 and RMSE.

2.5. Fire Severity Thresholds

We defined the following fire severity categories based on field fire severity indices: Extreme (>90), High (60–90), Moderate (30–60), Low (30–10), and Very low/unburned (<10). The proposed fire severity categories considered both CBI protocol categories [72] and previous research in post-fire pine mortality in Mexico [92,120] and in temperate pine ecosystems elsewhere [10]. The threshold for High fire severity was based on such previous studies, which observed higher post-fire pine mortality when approximately >2/3 of the tree crown volume was scorched. Conversely, lower fire-caused tree mortality is expected at the Moderate fire severity level, which can be in turn more susceptible to pest attack, in contrast with low mortality either by fire or pests expected at the *Low* fire severity level [10]. Finally, to support post-fire emergency soil erosion hazard assessments, we differentiated an Extreme fire severity level, corresponding to areas of full tree canopy consumption, due to crown fire. In these areas, the absence of scorched litter covering the bare mineral soil can result in higher soil erosion risk [7,8,91]. This is particularly marked when crown fire is accompanied by high soil organic matter consumption, consistent with high soil burn severity levels [9,90,121]. Based on these field fire severity categories, the best fit models of the best-performing spectral index and compositing technique were used to obtain the corresponding spectral index thresholds to map fire severity levels.

2.6. Section III: Comparison of GEE Composites for Mapping Burned Area Perimeter

In order to evaluate the burned/unburned (BUR/UNB) separability of the 24 analyzed composites for each wildfire (i.e., Sites 1–3), the *SI* values of the GEE composites were extracted to a sample of BUR/UNB points. We used filtered VIIRS active fires with $SI > 0$ inside the active fire perimeter of each wildfire as burned points (BUR). The same number of data sample points were randomly located inside the unburned buffer (UNB). Agricultural areas, agricultural–forest interface areas (downloaded from the SPPIF, accessed on 17 February 2021 http://forestales.ujed.mx/incendios2/#), bare soil, human settlements, water bodies, and ice/snow [122] were not considered for analysis. This resulted in a total of 5144, 634, and 1708 burned points for sites 1, 2, and 3, respectively, and an equal-sized sample of unburned points. For each wildfire, we tested the 99th, 95th, 90th, 85th, 80th, and 75th percentiles of *SI* values as candidates for the burned area threshold for the BUR sample, as well as the 1st, 5th, 10th, 15th, 20th, and 25th *SI* percentiles for the UNB sample. For every fire and for each candidate sampled composite *SI* threshold value, we calculated the overall accuracy, kappa coefficient, sensitivity (i.e., omission errors), and specificity (i.e., commission errors) [123] to classify burned/unburned data, using the *'caret'* package [124] (https://cran.r-project.org/web/packages/caret/, last accessed on 23 March 2022) from the *R* software [118]. The spectral index value with the highest kappa and overall accuracy was set as the optimal threshold to discriminate unburned from burned areas [23] for each composite and wildfire in question.

3. Results

3.1. Section I: Correspondence between Field-Based Severity Variables and Spectral Indices from Earth Explorer (Paired Images)

For the field severity variables and spectral indices analyzed, higher R^2 values were obtained with linear models, which are shown in Table S3. The highest correspondence with spectral indices was observed for *FSI* and *FSI$_2$*, which performed very similarly, followed by *wFSI$_2$*, generally outperforming *wFSI* (Table S3). *RBR5nc* and *RBR3c*, using bands 8a/12 and 8/12, respectively (Equations (14) and (16) [31]), had the first- and second-highest correspondence with the analyzed field measures of fire severity (an R^2 of approximatedly 0.85 for both *FSI* and *FSI$_2$*) and by individual stratum as well (Table S3). The third ranking spectral index was *dBAIS24nc* (Equation (10) [30]), with an R^2~0.83 for both *FSI* and *FSI$_2$* (Equations (1) and (2), respectively). *RBR5nc*, *RBR3c*, and *dBAIS24nc* also performed consistently, with the highest correspondence with overstory, canopy, and subcanopy severity and very similar performance between them for these strata ($R^2 = 0.80$–0.78) (Table S3). On the other hand, lower but significant correlations were observed for the understory and soil strata, with an R^2 of 0.69–0.67 (Table S3). For the understory stratum, *RBR5nc* and *RBR3c* were also the best-performing *SI*, with an R^2 of approximately 0.69, followed by *RBR1c* (using bands 11/12 [26], Equation (12), Table 4) and *dBAIS24nc*, with an R^2 of approximately 0.66 for both indices (Table S3). For soil burn severity, the highest correspondence was observed for *RBR5nc* (R^2 of 0.685), closely followed by *RBR3c* ($R^2 = 0.678$) (Table S3). The absolute fire severity spectral indices based on *NBR*, such as *dNBR5n* (8a/12 bands), consistently ranked higher than red and *red-edge* versions of *dNDVIc* for all strata but showed a lower performance compared to their relative versions [70] for all of the analyzed fire severity indices and strata.

3.2. Section II: Comparison of Google Earth Engine Composites with Varying Period, Technique, and Phenological Correction for Predicting Field-Observed Fire Severity

3.2.1. Correspondence between GEE Composites and Field-Based Severity Indices

Based on *Section I*, we selected the composite with the highest correspondence index, *RBR5nc* (hereafter, *RBR*) for further testing in GEE. We also selected for further testing its corresponding absolute version, *dNBR5nc* (hereafter, *dNBR*), in order to quantify the differences between absolute and relative indices with the same spectral bands and because of the wide utilization of this latter index [67]. Similar to *Section I*, higher R^2 values were again observed with linear models. The correspondences between field fire severity indices and the composites for the selected indices *RBR* and *dNBR* with varying composite lengths (1 or 3 months), techniques (*AA*, *AM*, or *MM*), and phenological correction (*c* and *rc*) are shown in Table 5. In agreement with the results in *Section I*, *RBR* composites showed better performance than *dNBR*. The greatest association levels were obtained for the relative phenological correction indices (i.e., *RBRrc* and *dNBRrc*) compared with their analogous constant phenological correction (**c*) (Table 5). These gains using *rc* composites were higher for the 3-month composites, where the R^2 values increased by 0.05–0.09 and 0.10–0.14 for *RBR* and *dNBR*, respectively (Table 5).

Table 5. Goodness of fit and coefficients of linear models between spectral indices from composite images and the field-based severity index FSI_2 for the three wildfires of Durango.

GEE Composite	β_0	β_1	θ	R^2	Diff. R^2 (rc-c)	Diff. R^2 (1-3)
RBRrc_AA1	−10.4	5.4	85.8	0.815	-	-
RBRrc_AM1	−16.7	5.8	87.1	0.830	-	-
RBRrc_MM1	−20.7	5.9	87.4	0.832	-	-
RBRrc_AA3	−5.9	4.4	77.6	0.781	-	0.034
RBRrc_AM3	−19.7	4.7	80.9	0.789	-	0.041
RBRrc_MM3	−21.5	5.1	87.5	0.788	-	0.044
RBRc_AA1	18.1	5.0	88.4	0.779	0.036	-
RBRc_AM1	6.2	5.5	85.8	0.817	0.013	-
RBRc_MM1	6.7	5.5	85.5	0.820	0.012	-
RBRc_AA3	36.1	4.3	86.5	0.727	**0.054**	**0.052**
RBRc_AM3	21.8	4.8	104.1	0.698	**0.091**	**0.119**
RBRc_MM3	15.1	4.9	101.3	0.721	**0.067**	**0.099**
dNBRrc_AA1	−15.4	6.5	113.0	0.785	-	-
dNBRrc_AM1	−24.6	7.0	117.8	0.793	-	-
dNBRrc_MM1	−28.9	6.9	116.3	0.796	-	-
dNBRrc_AA3	−10.6	5.5	103.3	0.759	-	0.026
dNBRrc_AM3	−25.8	5.8	106.5	0.765	-	0.028
dNBRrc_MM3	−24.8	5.8	108.7	0.755	-	0.041
dNBRc_AA1	25.2	5.9	118.8	0.731	**0.054**	-
dNBRc_AM1	8.4	6.5	116.5	0.773	0.020	-
dNBRc_MM1	8.6	6.4	114.3	0.778	0.018	-
dNBRc_AA3	51.5	5.2	125.9	0.655	**0.104**	**0.076**
dNBRc_AM3	38.9	5.8	149.6	0.623	**0.142**	**0.150**
dNBRc_MM3	27.7	5.5	136	0.642	**0.113**	**0.136**

β_0, β_1 = intercept and coefficient of linear regression model (Equation (34)), R^2 = coefficient of determination, θ = root mean square error, FSI_2 = Fire Severity Index 2 (Equation (2)), Suffix: c = phenological correction [73], rc = relative phenological correction (suggested in this study), composite technique: *AA* = average in pre- and post-fire composites; *AM* = average (pre) and minimum (post); *MM* = minimum (pre and post), time window: 1 and 3 months, Diff. R^2 (rc-c): observed difference in R^2 between rc and c composites for corresponding technique and index, and Diff. R^2 (1-3): observed difference in R^2 between 1-month and 3-month composites for corresponding technique and index. Changes in R^2 > 0.05 are marked in bold.

For all the analyzed compositing techniques and indices, the R^2 between field severity and composites was higher for the 1-month period than for the corresponding 3-month period (Table 5). The decrease in the R^2 with increasing composite length from 1 to 3 months was ≥ 0.05 for all of the *c* composites (0.05, 0.12, and 0.10 and 0.08, 0.15, and 0.14 decreases in the R^2 for *AA*, *AM*, and *MM* for *RBRc* and *dNBRc*, respectively). Interestingly, *rc* composites showed, in contrast, a ≤ 0.05 decrease in the R^2 (0.03–0.04) for all *RBRrc* and *dNBRrc* composites with increasing composite length from 1 to 3 months (Table 5). Regarding the compositing technique, the highest correspondence was obtained using the minimum in pre- and post-fire (*RBRrc_MM1* with $R^2 = 0.832$ and *RBRc_MM1* with $R^2 = 0.820$), closely followed by *AM* (R^2 of 0.830 and 0.817 for *RBR5nrc* and *RBR5nc*, respectively). Conversely, composites using the average in both pre- and post-fire images had the lowest R^2 for the 1-month composites, with an $R^2 = 0.815$ (*RBRrc*) and 0.779 (*RBRc*). *AA* composites also showed lower R^2 values than *AM* or *MM* composites for the 1-month *dNBR* composites (Table 5).

3.2.2. Fire Severity Thresholds for Each GEE Composite Technique

Based on the best-performance regression models obtained in Table 5, we determined composite thresholds categorizing FSI_2 (%) as follows: Extreme (>90), High (60–90), Moderate (30–60), Low (30–10), and Very low (<10) (Table S4). As expected, there were some variations in the thresholds between composites of different lengths, compositing techniques, and phenological corrections applied. Interestingly, the thresholds were higher for the 1-month composites compared to lower thresholds for the 3-month composites. For example, the *Extreme* fire severity thresholds for the *RBRrc* composites, using *AA* and *AM*, would be 478 and 506, respectively, for the 1-month composites, compared to values of 392 and 405 (approximately 100 *RBR* lower), respectively, for the 3-month composites. Similar magnitude differences were observed for all fire severity categories, for both *RBR* and *dNBR*, for all the compositing techniques applied.

3.2.3. Evaluating Phenological Variations for each GEE Composite Period and Technique

The differences between the average and minimum for 1- and 3-month *NBR* pre- and post-fire composites are illustrated in Figure 3 for two selected wildfires: Site 1 (Figure 3a–h) and Site 2 (Figure 3i–p). For the post-fire composites, lower *NBR* values, more clearly capturing the high-severity areas, can be observed for the minimum composites for both wildfires and compositing periods (Figure 3f,h,n,p). In contrast, the average post-composites showed a more diluted response, less clearly capturing high-severity areas in both wildfires (Figure 3b,d,j,l). This is particularly noteworthy for the post-composites in Site 2 (Figure 3j,l), where the average composites very weakly captured the burned surface, achieving positive (green and/or wet vegetation) post-*NBR* values, in contrast to a more distinct observation of burned areas in the corresponding minimum composite (Figure 3n,p).

The process of phenological correction, by sampling the unburned *dNBR*, is illustrated in Figure 4 for the 3-month *MM* composites of the three analyzed wildfires, Site 1 (Figure 4a–d), Site 2 (Figure 4e–h), and Site 3 (Figure 4i–l). The left column (Figure 4a,e,i) shows the uncorrected *dNBR*, together with the active fire perimeter buffer, used to sample the phenological changes in the *NBR* in the unburned area adjacent to the fire. An inspection of the *dNBR* within the unburned buffer suggests spatial variation in the phenological *NBR* changes, particularly for Site 1 and 2 (Figure 4a,e, respectively). For example, higher *dNBR* values can be observed in the southwest of Site 1 fire's unburned buffer (Figure 4a). Similarly, higher *dNBR* values can also be seen in the south of Site 2's unburned buffer (Figure 4e), in contrast with lower values in the northern part of the unburned area. A comparison against the corresponding *NBRpre* maps for these sites (Figure 3g,o, respectively) reveals that such areas with higher phenological change corresponded to the drying or senescence processes of areas with high *NBRpre* (i.e., dense, humid forests, shown in blue in Figure 3g,o), in contrast with lower phenological change (*dNBR*, Figure 4a,e) observed in the low-biomass (low-*NBR*) areas which were already dry in the pre-fire image (Figure 3g,o).

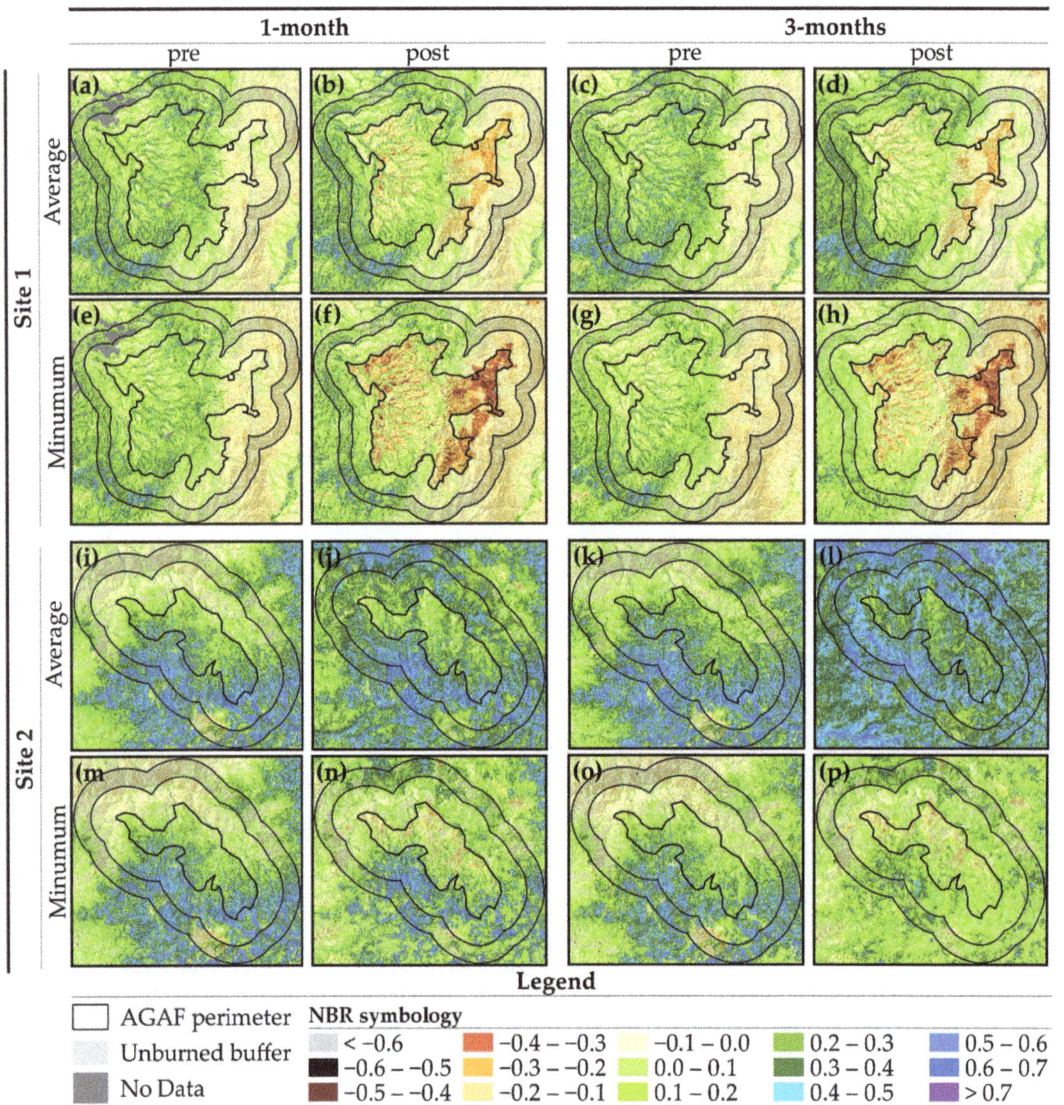

Figure 3. Pre- and post-fire *NBR* GEE composites with varying techniques average (**a–d**) and (**i–l**) and minimum (**e–h**) and (**m–p**) and different time windows: 1 month (left columns) and 3 months (right columns)) for Sites 1 (**a–h**) and 2 (**i–p**).

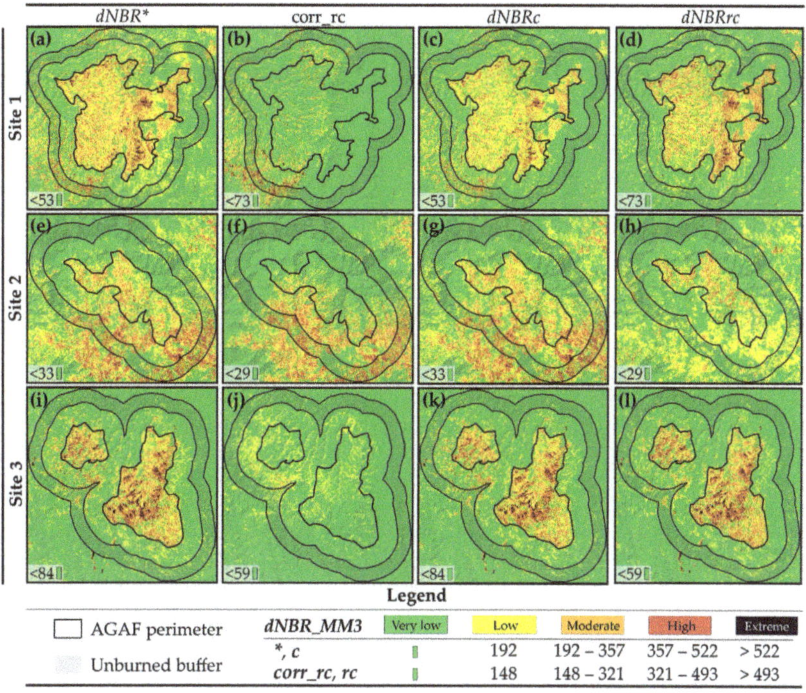

Figure 4. Phenological corrections on *dNBR* images, where *dNBR** = without any correction (**a,e,i**), corr_rc = relative correction map (**b,f,j**), *dNBRc* = constant-offset-corrected *dNBR* (**c,g,k**), and *dNBRrc* = relative-phenology-corrected *dNBR* (**d,h,l**) for sites 1 (**a–d**), 2 (**e–h**) and 3 (**i–l**). Very low/unburned thresholds (shown in green) were obtained for each fire in Section III (Table 6).

Table 6. Overall accuracy, kappa index, sensitivity, and specificity of the best-performing *RBR* burned area thresholds, for each analyzed composite technique in 1–3 months and different phenological corrections (*c* and *rc*).

| | | RBRc | | | | | | RBRrc | | | | | | Difference | | | |
Fire	Composite Technique	Perc	Threshold	OA	Kappa	Sens.	Spec.	Perc.	Threshold	OA	Kappa	Sens.	Spec.	Diff. OA (rc-c)	Diff. Kappa (rc-c)	Diff. Sens. (rc-c)	Diff. Spec. (rc-c)
	AA1	85 *	63	0.975	0.950	0.973	0.976	95 *	60	0.969	0.938	0.987	0.952	−0.01	−0.01	0.01	−0.02
	AM1	90 *	73	0.959	0.918	0.971	0.948	95 *	86	0.961	0.922	0.971	0.952	0.00	0.00	0.00	0.00
	MM1	10	73	0.959	0.918	0.966	0.952	95 *	70	0.960	0.920	0.971	0.950	0.00	0.00	0.01	0.00
Site 1	AA3	10	74	0.958	0.916	0.956	0.960	95 *	59	0.965	0.930	0.980	0.951	0.01	0.01	0.02	−0.01
	AM3	10	46	0.878	0.736	0.931	0.823	10	68	0.914	0.828	0.902	0.926	0.04	0.07	−0.04	0.10
	MM3	85 *	50	0.899	0.798	0.945	0.861	10	73	0.921	0.842	0.910	0.940	0.02	0.04	−0.04	0.08
	AA1	20	99	0.770	0.541	0.760	0.782	10	47	0.798	0.597	0.908	0.749	0.03	**0.06**	**0.15**	−0.03
	AM1	10	63	0.826	0.652	0.877	0.787	5	60	0.895	0.791	0.980	0.857	**0.07**	**0.14**	**0.10**	**0.07**
	MM1	80 *	92	0.826	0.652	0.830	0.822	85 *	57	0.893	0.786	0.993	0.860	**0.07**	**0.13**	**0.16**	0.04
Site 2	AA3	5	32	0.730	0.459	0.889	0.663	80 *	57	0.811	0.622	1.000	0.771	**0.08**	**0.16**	**0.11**	**0.11**
	AM3	75 *	81	0.789	0.578	0.813	0.768	10	48	0.865	0.730	0.902	0.842	**0.08**	**0.15**	**0.09**	**0.07**
	MM3	15	66	0.775	0.549	0.819	0.741	5	29	0.883	0.765	0.945	0.838	**0.11**	**0.22**	**0.13**	**0.10**
	AA1	10	76	0.977	0.955	0.971	0.984	1	67	0.974	0.948	0.989	0.961	0.00	−0.01	0.02	−0.02
	AM1	90 *	72	0.977	0.955	0.977	0.975	99 *	83	0.980	0.961	0.972	0.989	0.00	0.01	−0.01	0.01
	MM1	95 *	86	0.983	0.966	0.968	0.998	99 *	78	0.984	0.969	0.980	0.989	0.00	0.00	0.01	−0.01
Site 3	AA3	10	77	0.955	0.910	0.953	0.956	5	64	0.957	0.914	0.951	0.964	0.00	0.00	0.00	0.01
	AM3	10	76	0.953	0.905	0.952	0.953	95 *	53	0.946	0.892	0.943	0.950	−0.01	−0.01	−0.01	0.00
	MM3	10	78	0.964	0.928	0.954	0.975	5	59	0.960	0.920	0.951	0.970	0.00	−0.01	0.00	−0.01

Composite technique: *AA* = average in pre- and post-fire composites, *AM* = average (pre) and minimum (post) and *MM* = minimum (pre and post), Perc. = best-performing *RBR* percentile, * percentile of the unburned sample, OA = overall accuracy, Sens. = sensitivity, Spec. = specificity, time window: 1 and 3 months, phenological correction: *RBRc* = RBR with constant phenological correction [73], *RBRrc* = RBR with relative phenological correction (proposed in this study), Diff OA (rc-c) = difference in overall accuracy between *rc* and *c* composites, Diff kappa (rc-c) = difference in kappa coefficient between *rc* and *c* composites, Diff. Sens. (rc-c): difference in sensitivity between *rc* and *c* composites, and Diff. Spec. (rc-c): difference in specificity between *rc* and *c* composites. Differences in overall accuracy, kappa coefficient, sensitivity, or specificity between *rc* and *c* composites higher than 0.05 are shown in bold.

The average unburned *dNBR* values for every *NBRpre* value, used to calculate the relative correction, are shown in Figure 5 for the three analyzed wildfires, for every composite technique and period tested. Changes in the *NBR* in the unburned areas were substantially larger in the 3-month composites (Figure 5b,d,f), reaching values of up to 500–600 average *dNBR* in the unburned areas. In contrast, significantly lower unburned *dNBR*, of up to 100–200 values, were observed in the 1-month composites (Figure 5a,c,e), for all the wildfires and composite techniques analyzed. Both composites using the minimum on the post-*NBR* (*AM* and *MM*, showed in blue and green in Figure 5, respectively) behaved similarly, suggesting a higher influence of the post-fire image than the pre-fire image on the observed phenological change for the studied wildfires.

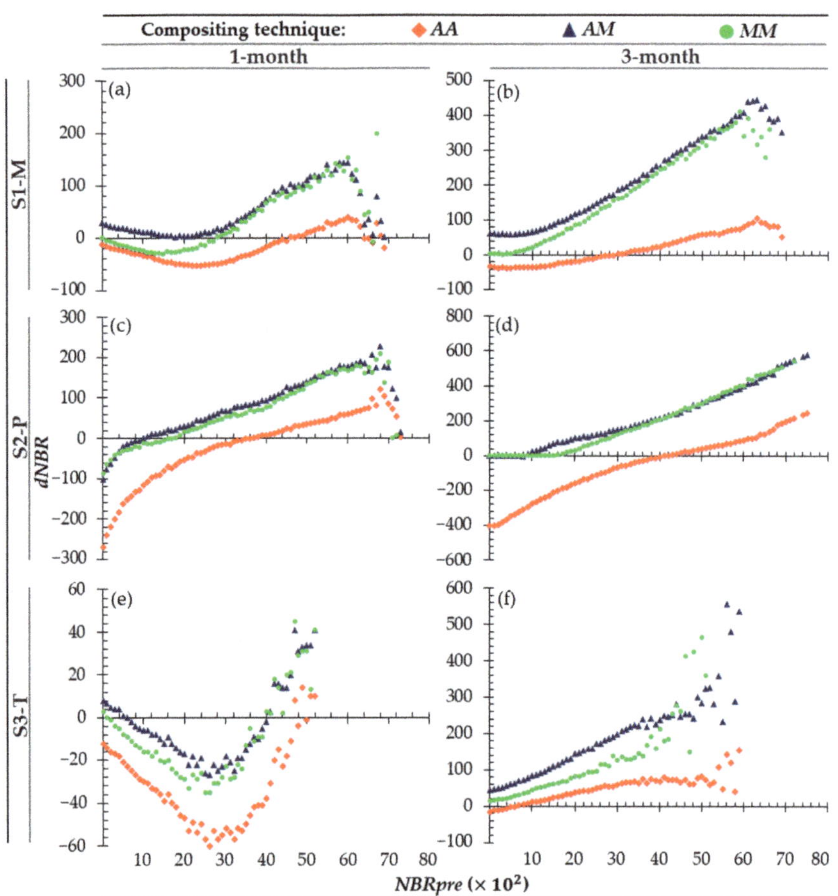

Figure 5. Average *dNBR* by *NBRpre* in the unburned buffer for sites 1 (**a,b**), 2 (**c,d**), and 3 (**e,f**) for 1-month and 3-month composites (**a,c,e** and **b,d,f**, respectively), where composite technique: *AA* = average in pre- and post-fire composites, *AM* = average (pre) and minimum (post), and *MM* = minimum (pre and post).

Interestingly, for both 1 and 3 months, unburned *dNBR* was mainly positive (i.e., drying) for *AM* and *MM* composites in the majority of fires and periods analyzed (shown in blue and green in Figure 5). In contrast, for composites using the average in both pre- and post-fire images (*AA*, shown in red in Figure 5), the unburned *dNBR* varied from negative (i.e., greening or increased moisture) in low-biomass (low-*NBR*) fuels to positive (i.e., drying) in higher-biomass (high-*NBR*) fuels. For those *AA* composites, averaging positive and negative *dNBR* values in a constant offset calculation could result in the underestimation of larger positive or negative phenological changes that occurred for each fuel type, not reflecting the spatial heterogeneity of phenological changes (Figure 5). In the case of composites with the minimum in the post-image *AM* and *AM*, a constant average offset could largely underestimate the higher drying or senescence present in unburned high-biomass fuels, particularly for the longer periods (Figure 5b,d,f). In addition, the smaller phenological changes that occurred in the low-biomass fuels (Figure 5) could also be overestimated by an average constant offset in this case.

By assigning these observed unburned average *dNBR* values (Figure 5) to each *NBRpre* value, we created a spatially variable relative correction (*rc*), shown in Figure 4b,f,j. The relative correction *rc* maps for the 3-month *MM* composites shown in Figure 4 use the patterns analyzed in the unburned buffer (Figure 5) to spatially represent the expected phenological change in the absence of fire for the entire area of study. Consequently, for Site 1 and Site 2 (Figure 4b,f), relative correction *rc* maps reflect that, on average, for those sites, pixels with a higher *NBRpre* experienced a higher phenological increase in the *dNBR* in the period analyzed. Conversely, observed lower phenological changes are mapped for the low-biomass (low-*NBR*) areas in those sites and periods of study (Figure 4b,f). Relative-phenology-corrected *dNBRrc* maps, calculated by subtracting *rc* rasters (second column, Figure 4b,f,j) from uncorrected *dNBRs* (first column, Figure 4a,e,i) are shown in the last column of Figure 4d,h,l. In contrast to the constant phenological *dNBRc* maps (Figure 4c,g,k), *dNBRrc* composites seem to minimize confusion (i.e., commission errors) in the unburned buffer (as evaluated in *Section III*), while also maintaining a clear distinction of the patches of higher severity (Figure 4d,h,l).

3.3. Section III: Evaluating the Effect of Composite Period Length, Technique, and Phenological Correction on the Accuraccy in Mapping the Burned Area Perimeter for each Wildfire

3.3.1. Burned Area Thresholds for Each Composite and Wildfire

We selected the best-performing *RBR* value (highest kappa and overall accuracy) from the candidate burned and unburned percentiles tested as the best burned area perimeter threshold for each wildfire and composite technique, period, and phenological correction (i.e., *c* and *rc*) (Table 6, Figure S1). Both the overall accuracy and kappa peaked at the same *RBR* values, which were in turn selected as burned area thresholds (Table 6). Burned area perimeters estimated with 1-month composites consistently showed higher kappa and overall accuracy values than their corresponding 3-month composites; this decrease in kappa with increasing composite period was higher for *c* than for *rc* composites (Table 6, Figure S1). Regarding the compositing technique, composites using the minimum (*AM* and *MM*) largely outperformed those using the average (*AA*) in Site 2 (Table 6 and Figure S1). For both Site 3 and 1-month composites in Site 1, relatively similar results were obtained for both the average and minimum techniques (Table 6 and Figure S1).

Remarkably, *rc* composites had generally higher kappa and overall accuracy values than their corresponding *c* composites (Table 6 and Figure S1). The highest increases in kappa were observed for Site 2 (where kappa values raised by 0.16–0.22 and 0.06–0.13 for 3 months and 1 month, respectively) and Site 1 (with kappa increases of up to 0.04–0.07 for the 3-month composites). For Site 2, those improvements in accuracy were generally a result of both gains in sensitivity of 0.09–0.16 (i.e., lower omission errors) and increases of up to 0.11 in specificity (lower commission errors) for the *rc* composites (Table 6). For the 3-month composites in Site 1, accuracy increases were mostly caused by improved specificity values of up to 0.10 (i.e., decreased commission errors). In contrast, for Site 3, both phenological corrections and all the composite techniques performed relatively similarly, together with a lower observed reduction in kappa values from 1 to 3 months.

The performance of *rc* and *c* composites for burned area mapping for the 1- and 3-month *MM* composites is shown in Figure 6, using the burned area thresholds obtained for each composite and wildfire analyzed (Table 6). Higher commission errors in the sampled unburned points (shown in yellow in Figure 6) can be observed for the *c* composites, particularly for the 3-month composites in Site 1 and 2 (Figure 6c,g), compared to a lower commission observed for their corresponding relative correction *rc* composites (Figure 6d,h). In addition, higher omission errors (shown in blue in Figure 6) for *c* composites in Site 2 can also be observed at both 1 (Figure 6e) and 3 months (Figure 6g), in contrast with lower omission errors for the *rc* composites during both periods (Figure 6f,h, respectively). Contrarily, slightly higher omission errors, mainly in low-severity areas, were observed with *rc* 3-month composites in Site 1 (Figure 6d), which nevertheless had a higher overall classification accuracy and kappa because of the large improvement in commission errors (Table 6).

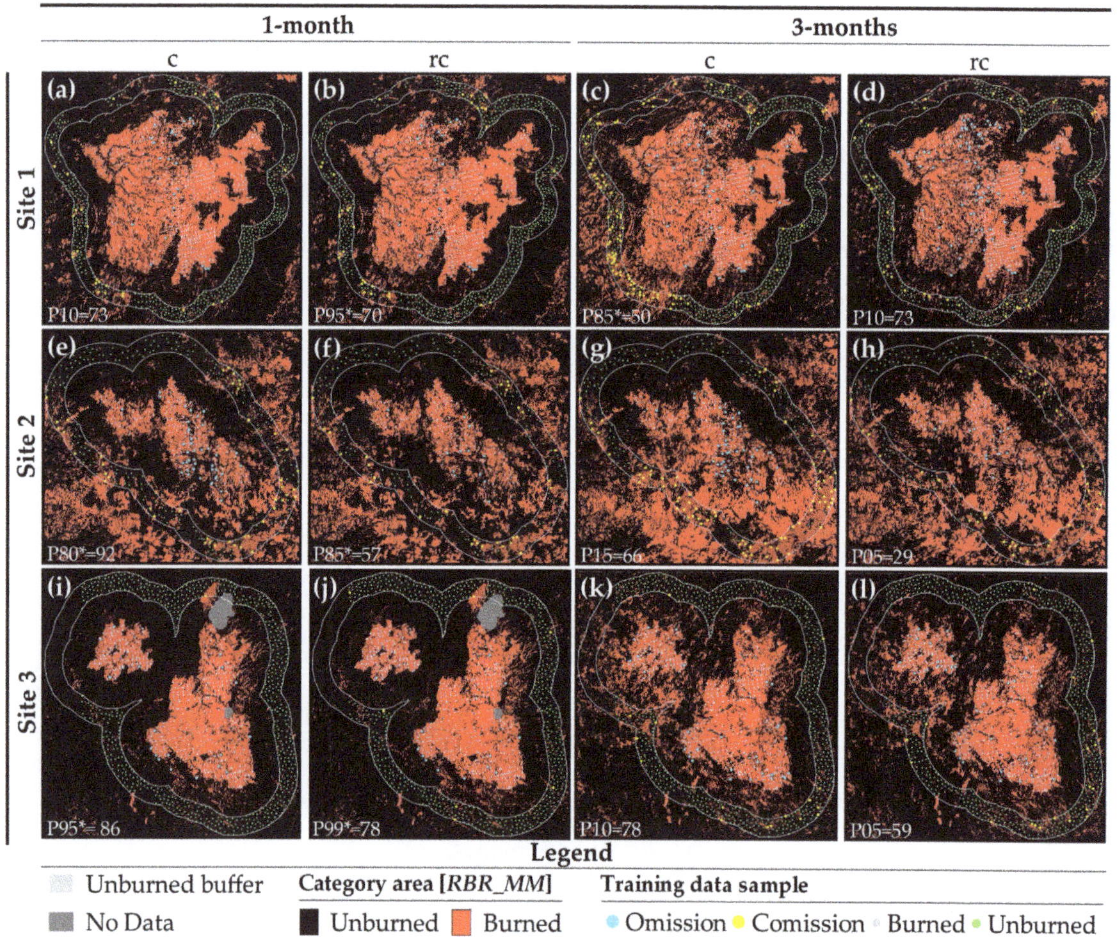

Figure 6. Burned area (shown in red) estimated for site 1 (**a–d**), 2 (**e–h**) and 3 (**i–l**), using the corresponding burned area threshold from Table 6, for 1-month (left columns) and 3-month (right columns) *MM* composites using constant phenology correction *c* [73] and relative phenological correction *rc* (proposed in this study). Sampled points for burned areas (active fires) and for unburned areas are shown in gray and green within the active fire perimeter and the unburned buffer, respectively. Omission and commission errors in the sampled burned and unburned point sample are shown in blue and yellow, respectively.

3.3.2. Fire Severity Mapping

We finally mapped fire severity (using FSI_2 thresholds from Table S4), also considering the wildfire-specific burned area thresholds for the Very low/unburned category (see Table 6). Fire severity maps for the three analyzed wildfires are shown in Figure 7 for the 1-month and 3-month *MM* composites, using constant and relative phenological corrections (*c* and *rc*). Notably, relative phenological corrections, in addition to allowing for a clearer burned/unburned delineation as previously shown in Figure 6 for these same composites, also maintained a clear distinction of high- and extreme-severity patches, even slightly improving the correspondence with field measured plots of fire severity (Table 5), which are shown in circles in Figure 7.

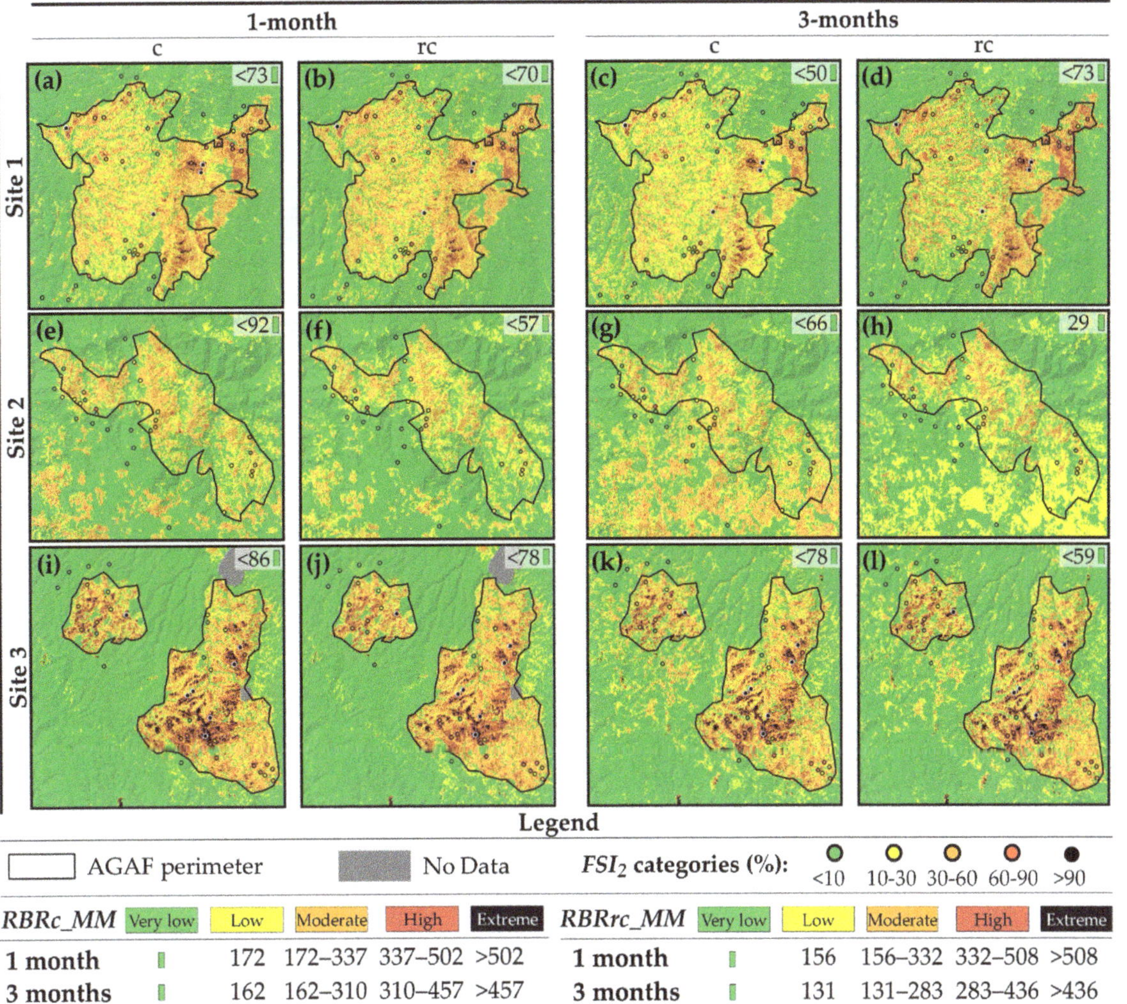

RBRc_MM	Very low	Low	Moderate	High	Extreme	RBRrc_MM	Very low	Low	Moderate	High	Extreme
1 month		172	172–337	337–502	>502	**1 month**		156	156–332	332–508	>508
3 months		162	162–310	310–457	>457	**3 months**		131	131–283	283–436	>436

Figure 7. Fire severity mapping with *MM* technique and different time windows (1 and 3 months) and constant (*c*) or relative phenological correction (*rc*) for wildfires in Site 1 (**a–d**), Site 2 (**e–h**), and Site 3 (**i–l**), using FSI_2 thresholds (see Table S4). The corresponding burned area threshold (Very low/unburned, shown in green) was assigned for each wildfire (see Table 6) according to the time window and composite technique.

4. Discussion

4.1. Field-Based Fire Severity Metrics and Paired Images S2 Spectral Indices Correspondence

The observed higher performance of *RBR5nc* and *RBR3c* (using bands 8a/12 and 8/12, respectively) as the best predictors of field-based fire severity (Table S3), agree with previous Sentinel-based studies of fire severity (e.g., [9,34,125]) and burned area (e.g., [35,126]). The superior performance of the relative index *RBR* compared to its absolute version *dNBR*, both for paired images and composites evaluated (Table 5 and Table S3), also supports the findings of previous studies such as Fernández-Manso [29], Botella-Martínez [127], Arellano et al. [128], Quintano et al. [129], Parks et al. [41,70], Holsinger et al. [19], and Fernández et al. [9], where relative indices showed a higher correlation against field-based severity data. In our study, the strong heterogeneity in fuel types (Figure 1) and pre-fire biomass (Figure 3) was probably related to this better performance of the relative spectral indexes.

The higher performance of fire severity indices (i.e., *FSI*, *FSI$_2$*) than individual strata (i.e., *OCS*, *UCS*, and *SSI*) supports previous observations where composite indices of fire/burn severity showed a higher R^2 with spectral indices than the individual strata evaluated [61,130]. For the temperate to semi-arid pine/oak study sites analyzed, our hypothesis that there could be correlations between canopy and subcanopy crown scorch volume and between understory and soil burn severity was met; in fact, we observed a significant correlation in both cases (R Pearsons = 0.96 and 0.77, respectively). This could explain our observed similar, even slightly higher performance of *FSI$_2$* against *FSI* and, more interestingly, our markedly higher R^2 for *wFSI$_2$* against *wFSI* (Table S3). This suggests some potential of *wFSI$_2$* to compensate for the overemphasis of soil strata in sparsely vegetated plots, observed using *wFSI*, with the latter being in line with previous observations [131]. Further research in a variety of forest ecosystems, analyzing the presence or absence of correlations between strata and also against satellite indices (e.g., [132]), might be useful in better understanding the limitations of both satellite and current integrated field indices of fire severity, and it could provide suggestions for potential improvements for simplified or more representative field indices.

Furthermore, while it is clear that integrated field fire severity indices can improve correlations with satellite data, the question remains as to whether such integrated indices are representative of ecosystem effects (e.g., [1,2,98,133]). In this sense, although they were lower than field fire severity indices, significant correlations were observed for all strata (Table S1). Such observed high correlations with overstory strata has been extensively documented elsewhere (e.g., [61,66,130]). This relationship can be potentially useful when coupled with mortality prediction models (e.g., [11,92,120]), while also accounting for forecasted post-fire pest risk (e.g., [10]), among other factors. Nevertheless, because of the complex interactions behind tree mortality, the performance of remotely sensed crown scorch volume as an input for predicting tree mortality is a subject of ongoing research (e.g., [12]).

Our observed relatively lower correlation for the understory and soil strata (R^2 of 0.691 and 0.685, respectively, Table S3) supports previous observations of lower accuracies for such strata against *SI* (e.g., [34,61,66]), with this limitation being potentially more marked in ecosystems with relatively high tree density. In this sense, for our study, the relatively low tree density in the majority of the pine/oak forests measured might have allowed for the signal of burned soil to be partially captured by the Sentinel spectral indices. This would agree with the observations of Allen and Sorbel [130], who also found significant correlations of substrate severity scores (R = 0.74), although they were also lower than for canopy (R = 0.83) in open-woodland canopies and tundra forest ecosystems. In addition, our observed outperformance of the *NBR* against *NDVI*-based indices for SBS prediction agrees with the results of Sobrino et al. [8] in temperate forests in NW Spain. Additionally, the best performance of *RBR* in the current study supports the best performance observed by Fernández et al. [9] for this index to predict post-fire physical soil properties.

While the relatively open canopies of our studied ecosystems seem to reinforce the concept that the ash and burned soil spectral response might be partially captured in relatively low-tree-density forests, as previously suggested by both some empirical [8,9,130] and radiative transfer modeling studies (e.g., [134]), this might not hold true under more dense canopy conditions (e.g., [66]). In this sense, several authors have stated that passive remote sensing imagery is more likely to detect fire effects on tree crowns than on the ground (e.g., [34,61,135]), together with its limitations to separate strata effects in ecosystems with complex structures [136]. This could be related to a relative inability of passive remote sensing to "pass through" over outer vegetation strata [34], this latter limitation being potentially more marked in ecosystems of relatively high tree density or cover. Accordingly, beyond using satellite reflectance alone, future studies could further explore the role of other ancillary variables such as land surface temperature from thermal bands, which has shown potential for predicting both soil burn severity [137,138] and field indices of fire severity such as CBI [139,140]. In addition, the role of topographic variables [141] or fire intensity and duration [142], with the latter possibly being inferred from active fire detections [143], could be also explored to improve our ability to predict and map fire effects on soil.

4.2. Evaluation of GEE Composites for Mapping Fire Severity and Perimeter

4.2.1. Effect of Compositing Technique on Fire Severity and Burned Area Mapping

Higher correspondence with field fire severity (Table 5) and increased burned area classification accuracy (Table 6) were observed for the composites using a minimum in the post-fire image, particularly for the *MM*. In addition, for the equivalent evaluation period and phenological correction, the fire severity thresholds for *MM* were higher than for their equivalent *AA* composites (Table S4). Both results are possibly related to their better potential to detect initial post-fire effects, by selecting the driest lower-biomass images both in pre- and post-fire periods (e.g., Figure 3h,p), and therefore being potentially less influenced by weather oscillations or regrowth effects in the composite period. In contrast, for average composites, the initial post-fire severity was more diluted, likely by forest recovery responses (e.g., Figure 3b,d) and possibly also by increased fuel and soil moisture with post-fire precipitation (Figure 3j,l), which was undesirable for burned area perimeter and initial fire severity detection. This supports studies that have proposed the use of the minimum or maximum values (depending on the spectral index) for burned area perimeter mapping (e.g., [27,51,144]). For example, Roteta et al. [26] reported that minimum *NBR* composites retained the pixel values when the burned signal was considered to be strongest, corroborated by our comparisons of post-fire minimum versus average *NBR* composites (Figure 3). Additionally, it confirms suggestions from previous studies that warned about the limitations of average composites for initial assessments in forests that revegetate rapidly after fire (e.g., [41,45]). In contrast, for a different goal, such as mapping extended assessments and long-term ecosystem responses, average composites of longer compositing periods might be valuable, particularly in forest ecosystems of slower post-fire recovery (e.g., [19,41,45]). These potential differences support the observations of Holsinger et al. [19], who suggested that the best approach for measuring fire severity may depend on the context and conditions of each fire, as well as the measurement objectives. Future studies in Mexico should focus on performing extended assessments of burn severity, potentially capturing delayed mortality, which might be particularly relevant for fire-sensitive species such as wet tropical forests [84].

4.2.2. Effect of Compositing Period on Fire Severity and Perimeter Mapping

Shorter-length (1-month) composites showed higher fire severity thresholds (Table S4) and higher correspondence with field fire severity (Table 5) than 3-month composites for the fires analyzed. Our observed decrease in the *dNBR* and *RBR* fire severity thresholds with increasing time confirms, for Sentinel-2 composites, previous observations from Landsat compositing studies (e.g., [19,41,45]) or from paired Landsat imagery (e.g., [54,58,61,130]), where longer evaluation periods also generally resulted in lower fire severity thresholds and lower correspondence with fire severity [19]. Furthermore, this better agreement of fire severity with imagery immediately after a fire seems to support the previous suggestions to, ideally, use an image captured within a few weeks of field plot measurements, so that the image reflects the conditions at the time of field measurements (e.g., [19,58,72]). In addition, the use of MODIS and VIIRS active fires to define the composite search window starting immediately post-fire (e.g., [19,47]) was confirmed to be useful, in the light of the fast decrease in the *dNBR* with increasing time since fire. In our study, an initial obscuration of fire effects, likely related to forest recovery, was already captured in the third month after fire, being particularly more marked when using average composites and constant offset corrections (Table S4, Figures 6 and 7). This likely agrees with similar rapid obscuration effects by regrowth in the first few months after a fire in fast-recovery ecosystems, such as semi-arid and tropical areas in the southwestern and southeastern United States and grasslands areas in the CONUS [61,71], or Mediterranean climate locations in southern Europe (e.g., [54]). We consequently recommend that, for the evaluation of initial fire severity in the semi-arid to temperate pine/oak ecosystems in Northern Mexico and similar fast-growing ecosystems, images taken immediately after fire and short compositing periods (<3 months) should be favored.

For our study area, images taken directly after a fire (1 month) were also particularly advantageous for fire perimeter mapping (Table 6, Figure 6). This supports various studies that have preferentially used early post-fire images for perimeter delineation, given the relatively clearer separation of burned/unburned areas in the fresh burn scar before ecosystem recovery (e.g., [54,62,68,145,146]). This need for short-term imagery has to be balanced with image availability, for which multi-sensor approaches [49,129,147] should be further explored [22]. Nevertheless, cloud-free, post-fire, phenologically matching images might not be readily available in the first month (e.g., [21,62]), particularly in in areas with persistent cloud cover. Consequently, in addition to defining the best season and time period for compositing, continued research in improved phenological correction techniques will likely continue to be necessary for sound burned area and fire severity evaluations.

Interestingly, the decrease in the R^2 for predicting field fire severity with increasing composite period was only higher than 0.05 for all the constant offset correction *c* composites. In contrast, *rc* composites using the proposed relative phenological correction showed two advantages for fire severity mapping: first, they provided a higher correspondence with field fire severity, particularly for the 3-month composites. Secondly, unlike *c* composites, *rc* composites experienced a much smaller decrease in the R^2 with increasing compositing periods to 3 months against corresponding 1-month *rc* composites. This suggests the better performance and potentially greater stability of the proposed relative phenological correction to minimize the observed decrease in correspondence against field fire severity with increasing observation periods. This should be corroborated in future studies testing this novel relative phenological correction technique in other locations.

4.2.3. Effect of Phenological Corrections on Fire Severity and Burned Area Mapping

The observed values in the unburned *dNBR* were much larger than the range of -100 and 100 *dNBR* proposed by Key and Benson [57] or Picotte [74] and Picotte et al. [67] to be sampled in the unburned buffer for a constant offset calculation, particularly for the 3-month period (Figure 5). Such observed strong spatial variability in the unburned *dNBR* would violate the assumptions for a reliable application of the constant offset method [57,74]. This limitation is not uncommon: in their evaluation of the 18,497 images manually processed by MTBS analysts in the 1984–2014 period, Picotte [74] found that close to one third of the calculated offset values did not meet the recommended range for a reliable constant offset application (offset average and standard deviation exceeding +/− 50). This common unsuitability of available imagery, limiting reliable constant phenology correction because of frequent strong, spatially variable phenological effects, reinforces the need to account for the temporal and spatial variability in phenology changes. In contrast to a constant offset calculation approach, where such strong variability in the range of the unburned *dNBR* can limit the average offset reliability, the relative correction method allows for the inclusion of the spatial variability in phenological changes in unburned areas, and therefore should not limit sampling values exceeding 100 or -100 *dNBR*. In fact, such variability can and should be incorporated into a spatially variable phenological correction to take into account the actual changes observed in different fuel types.

The relative phenological correction improved both the correspondence with field-observed fire severity (Figure 7) and burned area perimeter accuracy (Table 6 and Figure 6), mainly because of reduced commission, but also because of reduced omission errors in some sites. Improvements in specificity (reduced commission) in sites 1 and 2, which had large phenological effects in the unburned fuels, particularly higher during the longer compositing periods (Figure 5), were likely a consequence of the ability of *rc* to minimize the confusion of unburned stressed or senescent vegetation with burned areas by considering a fuel-specific (*NBRpre*) phenological change that better reproduces the spatial variability in those changes than a constant phenological correction.

Regarding omission errors, for Site 2, which had the strongest phenological effects, sensitivity was also largely improved using *rc*, possibly as a consequence of this clearer separability between burned and stressed vegetation for this site compared to the constant offset method. In contrast, the observed slight increase in omission errors for *rc* against *c* for the 3-month composites in Site 1 seems to suggest that, in some locations, in order to minimize commission errors, the *rc* perimeter thresholds might omit some of the lowest-severity areas. This is a common limitation using remote sensing imagery (e.g., [97]) and is relatively more acceptable than commission errors, since those low-severity areas are expected to be mildly impacted by fires and require fewer or no post-fire rehabilitation interventions (e.g., [10,90]).

In any case, omission errors should be treated with caution, given the spatial uncertainties inherent to the use of filtered active fire observations as proxies for burned area locations. This approach, already proposed by Roteta et al. [27], among others, for calibrating burned area thresholds for Sentinel and Landsat, has the advantage of reducing the need for extensive ground datasets of burned areas and of being susceptible to being automated in near real time using active fire observations to obtain fire-specific burned area thresholds [148]. Nonetheless, the relatively coarse resolution of active fires requires caution, particularly in smaller fires (with a lower number of observations than the large fires analyzed here), where the technique might have limited applicability because of this spatial uncertainty. Future studies could alternatively test the use of Landsat active fire detections [149] or aerial infrared imagery [150], where available, to calibrate locally adaptive burned area thresholds [148] from the percentile distribution of the spectral values of S2 indices, extracted from those detections at much finer spatial resolutions.

We observed a high degree of spatial and temporal variation in phenology, as measured by *rc*. In this sense, the most widely used method of phenological correction, the constant offset, had previously been documented to slightly improve correlations with CBI when analyzing several fires (e.g., [41,76]). Conversely, some studies reported decreased correlations with field CBI using constant offset correction methods compared to uncorrected images (e.g., [58]) and attributed this to the dynamic nature of the ecosystems studied, such that a single average *dNBR* value in unburned areas could not provide a reference point between pre- and post-fire images. This was also acknowledged by Parks et al. [41], who warned that, in their current formulations, constant offset methods would not be appropriate under heterogeneous fuel conditions, which can unfortunately be very common (e.g., [23,74]). Moreover, on an individual fire level, a constant offset correction is not expected to improve neither correlations with field fire severity (e.g., [130]) nor to change the individual fire perimeter delineation accuracy. It is obvious that simply subtracting a constant value to a raster image would not alter the relative differences between pixel values (e.g., burned/unburned) at an individual fire level. In contrast, spatially heterogeneous methods, such as the relative correction method, apply a different phenological correction to every fuel type, as defined by the *NBRpre*, by including the observed heterogeneity in phenological changes (Figures 4 and 5), resulting in the potential for improving accuracies for fire severity and burned area mapping (Table 6, Figure S1).

Interestingly, the higher increases in accuracy using *rc* composites for fire severity and fire perimeter mapping, observed for the 3-month composites, corresponded with the higher variation in the *dNBR* values observed in the unburned area for those longer periods (Figure 5). In addition, larger improvements using *rc* against c composites for burned area perimeter mapping also corresponded to sites where observed phenological changes were also larger, such as Site 1 and 2, while accuracies and fire severity and perimeter maps (Figures 6 and 7) were similar for sites with smaller observed phenological changes, such as Site 3. This seems to suggest that relative phenological correction methods could be routinely applied both for fire severity and fire perimeter delineation, resulting in a similar performance to the current constant offset method when phenological changes are small. More importantly, when large and spatially heterogeneous phenological changes are present, such spatially variable phenological correction could improve both fire severity and perimeter delineation accuracy. Because of the high number of images with large and/or spatially variable phenological effects (31% of all MTBS images analyzed by [74]) and the frequent lack of immediate post-fire imagery [62], we believe that the potential of the proposed *rc* method to improve both fire severity and perimeter delineation can be significant.

Previous approaches have used supervised methods that require a manual selection of control plots by subjectively interpreted fuel type based on expert knowledge (e.g., [77]) or the manual creation of phenology-matched composites [22], which currently require the large involvement of analyst manual supervision [23]. Unlike those approaches demanding manual supervision, the novel relative correction method proposed in this study uses a systematic approach by automatically stratifying by *NBRpre* as a proxy of fuel type and biomass and quantifying and correcting for the phenological changes for each of those spatially variable fuel categories. All the processes in the *rc* calculation can be easily automated and readily incorporated into existing operational GEE codes, very similar to the current automatization of constant offset [41,75], but only by automatically stratifying its calculation by the *NBRpre*, with this layer also already being calculated automatically in Parks et al. [41] GEE code. Including an automated *rc* calculation into existing GEE burned area tools could improve the accuracy of operational large-scale automated assessments of fire severity [19,41,44,45] and burned area [26,27,42]).

Additionally, another advantage of the proposed *rc* method is that, unlike methods that assume a baseline from previous years (e.g., internannual differencing [23]), this approach neither requires a long time series as a baseline, nor does it make any assumptions about the magnitude of the phenological changes from those previous periods. Instead, it quantifies the observed temporal and spatial variations in the unburned fuels and uses this observed change to normalize the image with respect to observed unburned phenological changes. In this sense, this method can capture phenological changes of varying magnitudes (Figure 5), potentially reflecting a variety of greening/drying conditions between different fuels, specific to each location and assessment time. The portability of the *rc* method suggests the potential for improved analysis in a variety of locations and assessment periods, as it minimizes non-fire-induced changes in vegetation dynamics. This could improve the transferability of fire severity thresholds, again, because it minimizes fuel-specific variations in phenological and weather influences. Further research in spatial and temporal variations of phenological corrections across a large variety of ecosystems should be encouraged and ultimately incorporated into operational burned area and fire severity mapping efforts, potentially improving their accuracy and robustness.

5. Conclusions

For the first time, the current study analyzed the accuracy of Sentinel-2 spectral indices to map fire severity and burned area perimeter in Mexico. It included an evaluation of

Google Earth Engine (GEE), composite length, and technique and tested a novel relative phenological correction (*rc*) against the constant offset (*c*) method. The *RBR* index, using bands 8a and 12, showed the highest correspondence with both field fire severity indexes and evaluated individual strata (i.e., canopy, subcanopy, understory, and soil), for a variety of sites with marked vegetation heterogeneity under a climatic gradient from semi-arid to temperate.

One-month GEE composites, using the minimum metric for post-fire images, showed a higher correspondence with field severity and higher fire perimeter delineation accuracies than 3-month composites or using the average. Interestingly, the decreases in fire severity and perimeter accuracy with increasing compositing period from 1 to 3 months were lower for the composites using the *rc* phenological correction, which showed greater stability compared to a constant offset correction. Therefore, the *rc* method could be potentially useful to minimize such effects when immediate post-fire, cloud-free imagery is not available.

Relative phenologically corrected composites showed both improved correspondence with field fire severity and higher accuracy for fire perimeter delineation for most sites and periods analyzed. This latter improvement was more significant in sites and periods where phenological effects (as quantified by changes in the *dNBR* in the unburned area) were stronger. Because strong and heterogeneous fuel phenology effects have been frequently observed, the *rc* method could have the strong potential to significantly improve the accuracy of fire severity and burned areas if routinely applied in post-fire evaluations.

The proposed *rc* approach, using the *NBRpre* to systematically stratify fuels and sample the spatially variable temporal variations in the spectral indices of the unburned fuels, goes beyond the constant offset method in incorporating fuel heterogeneity but shares its relative simplicity of being fully automated. Including such a spatially variable phenological correction into existing automated tools, such as *GEE*, could contribute to deriving improved, systematic large-scale evaluations of fire severity and perimeters in Mexico and elsewhere.

Supplementary Materials: The following supporting information can be downloaded at: https://www.mdpi.com/article/10.3390/rs14133122/s1, Figure S1: Kappa value across RBR values tested as candidate burned area thresholds for each wildfire (sites 1,2,3). Where: Phenological correction: c = constant offset phenological correction [73], rc = relative phenological correction (proposed in this study), Composite technique: AA = Average in pre and post fire composites (Figure 7a,b), AM = Average (pre) and Minimum (post) (Figure 7c,d), MM = Minimum (pre and post) (Figure 7e,f) and Time-window: 1 month (left column) and 3 months (right column).; Table S1: Acquired Sentinel 2 L1C level images from Earth Explorer summary; Table S2: Sentinel-L2A imagery period window for GEE composites indices calculation, Table S3: Coefficient of Determination (R2) and RMSE (θ) of univariate linear models between spectral indices (Table 3) (paired single day images) and field-based severity variables in the three wildfires of Durango, Mexico, Table S4: GEE Composites Index fire severity thresholds obtained using equations from table for FSI2 (%) levels where: Very Low/unburned (<10), Low (10–30), Moderate (30–60), High (60–90) and Extreme (>90).

Author Contributions: A.I.S.-C.: formal analysis, methodology; D.J.V.-N.: conceptualization, methodology, formal analysis, writing—original draft; J.B.-R.: programming; C.I.B.-H.: programming; P.M.L.-S.: methodology; J.J.C.-R.: methodology; S.A.P.: programming, writing—review and editing; L.M.H.: programming, writing—review and editing. All authors have read and agreed to the published version of the manuscript.

Funding: Funding for this study was provided by CONAFOR/CONACYT Project "CO-2018-2-A3-S-131553, Reforzamiento al Sistema Nacional de Predicción de Peligro de Incendios Forestales de México para el pronóstico de conglomerados y área quemada (2019-2022)", for the enhancement of the Forest Fire Danger Prediction System of Mexico to map and forecast active fire perimeters and burned area, funded by the Sectorial Fund for forest research, development and technological innovation "Fondo Sectorial para la investigación, el desarrollo y la innovación tecnológica forestal". Additional support was provided by the USDA Forest Service, Rocky Mountain Research Station, and Aldo Leopold Wilderness Research Institute. The findings and conclusions in this publication are those of the authors and should not be construed to represent any official USDA or U.S. Government determination or policy.

Institutional Review Board Statement: Not applicable.

Informed Consent Statement: Not applicable.

Data Availability Statement: MODIS and VIIRS active fire data used in the study can be accessed from FIRMS: https://firms.modaps.eosdis.nasa.gov/active_fire/ (accessed on 26 March 2022). Aggregated active fire perimeters for the study period analyzed are publicly available through the Forest Fire Danger Prediction System of Mexico: http://forestales.ujed.mx/incendios2/ (accessed on 26 March 2022). Sentinel-2 single-date L1C images can be freely downloaded from the United States Geological Survey (USGS) Earth Explorer data portal (EE, https://earthexplorer.usgs.gov/, accessed on 12 February 2022). Monthly composite indices of Sentinel-2 processing level 2A: bottom-of-atmosphere (GEE dataset ID: ee.ImageCollection ("COPERNICUS/S2_SR")) imagery can be freely accessed from *Google Earth Engine.*

Acknowledgments: We would like to thank CONAFOR's personnel for their support during the current study and their continuous feedback regarding the development of the *Sistema Nacional de Predicción de Peligro contra Incendios Forestales* de México.

Conflicts of Interest: The authors declare no conflict of interest. The funders had no role in the design of the study; in the collection, analyses, or interpretation of data; in the writing of the manuscript, or in the decision to publish the results.

References

1. Lentile, L.B.; Holden, Z.A.; Smith, A.M.; Falkowski, M.J.; Hudak, A.T.; Morgan, P.; Lewis, S.A.; Gessler, P.E.; Benson, N.C. Remote sensing techniques to assess active fire characteristics and post-fire effects. *Int. J. Wildland Fire* **2006**, *15*, 319–345. [CrossRef]
2. Morgan, P.; Keane, R.E.; Dillon, G.K.; Jain, T.B.; Hudak, A.T.; Karau, E.C.; Sikkink, P.G.; Holden, Z.A.; Strand, E.K. Challenges of assessing fire and burn severity using field measures, remote sensing and modelling. *Int. J. Wildland Fire* **2014**, *23*, 1045–1060. [CrossRef]
3. Chuvieco, E.; Mouillot, F.; Van der Werf, G.R.; San Miguel, J.; Tanase, M.; Koutsias, N.; García, M.; Yebra, M.; Padilla, M.; Gitas, I.; et al. Historical background and current developments for mapping burned area from satellite Earth observation. *Remote Sens. Environ.* **2019**, *225*, 45–64. [CrossRef]
4. Mouillot, F.; Schultz, M.G.; Yue, C.; Cadule, P.; Tansey, K.; Ciais, P.; Chuvieco, E. Ten years of global burned area products from spaceborne remote sensing-A review: Analysis of user needs and recommendations for future developments. *Int. J. Appl. Earth Obs.* **2014**, *26*, 64–79. [CrossRef]
5. French, N.H.F.; de Groot, W.J.; Jenkins, L.K.; Rogers, B.M.; Alvarado, E.; Amiro, B.; de Jong, B.; Goetz, S.; Hoy, E.; Hyer, E.; et al. Model comparisons for estimating carbon emissions from North American wildland fires. *J. Geophys. Res.* **2011**, *116*, G00K05. [CrossRef]
6. Prichard, S.; Larkin, N.; Ottmar, R.; French, N.; Baker, K.; Brown, T.; Clements, C.; Dickinson, M.; Hudak, A.; Kochanski, A.; et al. The Fire and Smoke Model Evaluation Experiment—A Plan for Integrated, Large Fire–Atmosphere Field Campaigns. *Atmosphere* **2019**, *10*, 66. [CrossRef]
7. Fernández, C.; Vega, J.A. Modelling the effect of soil burn severity on soil erosion at hillslope scale in the first year following wildfire in NW Spain. *Earth Surf. Processes Landf.* **2016**, *41*, 928–935. [CrossRef]
8. Sobrino, J.A.; Llorens, R.; Fernández, C.; Fernández-Alonso, J.M.; Vega, J.A. Relationship between soil burn severity in forest fires measured in situ and through spectral indices of remote detection. *Forests* **2019**, *10*, 457. [CrossRef]
9. Fernández, C.; Fernández-Alonso, J.M.; Vega, J.A.; Fontúrbel, T.; Llorens, R.; Sobrino, J.A. Exploring the use of spectral indices to assess alterations in soil properties in pine stands affected by crown fire in Spain. *Fire Ecol.* **2021**, *17*, 2. [CrossRef]
10. Vega, J.; Jimenez, E.; Vega, D.; Ortiz, L.; Pérez, J.R. Pinus pinaster Ait. tree mortality following wildfire in Spain. *For. Ecol. Manag.* **2011**, *261*, 2232–2242. [CrossRef]
11. Cansler, C.A.; Hood, S.M.; Varner, J.M.; van Mantgem, P.J.; Agne, M.C.; Andrus, R.A.; Ayres, M.P.; Ayres, B.D.; Bakker, J.D.; Battaglia, M.A.; et al. The Fire and Tree Mortality Database, for empirical modeling of individual tree mortality after fire. *Sci. Data* **2020**, *7*, 194. [CrossRef] [PubMed]
12. Furniss, T.J.; Kane, V.R.; Larson, A.J.; Lutz, J.A. Detecting tree mortality with Landsat-derived spectral indices: Improving ecological accuracy by examining uncertainty. *Remote Sens. Environ.* **2020**, *237*, 111497. [CrossRef]
13. Fernández-García, V.; Fulé, P.Z.; Marcos, E.; Calvo, L. The role of fire frequency and severity on the regeneration of Mediterranean serotinous pines under different environmental conditions. *Forest Ecol. Manag.* **2019**, *444*, 59–68. [CrossRef]
14. Viana-Soto, A.; Aguado, I.; Salas, J.; García, M. Identifying post-fire recovery trajectories and driving factors using landsat time series in fire-prone mediterranean pine forests. *Remote Sens.* **2020**, *12*, 1499. [CrossRef]

15. Fernández-Guisuraga, J.M.; Suárez-Seoane, S.; Calvo, L. Modeling Pinus pinaster forest structure after a large wildfire using remote sensing data at high spatial resolution. *For. Ecol. Manag.* **2019**, *446*, 257–271. [CrossRef]
16. Fernández-Guisuraga, J.M.; Calvo, L.; Fernandes, P.M.; Suárez-Seoane, S. Short-Term Recovery of the Aboveground Carbon Stock in Iberian Shrublands at the Extremes of an Environmental Gradient and as a Function of Burn Severity. *Forests* **2022**, *13*, 145. [CrossRef]
17. Fernández-Manso, A.; Quintano, C.; Roberts, D.A. Burn severity influence on post-fire vegetation cover resilience from Landsat MESMA fraction images time series in Mediterranean forest ecosystems. *Remote Sens. Environ.* **2016**, *184*, 112–123. [CrossRef]
18. Fernández-Guisuraga, J.M.; Suárez-Seoane, S.; Calvo, L. Radiative transfer modeling to measure fire impact and forest engineering resilience at short-term. *ISPRS J. Photogramm. Remote Sens.* **2021**, *176*, 30–41. [CrossRef]
19. Holsinger, L.M.; Parks, S.A.; Saperstein, L.B.; Loehman, R.A.; Whitman, E.; Barnes, J.; Parisien, M.A. Improved fire severity mapping in the North American boreal forest using a hybrid composite method. *Remote Sens. Ecol. Conserv.* **2021**, *8*, 222–235. [CrossRef]
20. Chen, D.; Loboda, T.V.; Hall, J.V. A systematic evaluation of influence of image selection process on remote sensing-based burn severity indices in North American boreal forest and tundra ecosystems. *ISPRS J. Photogramm. Remote Sens.* **2020**, *159*, 63–77. [CrossRef]
21. Chen, D.; Fu, C.; Hall, J.V.; Hoy, E.E.; Loboda, T.V. Spatio-temporal patterns of optimal Landsat data for burn severity index calculations: Implications for high northern latitudes wildfire research. *Remote Sens. Environ.* **2021**, *258*, 112393. [CrossRef]
22. Morresi, D.; Marzano, R.; Lingua, E.; Motta, R.; Garbarino, M. Mapping burn severity in the western Italian Alps through phenologically coherent reflectance composites derived from Sentinel-2 imagery. *Remote Sens. Environ.* **2022**, *269*, 112800. [CrossRef]
23. De Carvalho Júnior, O.A.; Guimarães, R.F.; Silva, C.R.; Gomes, R.A.T. Standardized time-series and interannual phenological deviation: New techniques for burned-area detection using long-term MODIS-NBR dataset. *Remote Sens.* **2015**, *7*, 6950–6985. [CrossRef]
24. Bastarrika, A.; Chuvieco, E.; Martín, M.P. Mapping burned areas from Landsat TM/ETM+ data with a two-phase algorithm: Balancing omission and commission errors. *Remote Sens. Environ.* **2011**, *115*, 1003–1012. [CrossRef]
25. Bastarrika, A.; Alvarado, M.; Artano, K.; Martinez, M.; Mesanza, A.; Torre, L.; Ramo, R.; Chuvieco, E. Bams: A tool for supervised burned area mapping using landsat data. *Remote Sens.* **2014**, *6*, 12360–12380. [CrossRef]
26. Roteta, E.; Bastarrika, A.; Franquesa, M.; Chuvieco, E. Landsat and Sentinel-2 Based Burned Area Mapping Tools in Google Earth Engine. *Remote Sens.* **2021**, *13*, 816. [CrossRef]
27. Roteta, E.; Bastarrika, A.; Ibisate, A.; Chuvieco, E. A Preliminary Global Automatic Burned-Area Algorithm at Medium Resolution in Google Earth Engine. *Remote Sens.* **2021**, *13*, 4298. [CrossRef]
28. Gitelson, A.A.; Gritz, Y.; Merzlyak, M.N. Relationships between leaf chlorophyll content and spectral reflectance and algorithms for non-destructive chlorophyll assessment in higher plant leaves. *J. Plant Physiol.* **2003**, *160*, 271–282. [CrossRef]
29. Fernández-Manso, A.; Fernández-Manso, O.; Quintano, C. SENTINEL-2A red-edge spectral indices suitability for discriminating burn severity. *Int. J. Appl. Earth Obs. Geoinf.* **2016**, *50*, 170–175. [CrossRef]
30. Filipponi, F. BAIS2: Burned Area Index for Sentinel-2. *Proceedings* **2018**, *2*, 364. [CrossRef]
31. Huang, H.; Roy, D.P.; Boschetti, L.; Zhang, H.K.; Yan, L.; Kumar, S.S.; Gomez-Dans, J.; Li, J. Separability analysis of Sentinel-2A multi-spectral instrument (MSI) data for burned area discrimination. *Remote Sens.* **2016**, *8*, 873. [CrossRef]
32. Navarro, G.; Caballero, I.; Silva, G.; Parra, P.C.; Vázquez, Á.; Caldeira, R. Evaluation of forest fire on Madeira Island using Sentinel-2A MSI imagery. *Int. J. Appl. Earth Obs. Geoinf.* **2017**, *58*, 97–106. [CrossRef]
33. Fernández-García, V.; Santamarta, M.; Fernández-Manso, A.; Quintano, C.; Marcos, E.; Calvo, L. Burn severity metrics in fire-prone pine ecosystems along a climatic gradient using Landsat imagery. *Remote Sens. Environ.* **2018**, *206*, 205–217. [CrossRef]
34. García-Llamas, P.; Suarez-Seoane, S.; Fernández-Guisuraga, J.M.; Fernández-García, V.; Fernández-Manso, A.; Quintano, C.; Taboada, A.; Marcos, E.; Calvo, L. Evaluation and comparison of Landsat 8, Sentinel-2 and Deimos-1 remote sensing indices for assessing burn severity in Mediterranean fire-prone ecosystems. *Int. J. Appl. Earth Obs. Geoinf.* **2019**, *80*, 137–144. [CrossRef]
35. Achour, H.; Toujani, A.; Trabelsi, H.; Jaouadi, W. Evaluation and comparison of Sentinel-2 MSI, Landsat 8 OLI, and EFFIS data for forest fires mapping. Illustrations from the summer 2017 fires in Tunisia. *Geocarto Int.* **2021**, *36*, 1–20. [CrossRef]
36. Vega-Nieva, D.J.; Briseño-Reyes, J.; Nava-Miranda, M.G.; Calleros-Flores, E.; López-Serrano, P.M.; Corral-Rivas, J.J.; Montiel-Antuna, E.; Cruz-López, M.I.; Cuahutle, M.; Ressl, R.; et al. Developing models to predict the number of fire hotspots from an accumulated fuel dryness index by vegetation type and region in Mexico. *Forests* **2018**, *9*, 190. [CrossRef]
37. Vega-Nieva, D.J.; Nava-Miranda, M.G.; Calleros-Flores, E.; López-Serrano, P.M.; Briseño-Reyes, J.; López-Sánchez, C.; Corral-Rivas, J.J.; Montiel-Antuna, E.; Cruz-Lopez, M.I.; Ressl, R.; et al. Temporal patterns of active fire density and its relationship with a satellite fuel greenness index by vegetation type and region in Mexico during 2003–2014. *Fire Ecol.* **2019**, *15*, 28. [CrossRef]
38. Briones-Herrera, C.I.; Vega-Nieva, D.J.; Monjarás-Vega, N.A.; Flores-Medina, F.; Lopez-Serrano, P.M.; Corral-Rivas, J.J.; Carrillo-Parra, A.; Pulgarin-Gámiz, M.A.; Alvarado-Celestino, E.; González-Cabán, A.; et al. Modeling and mapping forest fire occurrence from aboveground carbon density in Mexico. *Forests* **2019**, *10*, 402. [CrossRef]
39. Monjarás-Vega, N.A.; Briones-Herrera, C.I.; Vega-Nieva, D.J.; Calleros-Flores, E.; Corral-Rivas, J.J.; López-Serrano, P.M.; Pompa-García, M.; Rodríguez-Trejo, D.A.; Carrillo-Parra, A.; González-Cabán, A.; et al. Predicting forest fire kernel density at multiple scales with geographically weighted regression in Mexico. *Sci. Total Environ.* **2020**, *718*, 137313. [CrossRef]

40. White, J.C.; Wulder, M.A.; Hobart, G.W.; Luther, J.E.; Hermosilla, T.; Griffiths, P.; Coops, N.C.; Hall, R.J.; Hostert, P.; Dyk, A.; et al. Pixel-based image compositing for large-area dense time series applications and science. *Can. J. Remote Sens.* **2014**, *40*, 192–212. [CrossRef]

41. Parks, S.A.; Holsinger, L.M.; Voss, M.A.; Loehman, R.A.; Robinson, N.P. Mean composite fire severity metrics computed with Google Earth Engine offer improved accuracy and expanded mapping potential. *Remote Sens.* **2018**, *10*, 879. [CrossRef]

42. Long, T.; Zhang, Z.; He, G.; Jiao, W.; Tang, C.; Wu, B.; Zhang, X.; Wang, G.; Yin, R. 30 m resolution global annual burned area mapping based on Landsat Images and Google Earth Engine. *Remote Sens.* **2019**, *11*, 489. [CrossRef]

43. Guindon, L.; Bernier, P.; Gauthier, S.; Stinson, G.; Villemaire, P.; Beaudoin, A. Missing forest cover gains in boreal forests explained. *Ecosphere* **2018**, *9*, e02094. [CrossRef]

44. Parks, S.A.; Holsinger, L.M.; Koontz, M.J.; Collins, L.; Whitman, E.; Parisien, M.A.; Loehman, R.A.; Barnes, J.L.; Bourdon, J.F.; Boucher, J.; et al. Giving ecological meaning to satellite-derived fire severity metrics across North American forests. *Remote Sens.* **2019**, *11*, 1735. [CrossRef]

45. Whitman, E.; Parisien, M.A.; Holsinger, L.M.; Park, J.; Parks, S.A. A method for creating a burn severity atlas: An example from Alberta, Canada. *Int. J. Wildland Fire* **2020**, *29*, 995–1008. [CrossRef]

46. Flood, N. Seasonal composite Landsat TM/ETM+ Images using the medoid (a multi-dimensional median). *Remote Sens.* **2013**, *5*, 6481–6500. [CrossRef]

47. Hislop, S.; Haywood, A.; Jones, S.; Soto-Berelov, M.; Skidmore, A.; Nguyen, T.H. A satellite data driven approach to monitoring and reporting fire disturbance and recovery across boreal and temperate forests. *Int. J. Appl. Earth Obs. Geoinf.* **2020**, *87*, 102034. [CrossRef]

48. Soulard, C.E.; Albano, C.M.; Villarreal, M.L.; Walker, J.J. Continuous 1985–2012 Landsat Monitoring to Assess Fire Effects on Meadows in Yosemite National Park, California. *Remote Sens.* **2016**, *8*, 371. [CrossRef]

49. Roy, D.P.; Huang, H.; Boschetti, L.; Giglio, L.; Yan, L.; Zhang, H.H.; Li, Z. Landsat-8 and Sentinel-2 burned area mapping-A combined sensor multi-temporal change detection approach. *Remote Sens. Environ.* **2019**, *231*, 111254. [CrossRef]

50. Pinto, M.M.; Trigo, R.M.; Trigo, I.F.; DaCamara, C.C. A Practical Method for High-Resolution Burned Area Monitoring Using Sentinel-2 and VIIRS. *Remote Sens.* **2021**, *13*, 1608. [CrossRef]

51. Filipponi, F. Exploitation of sentinel-2 time series to map burned areas at the national level: A case study on the 2017 italy wildfires. *Remote Sens.* **2019**, *11*, 622. [CrossRef]

52. Barboza Castillo, E.; Turpo Cayo, E.Y.; de Almeida, C.M.; Salas López, R.; Rojas Briceño, N.B.; Silva López, J.O.; Barrena Gurbillón, M.Á.; Oliva, M.; Espinoza-Villar, R. Monitoring wildfires in the northeastern peruvian amazon using landsat-8 and sentinel-2 imagery in the GEE platform. *ISPRS Int. J. Geo-Inf.* **2020**, *9*, 564. [CrossRef]

53. Anaya, J.A.; Sione, W.F.; Rodríguez-Montellano, A.M. Identificación de áreas quemadas mediante el análisis de series de tiempo en el ámbito de computación en la nube. *Revista de Teledetección* **2018**, *51*, 61–73. [CrossRef]

54. Veraverbeke, S.; Lhermitte, S.; Verstraeten, W.W.; Goossens, R. The temporal dimension of differenced Normalized Burn Ratio (dNBR) fire/burn severity studies: The case of the large 2007 Peloponnese wildfires in Greece. *Remote Sens. Environ.* **2010**, *114*, 2548–2563. [CrossRef]

55. Kato, A.; Thau, D.; Hudak, A.T.; Meigs, G.W.; Moskal, L.M. Quantifying fire trends in boreal forests with Landsat time series and self-organized criticality. *Remote Sens. Environ.* **2020**, *237*, 111525. [CrossRef]

56. Lewis, S.A.; Robichaud, P.R.; Hudak, A.T.; Strand, E.K.; Eitel, J.U.H.; Brown, R.E. Evaluating the Persistence of Post-Wildfire Ash: A Multi-Platform Spatiotemporal Analysis. *Fire* **2021**, *4*, 68. [CrossRef]

57. Key, C.H.; Benson, N.C. Landscape Assessment (LA): Sampling and Analysis Methods. In *Firemon: Fire Effects Monitoring and Inventory System*; Lutes, D., Keane, R.E., Caratti, J.F., Key, C.H., Benson, N.C., Sutherland, S., Gangi, L., Eds.; RMRS-GTR-164; Rocky Mountain Research Station, US Department of Agriculture, Forest Service: Fort Collins, CO, USA, 2006; pp. LA-1–LA-51.

58. Picotte, J.J.; Robertson, K.M. Validation of remote sensing of burn severity in south-eastern US ecosystems. *Int. J. Wildland Fire* **2011**, *20*, 453–464. [CrossRef]

59. Wang, L.; Quan, X.; He, B.; Yebra, M.; Xing, M.; Liu, X. Assessment of the Dual Polarimetric Sentinel-1A Data for Forest Fuel Moisture Content Estimation. *Remote Sens.* **2019**, *11*, 1568. [CrossRef]

60. Yebra, M.; Dennison, P.E.; Chuvieco, E.; Riaño, D.; Zylstra, P.; Hunt, E.R., Jr.; Danson, F.M.; Qi, Y.; Jurdao, S. A global review of remote sensing of live fuel moisture content for fire danger assessment: Moving towards operational products. *Remote Sens. Environ.* **2013**, *136*, 455–468. [CrossRef]

61. Zhu, Z.; Key, C.; Ohlen, D.; Benson, N. *Evaluate Sensitivities of Burn-Severity Mapping Algorithms for Different Ecosystems and Fire Histories in the United States*; Final Report to the Joint Fire Science Program; JFSP 01-1-4-12; Joint Fire Science Program: Boise, ID, USA, 2006; 36p.

62. Picotte, J.J.; Robertson, K. Timing constraints on remote sensing of wildland fire burned area in the southeastern US. *Remote Sens.* **2011**, *3*, 1680–1690. [CrossRef]

63. Verbyla, D.L.; Kasischke, E.S.; Hoy, E.E. Seasonal and topographic effects on estimating fire severity from Landsat TM/ETM+ data. *Int. J. Wildland Fire* **2008**, *17*, 527–534. [CrossRef]

64. Saulino, L.; Rita, A.; Migliozzi, A.; Maffei, C.; Allevato, E.; Garonna, A.P.; Saracino, A. Detecting burn severity across mediterranean forest types by coupling medium-spatial resolution satellite imagery and field data. *Remote Sens.* **2020**, *12*, 741. [CrossRef]

65. Loboda, T.; O'neal, K.; Csiszar, I. Regionally adaptable dNBR-based algorithm for burned area mapping from MODIS data. *Remote Sens. Environ.* **2007**, *109*, 429–442. [CrossRef]
66. French, N.H.; Kasischke, E.S.; Hall, R.J.; Murphy, K.A.; Verbyla, D.L.; Hoy, E.E.; Allen, J.L. Using Landsat data to assess fire and burn severity in the North American boreal forest region: An overview and summary of results. *Int. J. Wildland Fire* **2008**, *17*, 443–462. [CrossRef]
67. Picotte, J.J.; Cansler, C.A.; Kolden, C.A.; Lutz, J.A.; Key, C.; Benson, N.C.; Robertson, K.M. Determination of burn severity models ranging from regional to national scales for the conterminous United States. *Remote Sens. Environ.* **2021**, *263*, 112569. [CrossRef]
68. Llorens, R.; Sobrino, J.A.; Fernández, C.; Fernández-Alonso, J.M.; Vega, J.A. A methodology to estimate forest fires burned areas and burn severity degrees using Sentinel-2 data. Application to the October 2017 fires in the Iberian Peninsula. *Int. J. Appl. Earth Obs. Geoinf.* **2021**, *95*, 102243. [CrossRef]
69. Kolden, C.A.; Smith, A.M.; Abatzoglou, J.T. Limitations and utilisation of Monitoring Trends in Burn Severity products for assessing wildfire severity in the USA. *Int. J. Wildland Fire* **2015**, *24*, 1023–1028. [CrossRef]
70. Parks, S.A.; Dillon, G.K.; Miller, C. A New Metric for Quantifying Burn Severity: The Relativized Burn Ratio. *Remote Sens.* **2014**, *6*, 1827–1844. [CrossRef]
71. Picotte, J.J.; Bhattarai, K.; Howard, D.; Lecker, J.; Epting, J.; Quayle, B.; Benson, N.; Nelson, K. Changes to the Monitoring Trends in Burn Severity program mapping production procedures and data products. *Fire Ecol.* **2020**, *16*, 16. [CrossRef]
72. Key, C.H. Ecological and sampling constraints on defining landscape fire severity. *Fire Ecol.* **2006**, *2*, 34–59. [CrossRef]
73. Miller, J.D.; Thode, A.E. Quantifying burn severity in a heterogeneous landscape with a relative version of the delta Normalized Burn Ratio (dNBR). *Remote Sens. Environ.* **2007**, *109*, 66–80. [CrossRef]
74. Picotte, J.J. Development of a new open-source tool to map burned area and burn severity. In *Proceedings RMRS-P-78*; Hood, S.M., Drury, S., Steelman, T., Steffens, R., Eds.; US Department of Agriculture, Forest Service, Rocky Mountain Research Station: Fort Collins, CO, USA, 2020; Volume 78, pp. 182–194.
75. Briones-Herrera, C.I.; Vega-Nieva, D.J.; Monjarás-Vega, N.A.; Briseño-Reyes, J.; López-Serrano, P.M.; Corral-Rivas, J.J.; Alvarado-Celestino, E.; Arellano-Pérez, S.; Álvarez-González, J.G.; Ruiz-González, A.D.; et al. Near Real-Time Automated Early Mapping of the Perimeter of Large Forest Fires from the Aggregation of VIIRS and MODIS Active Fires in Mexico. *Remote Sens.* **2020**, *12*, 2061. [CrossRef]
76. Meddens, A.J.; Kolden, C.A.; Lutz, J.A. Detecting unburned areas within wildfire perimeters using Landsat and ancillary data across the northwestern United States. *Remote Sens. Environ.* **2016**, *186*, 275–285. [CrossRef]
77. Lhermitte, S.; Verbesselt, J.; Verstraeten, W.W.; Coppin, P. A pixel based regeneration index using time series similarity and spatial context. *Photogramm. Eng. Remote Sens.* **2010**, *76*, 673–682. [CrossRef]
78. García, M.L.; Caselles, V. Mapping burns and natural reforestation using Thematic Mapper data. *Geocarto Int.* **1991**, *6*, 31–37. [CrossRef]
79. González-Elizondo, M.S.; González-Elizondo, M.; Tena-Flores, J.A.; Ruacho-González, L.; López-Enríquez, I.L. Vegetación de la Sierra Madre Occidental, México: Una síntesis. *Acta Botánica Mex.* **2012**, *100*, 351–403. [CrossRef]
80. INEGI (Instituto Nacional de Estadística, Geografía e Informática). Conjunto de datos vectoriales de Uso del Suelo y Vegetación, Serie VI, Escala 1:250,000. México. 2017. Available online: http://www.beta.inegi.org.mx/app/biblioteca/ficha.html?upc=889463598459 (accessed on 2 July 2019).
81. García, E. *Modificaciones al Sistema de Clasificación Climática de Köppen*, 5th ed.; Instituto de Geografía, Universidad Nacional Autónoma de México: CDMX, Mexico, 2004; p. 98.
82. INEGI. Conjunto de Datos Vectoriales, Escala 1:1,000,000. Unidades Climatológicas. Instituto Nacional de Estadística, Geografía e Informática, México. 2008. Available online: http://www.beta.inegi.org.mx/app/biblioteca/ficha.html?upc=702825267568 (accessed on 1 December 2019).
83. INEGI. Conjunto de Datos Vectoriales Fisiográficos, Continuo Nacional Escala 1:1,000,000, Serie, I. Instituto Nacional de Estadística, Geografía e Informática, México. 2001. Available online: http://www.beta.inegi.org.mx/app/biblioteca/ficha.html?upc=702825267582 (accessed on 1 December 2019).
84. Jardel, E.J.; Quintero Gardilla, S.D.; Lomelí Jiménez, A.J.; Graf Pérez, J.D.; Rodríguez Gómez, J.N.; Pérez Salicrup, D.P. Guía técnica Divulgativa Para el Uso de Modelos de Comportamiento del Fuego Para los Tipos de Combustibles Forestales de México. Universidad de Guadalajara. Tech. Rep. "Caracterización y Clasificación de Combustibles Para Generar y Validar Modelos de Combustibles Forestales Para México" CONAFOR-CONACyT 2014- CO2-251694, México. 2018. Available online: https://snigf.cnf.gob.mx/wp-content/uploads/Incendios/Insumos%20Manejo%20Fuego/Modelos%20combustibles/Guia%20modelos.pdf (accessed on 30 October 2020).
85. CONAFOR (Comisión Nacional Forestal). Polígonos de Incendios Forestales: 2019 Concentrado Nacional. Sistema Nacional y Gestión Forestal, Mexico. 2019. Available online: https://snigf.cnf.gob.mx/incendios-forestales/ (accessed on 30 October 2020).
86. Silva Cardoza, A.I. Evaluation and mapping of forest fires severity in the Western Sierra Madre, Mexico. In Proceedings of the XIV Congreso Mexicano de Recursos Forestales, Durango, Mexico, 6 November 2019.

87. Silva-Cardoza, A.I.; Vega-Nieva, D.J.; López-Serrano, P.M.; Corral-Rivas, J.J.; Briseño Reyes, J.; Briones-Herrera, C.I.; Loera Medina, J.C.; Parra Aguirre, E.; Rodríguez-Trejo, D.A.; Jardel-Peláez, E. Metodología para la evaluación de la severidad de incendios forestales en campo, en ecosistemas de bosque templado de México. Universidad Juárez del Estado de Durango, Universidad Autónoma Chapingo, Universidad de Guadalajara. Tech. Rep. "Reforzamiento al sistema nacional de predicción de peligro de incendios forestales de México para el pronóstico de conglomerados y área Quemada" CONAFOR-CONACYT-2018-C02-B-S-131553, México. 2021. Available online: http://forestales.ujed.mx/incendios2/php/publicaciones_documentos/7_3_Silva%20et%20al%2020211201_Metodologia_severidad_v1.pdf (accessed on 30 October 2021).

88. Trimble. *Spectra Geospatial MobileMapper 60 Handheld User Guide*; Trimble Inc.: Sunnyvale, CA, USA, 2019; Available online: http://trl.trimble.com/docushare/dsweb/Get/Document-914012/MM60-v2-Datasheet-EN.pdf (accessed on 27 October 2021).

89. Parson, A.; Robichaud, P.R.; Lewis, S.A.; Napper, C.; Clark, J.T. *Field Guide for Mapping Post-Fire Soil Burn Severity*; Gen. Tech. Rep. RMRS-GTR-243; U.S. Department of Agriculture, Forest Service, Rocky Mountain Research Station: Fort Collins, CO, USA, 2010; p. 49. Available online: https://www.fs.fed.us/rm/pubs/rmrs_gtr243.pdf (accessed on 7 July 2019).

90. Vega, J.A.; Fontúrbel, T.; Fernández, C.; Díaz-Raviña, M.; Carballas, T.; Martín, A.; González-Prieto, S.; Merino, A.; Benito, E. *Acciones Urgentes Contra la Erosión en Áreas Forestales Quemadas: Guía Para su Planificación en Galicia*; Andavira: Galicia, Spain, 2013; p. 140.

91. Vega, J.A.; Fontúrbel, T.; Merino, A.; Fernández, C.; Ferreiro, A.; Jiménez, E. Testing the ability of visual indicators of soil burn severity to reflect changes in soil chemical and microbial properties in pine forests and shrubland. *Plant Soil* **2013**, *369*, 73–91. [CrossRef]

92. Rodríguez-Trejo, D.A.; Martínez-Muñoz, P.; Pulido-Luna, J.A.; Martínez-Lara, P.J.; Cruz-López, J.D. Instructivo de Quemas Prescritas para el Manejo Integral del Fuego en el municipio de Villaflores y la Reserva de la Biosfera La Sepultura, Chiapas, México. In *Fondo Mexicano para la Conservación de la Naturaleza, USDA FS, USAID, BIOMASA, A. C., Universidad Autónoma Chapingo, Ayuntamiento de Villaflores, SEMARNAT, CONAFOR, CONANP, Gobierno del Estado de Chiapas*; Fondo Mexicano para la Conservación de la Naturaleza: CDMX, Mexico, 2019; p. 183.

93. Varner, J.M.; Hood, S.M.; Aubrey, D.P.; Yedinak, K.; Hiers, J.K.; Jolly, W.M.; Shearman, T.M.; McDaniel, J.K.; O'Brien, J.J.; Rowell, E.M. Tree crown injury from wildland fires: Causes, measurement and ecological and physiological consequences. *New Phytol.* **2021**, *231*, 1676–1685. [CrossRef]

94. CONAFOR. Inventario Nacional Forestal y de Suelos: Procedimientos de muestreo. Jalisco, Mexico. 2017. Available online: https://snigf.cnf.gob.mx/wp-content/uploads/Documentos%20metodologicos/2019/ANEXO%20Procedimientos%20de%20muestreo%20V%2019.0.pdf (accessed on 30 October 2020).

95. Lujan-Soto, J.E.; Corral-Rivas, J.J.; Aguirre-Calderón, O.A.; Gadow, K.V. Grouping forest tree species on the Sierra Madre Occidental, Mexico. *Allg. Forst Und Jagdztg.* **2015**, *186*, 63–71.

96. Flores Medina, F.; Vega-Nieva, D.J.; Corral-Rivas, J.J.; Álvarez-González, J.G.; Ruiz-González, A.D.; López-Sánchez, C.A.; Carillo Parra, A. Development of biomass allometric equations for the regeneration of four species in Durango, Mexico. *Rev. Mex. De Cienc. For.* **2018**, *9*, 157–185. [CrossRef]

97. Stambaugh, M.C.; Hammer, L.D.; Godfrey, R. Performance of burn-severity metrics and classification in oak woodlands and grasslands. *Remote Sens.* **2015**, *7*, 10501–10522. [CrossRef]

98. Keeley, J.E. Fire intensity, fire severity and burn severity: A brief review and suggested usage. *Int. J. Wildland Fire* **2009**, *18*, 116–126. [CrossRef]

99. De Santis, A.; Chuvieco, E. GeoCBI: A modified version of the Composite Burn Index for the initial assessment of the short-term burn severity from remotely sensed data. *Remote Sens. Environ.* **2009**, *113*, 554–562. [CrossRef]

100. Vega-Nieva, D.J.; Nava-Miranda, M.G.; Calleros-Flores, E.; López-Serrano, P.M.; Briseño-Reyes, J.; Corral-Rivas, J.J.; Cruz-López, M.I.; Cuahutle, M.; Ressl, R.; Alvarado-Celestino, E.; et al. The Forest Fire Danger Prediction System of Mexico. In Proceedings of the Remote Sensing Early Warning Systems, Virtual Wildfire Workshop, Online, 6–8 October 2020; Commission for Environmental Cooperation (CEC): Montreal, QC, Canada, 2020. Available online: http://forestales.ujed.mx/incendios2/php/publicaciones_documentos/Vega20_v2_The_Forest_Fire_Danger_Prediction_System_of_Mexico.pdf (accessed on 26 March 2022).

101. Vega-Nieva, D.J.; Briseño-Reyes, J.; Briones-Herrera, C.I.; Monjarás, N.; Silva-Cardoza, A.; Nava, M.G.; Calleros, E.; Flores, F.; López-Serrano, P.M.; Corral-Rivas, J.J.; et al. User Manual of the Forest Fire Danger Forecast System of Mexico [Manual de Usuario del Sistema de Predicción de Peligro de Incendios Forestales de México]. 2020. Available online: http://forestales.ujed.mx/incendios2/php/publicaciones_documentos/1_1-%20MANUAL_DE_USUARIO_SPPIF_v15_DV290820.pdf (accessed on 26 March 2022). (In Spanish).

102. Giglio, L.; Schroeder, W.; Justice, C.O. The collection 6 MODIS active fire detection algorithm and fire products. *Remote Sens. Environ.* **2016**, *178*, 31–41. [CrossRef] [PubMed]

103. Schroeder, W.; Oliva, P.; Giglio, L.; Csiszar, I. The New VIIRS 375 m active fire detection data product: Algorithm description and initial assessment. *Remote Sens. Environ.* **2014**, *143*, 85–96. [CrossRef]

104. Briones-Herrera, C.I.; Vega-Nieva, D.J.; Briseño-Reyes, J.; Monjarás-Vega, N.A.; López-Serrano, P.M.; Corral-Rivas, J.J.; Alvarado, E.; Arellano-Pérez, S.; Jardel Peláez, E.J.; Pérez Salicrup, D.R.; et al. Fuel-Specific Aggregation of Active Fire Detections for Rapid Mapping of Forest Fire Perimeters in Mexico. *Forests* **2022**, *13*, 124. [CrossRef]

105. Cruz-López, M.I. Sistema de alerta temprana, monitoreo e impacto de los incendios forestales en México y Centroamérica. In Proceedings of the 4th Wildland Fire International Conference, Seville, Spain, 14–17 May 2007; Available online: https://gfmc.online/doc/cd/REGIONALES/A_IBEROAMERICA/Cruz_MEJICO.pdf (accessed on 17 October 2021).

106. Chávez, P.S. Image-based atmospheric corrections-revisited and improved. *Photogramm. Eng. Remote Sens.* **1996**, *62*, 1025–1035.

107. Congedo, L. Semi-automatic classification plugin documentation. *Release* **2016**, *4*, 29.

108. QGIS Development Team. QGIS Geographic Information System. Open-Source Geospatial Foundation Project, 2020. Hannover, Germany. Available online: http://qgis.osgeo.org (accessed on 30 October 2020).

109. Key, C.H.; Benson, N.C. Measuring and remote sensing of burn severity: The CBI and NBR. Poster abstract/Eds. LF Neuenschwander, KC Ryan. In Proceedings of the Joint Fire Science Conference and Workshop, Boise, ID, USA, 15–17 June 1999; p. 284.

110. Sims, D.A.; Gamon, J.A. Relationships between leaf pigment content and spectral reflectance across a wide range of species, leaf structures and developmental stages. *Remote Sens. Environ.* **2002**, *81*, 337–354. [CrossRef]

111. Gitelson, A.; Merzlyak, M.N. Spectral reflectance changes associated with autumn senescence of Aesculus hippocastanum L. and Acer platanoides L. leaves. Spectral features and relation to chlorophyll estimation. *J. Plant Physiol.* **1994**, *143*, 286–292. [CrossRef]

112. Barnes, E.M.; Clarke, T.R.; Richards, S.E.; Colaizzi, P.D.; Haberland, J.; Kostrzewski, M.; Waller, P.; Choi, C.; Riley, E.; Thompson, T.; et al. Coincident detection of crop water stress, nitrogen status and canopy density using ground based multispectral data. In Proceedings of the Fifth International Conference on Precision Agriculture, Bloomington, MN, USA, 16–19 July 2000; p. 1619.

113. Tucker, C.J. Red and photographic infrared linear combinations for monitoring vegetation. *Remote Sens. Environ.* **1979**, *8*, 127–150. [CrossRef]

114. Hamada, M.A.; Kanat, Y.; Abiche, A.E. Multi-spectral image segmentation based on the K-means clustering. *Int. J. Innov. Technol. Explor. Eng.* **2019**, *9*, 1016–1019. [CrossRef]

115. Chen, J.M. Evaluation of vegetation indices and a modified simple ratio for boreal applications. *Can. J. Remote Sens.* **1996**, *22*, 229–242. [CrossRef]

116. ESA (European Space Agency, Paris). Sentinel-2: Satellite Description: Resolution and Swath. 2015. Available online: https://sentinel.esa.int/web/sentinel/missions/sentinel-2/instrument-payload/resolution-and-swath (accessed on 4 April 2019).

117. Gorelick, N.; Hancher, M.; Dixon, M.; Ilyushchenko, S.; Thau, D.; Moore, R. Google Earth Engine: Planetary-scale geospatial analysis for everyone. *Remote Sens. Environ.* **2017**, *202*, 18–27. [CrossRef]

118. RStudio Team. *RStudio: Integrated Development for R. v2021.09.3*; RStudio. Inc.: Boston, MA, USA, 2021; Available online: http://www.rstudio.com (accessed on 4 April 2021).

119. Ryan, T.P. *Modern Regression Methods*, 2nd ed.; Wiley-Interscience: Hoboken, NJ, USA, 2008; p. 672.

120. Rodríguez Trejo, D.A. *Incendios de Vegetación: Su Ecología, Manejo e Historia*, 1st ed.; UACH, Semarnat, PPCIF, Conafor, Conanp, PNIP, ANCF, AMPF. UACh: Texcoco, Mexico, 2015; Volume 1, pp. 893–1705.

121. Chafer, C.J. A comparison of fire severity measures: An Australian example and implications for predicting major areas of soil erosion. *Catena* **2008**, *74*, 235–245. [CrossRef]

122. CONABIO (Comisión Nacional para el Conocimiento y Uso de la Biodiversidad). Mapa de Cobertura de Tipos de Vegetación y uso de Suelo MAD-Mex 31 Clases, Sentinel 2 (2018), Resolución 30 m. Sistema Integral de Monitoreo de Biodiversidad y Degradación en Áreas Naturales Protegidas, México. 2018. Available online: https://monitoreo.conabio.gob.mx/snmb_charts/descarga_datos_madmex.html (accessed on 7 July 2019).

123. Congalton, R.; Green, K. *Assessing the Accuracy of Remotely Sensed Data: Principles and Practices*, 3rd ed.; CRC Press: Boca Raton, FL, USA, 2019. [CrossRef]

124. Kuhn, M. Caret package. *J. Stat. Softw.* **2008**, *28*, 1–26.

125. Mallinis, G.; Mitsopoulos, I.; Chrysafi, I. Evaluating and comparing Sentinel 2A and Landsat-8 Operational Land Imager (OLI) spectral indices for estimating fire severity in a Mediterranean pine ecosystem of Greece. *GIScience Remote Sens.* **2018**, *55*, 1–18. [CrossRef]

126. Amos, C.; Petropoulos, G.P.; Ferentinos, K.P. Determining the use of Sentinel-2A MSI for wildfire burning & severity detection. *Int. J. Remote Sens.* **2019**, *40*, 905–930. [CrossRef]

127. Botella-Martínez, M.A.; Fernández-Manso, A. Estudio de la severidad post-incendio en la Comunidad Valenciana comparando los índices dNBR, RdNBR y RBR a partir de imágenes Landsat 8. *Rev. De Teledetección* **2017**, *49*, 33–47. [CrossRef]

128. Arellano, S.; Vega, J.A.; Rodríguez y Silva, F.; Fernández, C.; Vega-Nieva, D.; Álvarez-González, J.G.; Ruiz-González, A.D. Validación de los índices de teledetección dNBR y RdNBR para determinar la severidad del fuego en el incendio forestal de Oia-O Rosal (Pontevedra) en 2013. *Rev. De Teledetección* **2017**, *49*, 49–61. [CrossRef]

129. Quintano, C.; Fernández-Manso, A.; Fernández-Manso, O. Combination of Landsat and Sentinel-2 MSI data for initial assessing of burn severity. *Int. J. Appl. Earth Obs. Geoinf.* **2018**, *64*, 221–225. [CrossRef]

130. Allen, J.L.; Sorbel, B. Assessing the differenced Normalized Burn Ratio's ability to map burn severity in the boreal forest and tundra ecosystems of Alaska's national parks. *Int. J. Wildland Fire* **2008**, *17*, 463–475. [CrossRef]

131. Cansler, C.A.; McKenzie, D. How robust are burn severity indices when applied in a new region? Evaluation of alternate field-based and remote-sensing methods. *Remote Sens.* **2012**, *4*, 456–483. [CrossRef]

132. French, N.H.F.; Graham, J.; Whitman, E.; Bourgeau-Chavez, L.L. Quantifying surface severity of the 2014 and 2015 fires in the Great Slave Lake area of Canada. *Int. J. Wildland Fire* **2020**, *29*, 892–906. [CrossRef]
133. Smith, A.M.; Sparks, A.M.; Kolden, C.A.; Abatzoglou, J.T.; Talhelm, A.F.; Johnson, D.M.; Boschetti, L.; Lutz, J.A.; Apostol, K.G.; Yedinak, K.M.; et al. Towards a new paradigm in fire severity research using dose–response experiments. *Int. J. Wildland Fire* **2016**, *25*, 158–166. [CrossRef]
134. Chuvieco, E.; De Santis, A.; Riaño, D.; Halligan, K. Halligan. Simulation Approaches for Burn Severity Estimation Using Remotely Sensed Images. *Fire Ecol.* **2007**, *3*, 129–150. [CrossRef]
135. Hoy, E.E.; French, N.H.; Turetsky, M.R.; Trigg, S.N.; Kasischke, E.S. Evaluating the potential of Landsat TM/ETM+ imagery for assessing fire severity in Alaskan black spruce forests. *Int. J. Wildland Fire* **2008**, *17*, 500–514. [CrossRef]
136. Veraverbeke, S.; Hook, S.J. Evaluating spectral indices and spectral mixture analysis for assessing fire severity, combustion completeness and carbon emissions. *Int. J. Wildl. Fire* **2013**, *22*, 707. [CrossRef]
137. Vlassova, L.; Pérez-Cabello, F.; Mimbrero, M.; Llovería, R.; García-Martín, A. Analysis of the Relationship between Land Surface Temperature and Wildfire Severity in a Series of Landsat Images. *Remote Sens.* **2014**, *6*, 6136–6162. [CrossRef]
138. Marcos, E.; Fernández-García, V.; Fernández-Manso, A.; Quintano, C.; Valbuena, L.; Tárrega, R.; Luis-Calabuig, E.; Calvo, L. Evaluation of composite burn index and land surface temperature for assessing soil burn severity in mediterranean fire-prone pine ecosystems. *Forests* **2019**, *9*, 494. [CrossRef]
139. Quintano, C.; Fernández-Manso, A.; Calvo, L.; Marcos, E.; Valbuena, L. Land surface temperature as potential indicator of burn severity in forest Mediterranean ecosystems. *Int. J. Appl. Earth Obs. Geoinf.* **2015**, *36*, 1–12. [CrossRef]
140. Zheng, Z.; Zeng, Y.; Li, S.; Huang, W. A new burn severity index based on land surface temperature and enhanced vegetation index. *Int. J. Appl. Earth Obs. Geoinf.* **2015**, *45*, 84–94. [CrossRef]
141. Veraverbeke, S.; Rogers, B.M.; Randerson, J.T. Daily burned area and carbon emissions from boreal fires in Alaska. *Biogeosciences* **2015**, *12*, 3579–3601. [CrossRef]
142. Hudspith, V.A.; Belcher, C.M.; Barnes, J.; Dash, C.B.; Kelly, R.; Hu, F.S. Charcoal reflectance suggests heating duration and fuel moisture affected burn severity in four Alaskan tundra wildfires. *Int. J. Wildland Fire* **2017**, *26*, 306. [CrossRef]
143. Barrett, K.; McGuire, A.D.; Hoy, E.E.; Kasischke, E.S. Potential shifts in dominant forest cover in interior Alaska driven by variations in fire severity. *Ecol. Appl.* **2011**, *21*, 2380–2396. [CrossRef]
144. Macauley, K.A.; McLoughlin, N.; Beverly, J.L. Modelling fire perimeter formation in the Canadian Rocky Mountains. *Forest Ecol. Manag.* **2022**, *506*, 119958. [CrossRef]
145. Soverel, N.O.; Perrakis, D.D.; Coops, N.C. Estimating burn severity from Landsat dNBR and RdNBR indices across western Canada. *Remote Sens. Environ.* **2010**, *114*, 1896–1909. [CrossRef]
146. Fornacca, D.; Ren, G.; Xiao, W. Evaluating the best spectral indices for the detection of burn scars at several post-fire dates in a mountainous region of Northwest Yunnan, China. *Remote Sens.* **2018**, *10*, 1196. [CrossRef]
147. Claverie, M.; Ju, J.; Masek, J.G.; Dungan, J.L.; Vermote, E.F.; Roger, J.C.; Skakun, S.V.; Justice, C. The Harmonized Landsat and Sentinel-2 surface reflectance data set. *Remote Sens. Environ.* **2018**, *219*, 145–161. [CrossRef]
148. Lizundia-Loiola, J.; Otón, G.; Ramo, R.; Chuvieco, E. A spatio-temporal active-fire clustering approach for global burned area mapping at 250 m from MODIS data. *Remote Sens. Environ.* **2020**, *236*, 111493. [CrossRef]
149. Schroeder, W.; Oliva, P.; Giglio, L.; Quayle, B.; Lorenz, E.; Morelli, F. Active fire detection using Landsat-8/OLI data. *Remote Sens. Environ.* **2016**, *185*, 210–220. [CrossRef]
150. Scaduto, E.; Chen, B.; Jin, Y. Satellite-Based Fire Progression Mapping: A Comprehensive Assessment for Large Fires in Northern California. *IEEE J. Sel. Top. Appl. Earth Obs. Remote Sens.* **2020**, *13*, 5102–5114. [CrossRef]

Article

Satellite Observations of Fire Activity in Relation to Biophysical Forcing Effect of Land Surface Temperature in Mediterranean Climate

Julia S. Stoyanova *, Christo G. Georgiev and Plamen N. Neytchev

National Institute of Meteorology and Hydrology, 1784 Sofia, Bulgaria; christo.georgiev@meteo.bg (C.G.G.); plamen.neytchev@meteo.bg (P.N.N.)
* Correspondence: julia.stoyanova@meteo.bg; Tel.: +359-2-462-4603

Abstract: The present work is aimed at gaining more knowledge on the nature of the relation between land surface temperature (LST) as a biophysical parameter, which is related to the coupled effect of the energy and water cycles, and fire activity over Bulgaria, in the Eastern Mediterranean. In the ecosystems of this area, prolonged droughts and heat waves create preconditions in the land surface state that increase the frequency and intensity of landscape fires. The relationships between the spatial–temporal variability of LST and fire activity modulated by land cover types and Soil Moisture Availability (SMA) are quantified. Long-term (2007–2018) datasets derived from geostationary MSG satellite observations are used: LST retrieved by the LSASAF LST product; fire activity assessed by the LSASAF FRP-Pixel product. All fires in the period of July–September occur in days associated with positive LST anomalies. Exponential regression models fit the link between LST monthly means, LST positive anomalies, LST-T2 (as a first proxy of sensible heat exchange with atmosphere), and FRP fire characteristics (number of detections; released energy FRP, MW) at high correlations. The values of biophysical drivers, at which the maximum FRP (MW) might be expected at the corresponding probability level, are identified. Results suggest that the biophysical index LST is sensitive to the changes in the dynamics of vegetation fire occurrence and severity. Dependences are found for forest, shrubs, and cultivated LCs, which indicate that satellite IR retrievals of radiative temperature is a reliable source of information for vegetation dryness and fire activity.

Keywords: geostationary satellite observations; wildfire regime; biophysical drivers; land surface temperature; land cover type; trends

Citation: Stoyanova, J.S.; Georgiev, C.G.; Neytchev, P.N. Satellite Observations of Fire Activity in Relation to Biophysical Forcing Effect of Land Surface Temperature in Mediterranean Climate. *Remote Sens.* **2022**, *14*, 1747. https://doi.org/10.3390/rs14071747

Academic Editors: Elena Marcos, Leonor Calvo, Susana Suarez-Seoane and Víctor Fernández-García

Received: 28 February 2022
Accepted: 1 April 2022
Published: 5 April 2022

Publisher's Note: MDPI stays neutral with regard to jurisdictional claims in published maps and institutional affiliations.

1. Introduction

Fire is a global phenomenon closely linked to climate variability and human practices with critical regional implications [1,2]. It is an important process in the modulation of the Earth system through the links between weather, climate, and vegetation and has the potential to impact on the global climate system by changing the ability of the surface to absorb and emit energy [3,4].

Climatic variability is a major driver of fire in many terrestrial ecosystems, as reflected in Bradstock's conceptual model of four climatic "switches" that influence fire regimes by controlling fuel amount, fuel moisture, and fire weather at contrasting temporal scales [1]. Fire regimes are also affected by other controls such as landscape-scale patterns of vegetation, topography, and human activities [5]. Climate is connected to fires at two distinct temporal scales [6]. Short-term climatic anomalies (from months to years) affect fires by modifying vegetation growth and fuel moisture before the fire and by influencing weather during the period of fire spread. In addition, climate has more indirect, long-term (decadal or longer) effects on the distribution of major vegetation types, which in turn constrain the landscape-scale mosaics of fuels and vegetation.

Soil moisture (SM) has been identified as a key variable in understanding and predicting wildfire hazards [7–11]. Soil moisture, defined as the water contained in the unsaturated soil zone [12], not only influences vegetation growth conditions and consequently the accumulation of wildfire fuel, but also determines the vegetation moisture content and hence the flammability of the vegetation [7,9,13].

Land surface temperature (LST) is one of the most important parameters controlling physical processes responsible for the land surface balance of water, energy, and CO_2 [14,15]. In the context of wildfire studies, the LST consideration has two main aspects: first, as a pre-fire indicator of land surface state (SM deficit in root zone and vegetation dryness, evapotranspiration, leaf temperature), and second, as a post-fire characteristic of fire-induced environmental changes. The variations in the spatial distribution of LST as a result of fires are usually the focus of research works that characterize fire intensity and burn severity. Higher post-fire LST of burned areas has been reported in remotely-sensed images [3,4,16–18], that is accounted mainly due to a decrease in transpiration and an increase in the Bowen ratio (β = sensible heating/latent heating) [19].

Although relationships between drought and fire seem quite interrelated, only a few studies have explored LST as a pre-fire indicator [20,21]. Short-term analyses based on LST daily anomalies have been performed to predict fire occurrence [20].

The vegetation cover partitions the incoming solar radiation into sensible and latent heat fluxes depending on its structure as well as affects the surface roughness, which can in turn alter heat and moisture transport; the type of vegetation and its seasonal dynamics affect the land's vulnerability to fire ignition and spread. That is why spatial–temporal patterns of LST can contribute to the monitoring of processes that structure ecosystem development and may be associated with fire occurrences to help fire management. Satellite measurements are a source of information for the accurate estimation of LST at the global and regional level, thus helping to evaluate the land surface–atmosphere exchange processes and can serve as a valuable metric of surface state [4,14].

In this study, we consider data for skin surface temperature retrieved by satellite measurements to explore the significance of LST as a biophysical driver, which (in combination with SM, air temperature, and humidity) controlled long-term wildfire activity during the study period of 2007–2018 over the Eastern Mediterranean region, as seen in coherent satellite active fire observations.

Studies have used satellite remote sensing for fire monitoring from the early 1990s. With the rapid development of remote sensing technology, satellites are increasingly used for fire monitoring [22–26] and applied at large scales. In recent years, many studies have been committed to identifying the spatial and temporal characteristics of fires from the global and regional perspectives with the use of satellite data [27–31]. All these studies were performed by using observations by low earth orbit (LEO) satellite sensors with the Very High Resolution Radiometer (AVHRR), Moderate Resolution Imaging Spectroradiometer (MODIS), and Advanced Spaceborne Thermal Emission and Reflection Radiometer (ASTER).

Geostationary (GEO) fire products were first generated over the Americas using the Geostationary Operational Environmental Satellite Visible Infrared Spin Scan Radiometer Atmospheric Sounder (GOES-VAS), e.g., in [32], and this led to the development of the long-standing GOES WildFire Automated Biomass Burning Algorithm (GOES WFABBA) product [33]. Roberts et al. [34] first demonstrated the retrieval of Fire Radiative Power (FRP) from GEO data and developed a full "fire thermal anomaly" active fire detection and FRP retrieval algorithm for GEO systems. This was initially applied to data from Meteosat Second Generation [35], and an operational version is now used to generate a series of geostationary active fire detection and FRP retrieval products that span much of the globe, including Meteosat over Africa and Europe [36,37], GOES-East and West over the Americas [38], and Himawari over Asia [39]. Wooster et al. [40] summarized the history of achievements in the field of active fire remote sensing, reviewed the physical basis of the approaches used, the nature of the active fire detection and characterization techniques deployed, and highlighted some of the key current capabilities and applications.

However, since every observation by the satellite sensors lasts only an instant, what they actually measured was the rate of release of FRE per unit of time or Fire Radiative Power (FRP) in MW [24,41,42]. It was shown that the FRE released by a fire is directly proportional to the biomass consumed as well as the smoke emitted [43]. Measurement of fire radiative energy (FRE) release rates or power (FRP) from space offers opportunities for the characterization of biomass burning and emissions in a quantitative manner. In essence, it has enabled the rating of fires as well as an estimation of regional FRP fluxes, which reflect the relative concentrations of biomass burning activities.

Mediterranean regions are some of the most affected by wildfires, and remote-sensed information about fire activity as provided by the SEVIRI instrument on board Meteosat-8 is especially valuable for forest and civil protection activities [44]. On the other hand, the Mediterranean is one of the most responsive regions to climate change, as evidenced by "pronounced warming" and significant decreases in spring and summer precipitation, which have led to regime shifts toward more arid climates [45,46]. Triggered by large-scale atmospheric forcing, Mediterranean regional heat waves are often amplified by surface preconditioning such as negative soil moisture anomalies and vegetation stress [47].

The research works of Sifakis et al. [48], Amraoui et al. [44], and Di Biase and Laneve [49] were aimed at understanding the spatial and temporal patterns of landscape fires in the Mediterranean, applying different remote sensing data from GEO sensors. Compared to LEO systems, GEO products offer higher temporal resolutions but coarser spatial resolutions, and each sensor only provides data over a specific region of the Earth. The observations from the geostationary meteorological satellite MSG provide a valuable source of information about fire occurrences in the Mediterranean region. Sifakis et al. [48] reported that during the summer period of 2007, MSG-SEVIRI data successfully detected 82% of the fire events in Greek territory, with less than 1% false alarms. Using data from MSG-1 satellite, Amraoui et al. [44] performed an analysis of the spatial distribution of fire events in the months of July and August during the period of 2007–2009. Around half of the fire pixels were detected in croplands, and the remaining half was evenly distributed between forest and shrub. Based on the analysis of the low, mid, and upper atmospheric fields of geopotential, temperature, relative humidity, and wind in two extreme events of fire activity that struck Greece and Italy on 24–25 July and 22–27 August 2007, they suggested a conceptual model for meteorological conditions that favor the occurrence of severe wildfire episodes in Italy and the Balkan Peninsula.

The current study is the first one to use the FRP product retrieved by geostationary satellite observations to investigate fire activity in a recent long-term period for the Eastern Mediterranean.

In Mediterranean ecosystems, prolonged droughts and heat waves have created the preconditions for the increasing frequency and intensity of forest fires [50], the underlying mechanism being the reduction of live and dead fuel moisture content as a response of the soil–plant system to the increased vapor pressure deficit [51]. Vegetation response varies with species as well as with forest structure and soil/terrain characteristics, and it is determined by evapotranspiration. The moisture of dead fuels is affected by weather variations as well, and it is regulated through evaporation [52]. The climate and weather patterns of the Mediterranean region highlight the value of having an improved understanding of the relationships between drought and wildfire; more specifically, an understanding of how drought is related to fire danger outputs. A number of studies have examined the link between drought indicators and wildfire occurrence [53–55]. Many drought indices (see [56]) are driven by stand and climate variables of precipitation and/or temperature, but more recent developments include variables that express conditions at the land surface–atmosphere interface such as vegetation health [57], soil moisture deficit [58–62], actual evapotranspiration [63], and evaporative demand [56,64,65]. These last indexes are important because they can reflect the outcome of the coupled physical processes of energy–water cycles on the land surface.

In our previous work [66], long-term LST anomalies retrieved by MSG SIVIRI were used to identify drought-prone areas in the Eastern Mediterranean. Song [21] showed that there is no linear relationship between LST anomaly and fire occurrences, but the time series of LST anomaly before fire events show a significant trend. Resolving LST–fire interactions facilitate fire predictions and the issuance of early warnings.

We use long-term spatially and temporally consistent satellite observations from the SEVIRI sensor of geostationary MSG satellite to assess the biophysical forcing effect of vegetation fires on a climatic basis at a regional scale. The study is focused on two key aspects of wild fire problems that are not systematically studied in the literature: (i) an evaluation of the importance of LST (monthly mean values, anomalies, and difference with air temperature, T2m) as precursor/s of vegetated land surface dryness and the related pre-fire signals of vegetation stress and fire occurrence; (ii) an exploration of the variability of these biophysical drivers of vulnerability in biomass burning across the climatic gradient, delineated into main land cover types, and the quantification of relations with fire activity using long-term records of satellite information from Meteosat, ground observations, and SVAT model data of Soil Moisture Availability (SMA) to the land cover as reference data over the Eastern Mediterranean region (Bulgaria). The specific objectives to be met are the following:

- To characterize spatial–temporal patterns of fire activity (July–September) using long-term satellite data records from SEVIRI observations (2004–2019) in terms of the number of biomass burning detections and the severity of burning (FRP, MW) according to the LSASAF FRP-Pixel product;
- To use LSASAF LST product data to statistically investigate and evaluate the relationship of the biophysical parameters of LST, LST anomaly, and the difference between skin and air temperatures (LST-T2m) to the occurrence and severity of wild fires on a short-term climatic basis (2007–2018);
- To characterize the wild fire vulnerability of the main vegetation types (forest, shrublands, cultivated) in relation to the LST and SMA warm and dry anomalies.

2. Materials and Methods

This work is based on the methodology developed in our work [66], adapted to assess environmental control on fire activity and its spatial–temporal variability by using satellite observations from geostationary Meteosat. As a biophysical index related to the forcing of fire activity, the satellite-derived IR skin temperature from SEVIRI observations with the EUMETSAT LSASAF LST product is proposed. Based on long-term data records, LST mean summer monthly values, and their anomalies and deviation from the air temperature, T2m were applied as indicators of the coupled energy and water cycles. Fire-promoting SM anomaly patterns derived by the SVAT modeling approach were used as a reference for the evaluation of the LST–fire intertwining. The pattern of fire activity over Bulgaria was characterized by the evaluation of SEVIRI-based satellite detections of biomass burning according to the LSASAF FRP-Pixel product. Benefiting from the geostationary satellites' high temporal resolution, the relationship between LST and fire regimes was studied and the underlying biophysical processes were explored.

Statistical comparative analyses of long-term data records of the LSASAF product, SVAT meteorological model outputs, and ground observations (meteorological parameters and actual fires) were applied following the methodology in [66], which was further extended (in Section 2.5). Overlaid graphical analyses of the summer season (June–September 2007–2018) dynamics of biophysical indexes in relation to fire characteristics (number of detections and severity) were performed to study the sensitivity of the approach. All datasets taken from information sources with different spatial resolutions were placed over the grid of the ECMWF global NWP model, IFS version O1280 (about 9 km spatial resolution). Only values with 1400 points that cover the region of Bulgaria were considered.

2.1. The LSASAF FRP-PIXEL Product

In order to characterize fire activity in this study, radiative energy released by fire events, as provided by the LSA SAF Fire Radiative Power-Pixel (FRP) product [36,67], was used. The LSASAF FRP-PIXEL product provided information on the location, timing, and fire radiative power (FRP, in MW) output of landscape fires ("wildfires") detected every 15 min across the full Meteosat disk at the native spatial resolution of the SEVIRI sensor. The FRP provided information on the measured radiant heat output of detected fires; the heat output was a direct result of the combustion process whereby carbon-based fuel is oxidized to CO_2 with the release of a certain "heat yield" [24]. FRP calculation relies on the Middle Infrared (MIR) method and assumes FRP to be proportional to the difference between the observed fire pixel radiance measured from the MSG SEVIRI radiometer and the "background" radiance that would have been observed at the same location in the absence of fire. The LSASAF algorithm for deriving FRP from SEVIRI radiance measurements is fully described in [37]. The type of fire detection algorithm that is proposed for use with SEVIRI is based on the principles applied to generate active fire detections within the MODIS Fire Products [24].

While this FRP methodology offers many advantages, among the uncertainness reported, two factors are likely to reduce the satellite-measured FRP that is emitted by the fire:

- The first concerns surface fires in forests, where an unknown amount of radiant energy may be intercepted (scattered and absorbed) by the forest canopy.
- Second, although atmospheric effects perturb MIR wavelength observation far less than those in shorter wavelengths, allowing FRP retrieval through even dense smoke and plumes, the radiative impact of the absorptive black carbon released during combustion may result in some underestimation of the FRP by the satellite measurement.

The FRP algorithm is subject to other sources of uncertainties, essentially due to the characteristics of SEVIRI, as described in [37]. Among these limitations, one can state that the FRP-derived value is quite sensitive to the fire location within the pixel. A fire located at the center of the instrument's instantaneous field-of-view will elevate the pixel temperature much more than a fire located far away from its center. Additionally, abnormally low radiances might be observed surrounding a fire pixel due to the negative lobes of the point spread function. As a result, the background temperature might be colder, thereby increasing the estimated FRP value. The rather coarse spatial resolution of SEVIRI may cause the non-detection of smaller/less intense fires that MODIS can detect, for instance, but SEVIRI misses. The non-linearity of the SEVIRI 3.9 channel above 310 K and the saturation of that band above 335 K are responsible for the error in the FRP estimation. As a result, FRP values derived from SEVIRI underestimate those derived from MODIS by about 40% [34].

A significant remote sensing constraint for fires is the presence of cloudiness in fire weather situations, as all satellite methods used for monitoring fires have limitations that tend to cause important biases in the final product. The algorithms cannot detect existing fires due to the elimination of cloudy pixels or the assumption of some detection in partially cloudy pixels as "false" fire detections based on "contextual" tests.

2.2. Biophysical Indexes

In our work [66], the spatial–temporal variability of land surface state dry anomalies was characterized by the land surface temperature with the use of LST retrievals from the MSG satellite measurements and SMA. Using long-term data records, the same biophysical drivers of drought occurrence have been further explored here in terms of their forcing effects on fire activity through the related biophysical processes. In addition to land surface radiometric temperature characterized by its monthly values and calculated anomalies, the blended parameter, which is the difference between skin surface temperature and air temperature at 2 m above the earth's surface (LST-T2m), is proposed to be used as a first proxy of sensible heat flux exchange. The patterns of these biophysical indexes were defined as the mean monthly values for June, July, August, and September. The SMA anomalies during the summer months were considered in parallel to explore the fire forcing effect of soil moisture deficit.

To quantify the relations, statistical analyses and modeling were applied. Since the considered biophysical parameters and the procedure of their application have been explained in detail in [66], only the main points are marked here, and the new elements are further explained in Section 2.5.

2.2.1. The LSASAF LST Product

For the purposes of the current study, the LSASAF LST product (MLST, LSA-001) has been applied [68]. The LST retrieval was based on clear-sky measurements from the MSG system in the thermal infrared window (MSG/SEVIRI channels IR10.8 and IR12.0), every 15 min within the area covered by the MSG disk. Theoretically, LST values could be determined from MSG 96 times per day, but in practice, fewer observations were available due to cloud cover.

The 12-year time series for the region of Bulgaria for the period of June–September (2007–2018), which included only days/locations with satellite fire detections, was constructed. Datasets for LSASAF LST at 0900 UTC and at 1200 UTC (averaged over 5 measurements within ±30 min of these two time slots to avoid the limitations of cloudy pixels) were inferred at MSG pixel bases. For some climatic assessments, the averaged 3×3 pixels LST values around the grid points were also processed. The difference between LST and the air temperature, T2m (observations at the SYNOP network at 0900 UTC and 1200 UTC) was also constructed for the same studied period.

2.2.2. SVAT Model and SMAI

For the identification of fire-promoting SM anomaly patterns, the "SVAT_bg" model developed at the National Institute of Meteorology and Hydrology (NIMH) of Bulgaria was used [61,69,70]. This is a simple 1D site-scale meteorological model that exploits the concept of one layer of vegetated land surface and two levels of moisture availability along the root zone depth. One of the main "SVAT_bg" model output parameters is soil moisture (SM) for different root zone soil depths (20, 50, 100 cm). For assessing land surface state, the SMA concept was adopted to serve as an information source for "warnings" of environmental constraints. Based on the "SVAT_bg" model-derived SM, a quantitative SMA index, SMAI was developed and operationally calculated at a site scale for the region of each NIMH synoptic station. At each point of the used grid, the SMAI values at the nearest synoptic station were considered for the calculation of the anomalous water content in the unsaturated root zone. SMAI was designed as a 6-level threshold scheme to account for moistening conditions [66]. The site-scale assessment of mean monthly SMAI anomalies at synoptic stations was compared with the overlaid FRP Pixel detections from corresponding MSG pixels covering the region of the station applying the statistical analyses.

2.3. Ground Observations of Forest Fires

Ground observations of the number of actual forest fires were considered according to the database of the State Forest Agency of Bulgaria (SFA), the responsible national institution, and were used for comparison with satellite detections.

2.4. Target Region and Land Cover

The study area includes the whole territory of Bulgaria, Southeastern Europe, located within latitude circles of 40.25 N and 45.0 N and meridians of 20.5 N and 29.5 E. The region falls under Mediterranean climate influences, characterized by dry summers and mild, wet winters, both with irregular precipitation distribution. Information about land covers (LC) provided by the ESA-CCI Land Cover Map product at a spatial resolution of 300 m [71] was used. Maps were updated on an annual basis. The LC classification system from ESA-CCI includes 24 vegetation types. For the purposes of this study, we reclassified these types into the following three main vegetation groups (Figure 1): forests (classes 50, 60,

61, 62, 71, 72, 80, 81, 82, 90), scrubland (11, 12, 40, 100, 110, 120, 121, 122), and cultivated areas (10, 20, 30, 130). Other land cover types, including permanent snow and ice (220) and barren and water bodies (200, 210), where there are limited fires, were not analyzed in our study. Urban areas (190) were also excluded.

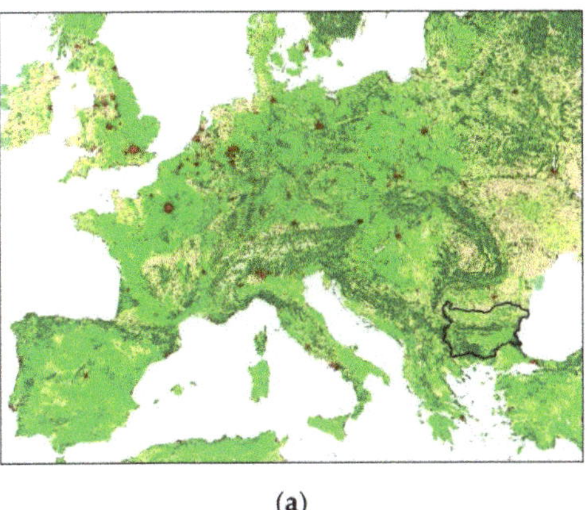

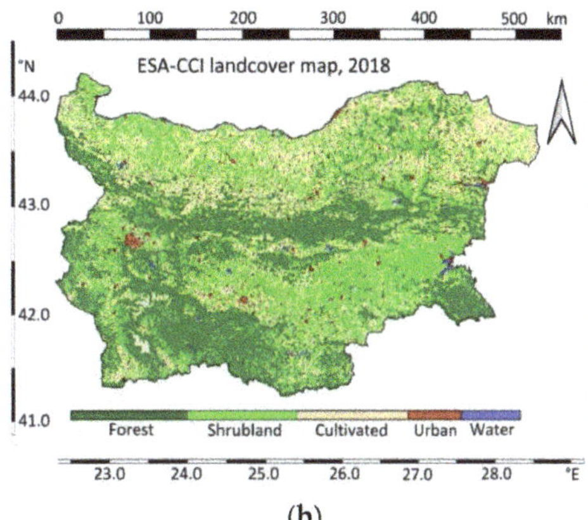

(**a**)   (**b**)

Figure 1. Geographical distribution of the vegetation cover types according to the ESA-CCI Land Cover database 2018 over (**a**) Europe; (**b**) Bulgaria, SE Europe, reclassified for the purposes of the study into three categories: forest, shrubland, grassland.

2.5. Numerical Analyses

To characterize the fire distribution pattern, a dataset of FRP detections (location, timing, and energy) for the period of June–September (2004–2019) was constructed [72,73]. All pixels with at least one fire detection were considered. Fire regime was characterized by the frequency of detections and severity according to the released energy from biomass burning (FRP fire radiative energy, MW). Each fire pixel was associated with the coordinates of the nearest point on the grid, where LST, LST anomalies, and (LST-T2m) had been derived. The procedure was performed for monthly (June, July, August, and September) accumulated energy released from fires in the corresponding grid point.

In order to validate the adopted methodology, monthly mean anomalies of LST and SMA that had been inferred in [66] and visualized in color-coded maps were evaluated for consistency with FRP fire pixel spatial distribution over Bulgaria by using qualitative comparative analysis. Examples for July 2007 are presented in Figure 2. Figure 2a shows that the higher positive LST anomalies (in red) correspond to a higher number of FRP-detected fire pixels. The stronger negative SMA anomalies (indicated by reddish colors) are related to a higher number of fire detections (Figure 2b). The consistency of the data allows quantitative studies to be performed further.

Based on long-term records (June–September 2007–2018), stochastic graphical analysis was performed. The consistency in the behavior of the fire activity and the biophysical drivers LST, LST anomalies, (LST-T2m), and SMA anomalies was analyzed in terms of their mean, spatiotemporal variability on a monthly and annual basis, as well as their anomalous distribution and relations. The summer seasonal dynamics of biophysical conditions for different LC vegetation types were studied and their relation to vulnerability of fire ignition/spread was evaluated.

A graphical description of the locality, the spread and skewness groups of numerical data of FRP detections, released energy from biomass burning, LST, LST anomalies, and the (LST-T2m) temperature difference was performed through their boxplot quartiles. Two types of regression analyses were performed; the first one made use of estimates by the method of least squares of the conditional mean of the response variable across values of the predictor variables. The second method used Quantile Regression (QR) estimates of the conditional median of the response variable, which has received increasing attention in recent years and is applied in many areas, including data analysis in the natural sciences.

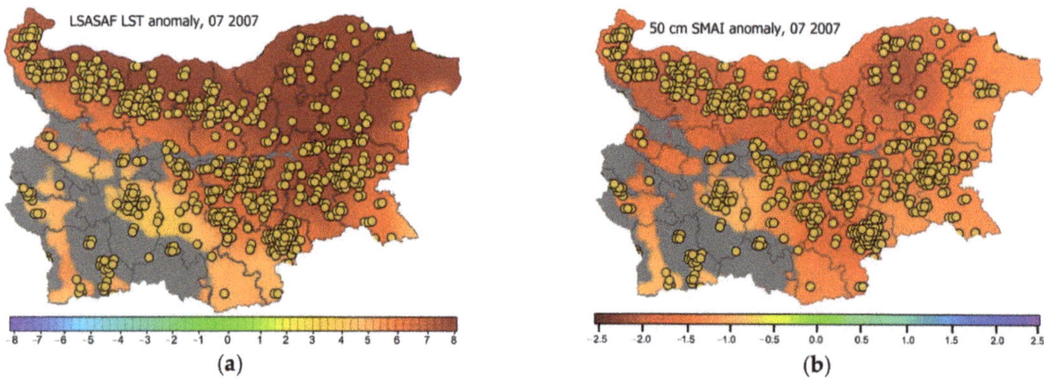

Figure 2. Spatial distribution of LSASAF FRP fire pixel detections superimposed over: (**a**) LSASAF LST 0900 UTC anomalies and (**b**) monthly mean SMA anomalies (50 cm soil depth). Examples for July 2007, Bulgaria: The LST and SMA anomalies were calculated for the 2007–2018 period [66].

To identify regions most vulnerable to fires in Bulgaria, the spatial pattern of the accumulated fire numbers detected by the LSASAF FRP-Pixel product for the period of June–September 2004–2019 was considered. A statistical evaluation of the spatial distribution of fire pixels was performed by applying the Mann–Kendall statistic test. To perform all these comparisons, the R language for statistical computing was used [74].

All numerical evaluations were performed for the three main vegetation groups: forest, shrubland, and cultivated, as well as considered all together in a sample "all land cover". Such a reclassification into three LC categories has been used in other studies of fire activity over the Mediterranean using satellite data, e.g., in [44].

A summary of all data used for comparative analyses in the current study are presented in Table 1.

Table 1. Summary of data used and their characteristics.

Data	Temporal Resolution	Spatial Resolution
Fire Radiative Power-Pixel product	15 min	SEVIRI, about 5 km over Bulgaria
Land Surface Temperature (LST)	15 min	SEVIRI, about 5 km over Bulgaria
Temperature difference between LST and air temperature at 2 m (LST-T2m)	0900 and 1200 UTC	NIMH synoptic station network
Soil Moisture Availability Index (SMAI)	Daily, 0600 UTC	NIMH synoptic station network

3. Results

Fire regimes were described through statistical distributions of frequency and severity over the studied area in the summer season during a 16-year time period. Using 12 years of data for LST and related biophysical parameters, the environmental determinants of fire regimes were assessed by exploring how environmental drivers operating over a range of scales affected the spatial and temporal patterns of these fires.

3.1. Active Fire Monitoring from Space

The results from the study of fire activity derived from satellite observations over Bulgaria are presented in maps of accumulated fire detections for each month of the July–September period and each year of the whole 2004–2019 period. Examples for selected years are shown in Figure 3 (see also [72,73]). The fire severity was assessed following the color-indicated released energy of burning biomass FRP (MW). The fire activity exhibited various seasonal distributions from year to year, e.g., maximum activity and severity was observed in July 2007 (also in 2006, 2016, 2017, not presented), in August 2004 (also 2006, 2008), and in September 2019 (2011, 2012). There are repetitions of spots with higher fire activity, suggesting that the spatial distribution varies depending on the evolution of the horizontal patterns of the biophysical drivers (to be further considered in Sections 3.2 and 3.3).

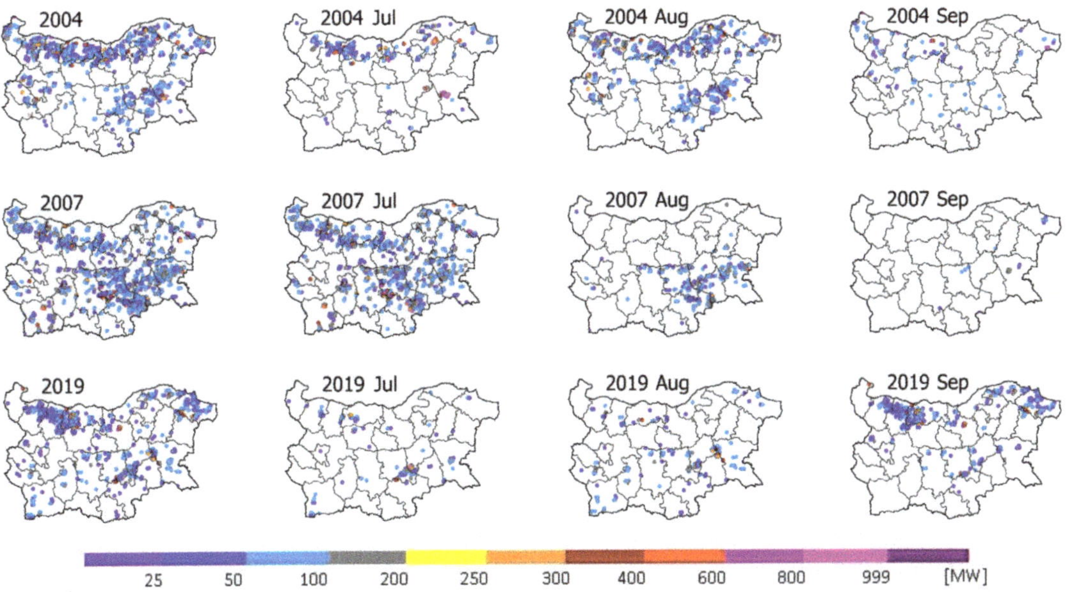

Figure 3. Maps of the spatial–temporal (monthly/annual) distribution of fire activity over Bulgaria. Color-coded severity of biomass burning according LSASAF FRP-Pixel (MW) is indicated. Examples for July, August, September and annually for 2004, 2007, and 2019 are presented.

Results from the validation of the MSG FRP-Pixel product comparing the number of detected fire pixels in forested areas with the course of actual forest fires reported by ground observations of the State Forest Agency (SFA) of Bulgaria are shown in Figure 4. Comparisons are performed for the period of July–September 2007–2018. A good agreement between the two independent sources of information for forest fire dynamics during different months and years is observed: the increase in actual fires corresponds to the increase in satellite fire pixel detections; the highest forest fire activity is in July 2007; for July and August, the two lines are very close, indicating an almost similar number of fire detections; for September, the courses are similar but the number of satellite detections are lower, probably due to the cloudier conditions in September.

3.2. Biophysical Drivers and Fire Activity

3.2.1. Annual Trends in Fire Activity along with LST

Figure 5 shows a comparison of LST (at 0900 UTC) with the fire energy released according to the FRP-Pixel product (MW). The course of LST over time (red lines) shows behavior synchronized with fire energy (blue lines) for all vegetation types. The sample size allows a regression analysis to be applied, and the results show an existing linear relationship between the parameters: Trend lines of both parameters, LST (red dashed line) and FRP, MW (blue dashed line), for the considered LC types were obtained (Figure 5a–d). The coefficient of determination R^2, which is the square of the correlation, shows how much of the observed scatter in the data is due to the hypothetical linear component as opposed to the unexplained random error. Despite the rather complex nature of the curves, the resulting R^2 values confirm a significant linear component (values bottom right in the panels).

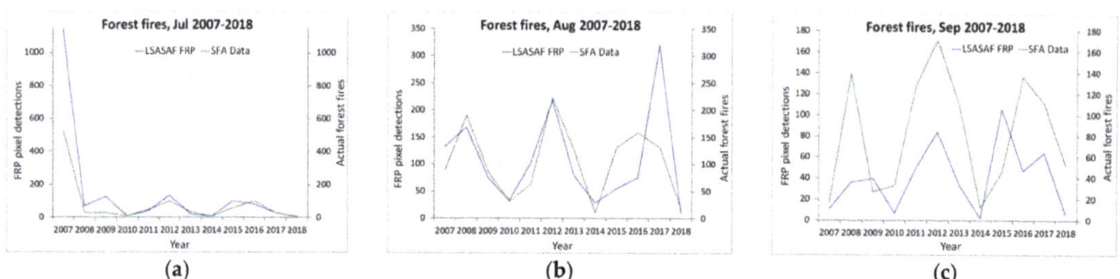

Figure 4. Inter-annual dynamics of forest fires over Bulgaria according the LSASAF FRP-Pixel product detections (blue line) and ground observations of actual forest fires by the national SFA (green line) for (**a**) July, (**b**) August, (**c**) September 2007–2018.

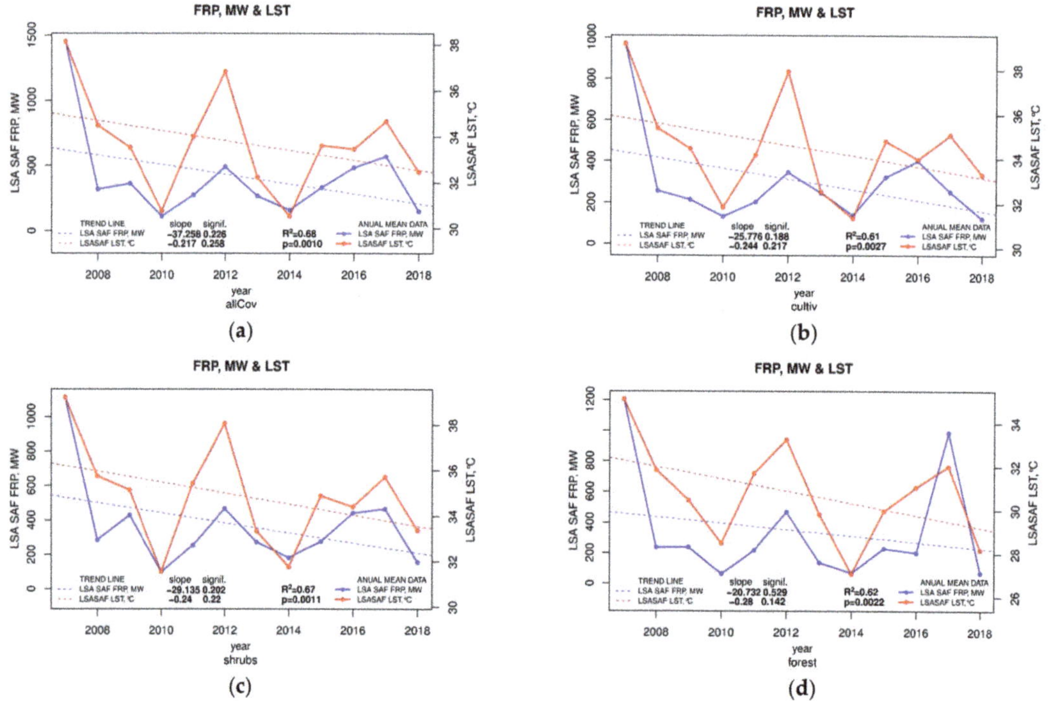

Figure 5. Time series of fire activity characterized by the total energy released from biomass burning per year FRP, MW (blue line) along with LST (red line) (June–September, 2007–2018) for: (**a**) All LC samples; (**b**) Cultivated LC; (**c**) Shrubs LC; (**d**) Forest LC. Each time series is fitted with a trend line using the linear regression technique (least squares method).

Figure 5a shows the trend lines for LST and the energy released from biomass burning for a sample of all land cover types without classifying them ("all LC"); the linear regression line fits the relation between the parameters at high R^2 values of 0.68. The values of R^2 for forest-shrubs-cultivated types vary between 0.61 and 0.68, and the validity of these conclusions passed the significant test from 0.001 up to 0.003 levels (Figure 5b–d). The FRP (blue lines) decreases with the decrease in LST (red lines) within individual diapason for biomes (e.g., 32–29 °C for forest and 36–34 °C for shrubs/cultivated, as seen in

Figure 5c,d). A high level of fit between the trends in the number of fire pixels and the LST as a biophysical driver is also experienced (Supplementary Material). In this case R^2 varies within 0.67–0.62 for shrubs and forest vegetation types, at high significant levels, indicating a descending trend for the fire accuracy along with the LST decrease.

To clarify the nature of this linear structure, it is examined by linear regression on the individual components in each data pair. Again, the results are similar (values in the bottom left panel in Figure 5), although at the limit of reliability. For at least one group in each pair, a linear model with a similar slope of about −0.25 is obtained, and its significance level is slightly above 0.10. Therefore, for at least one pair, a significant confirmation of the estimated downward trend line was obtained.

3.3. Statistical Analyses

Monthly wildfire activity (2007–2018) characterized by the number of fire pixel according to the LSASAF FRP-Pixel product, and the ratio between the total energy released (MW) and the total number of fire pixels detected for a specific month and LC type are shown in Table 2. The number of fire pixels in forest and shrubs in June represents 6.5–7.6% percent of the fire pixels detected in July. Fires in cultivated LC in June seem lower by about 5% than their amount in July due to the specific vegetation in the managed LC areas still being in the growing phase. For September, the wildfires (forest, shrubs) from the forest fires in August are at 37% and 50% from the shrub fires in the same month. However, in the cultivated LC, the relatively large number of fire pixels are detected in September (75% from their number in August), which is likely due to some controlled agriculture burning activities.

Table 2. Monthly accumulated FRP-Pixel detections for the period of 2007–2018 and the total FRP energy released per pixel (MW), DFI. Examples for shrubs, forest, and cultivated LCs for June, July, August, and September are presented.

Monthly Accumulated (2007–2018) Fire Characteristics	June	July	August	September
Shrubs LC, FirePixelDetections	232	3572	2819	1429
DFI (MW/pixel)	72.63	99.78	101.28	85.66
Forest LC, FirePixelDetections	137	1795	1306	486
DFI (MW/pixel)	35.06	72.97	96.05	76.29
Cultivated LC Detections	130	2533	1928	1441
DFI (MW/pixel)	69.82	98.10	107.18	83.42

The ratio between the total FRP energy released and the accumulated number of fire pixels at a specific LC type was introduced to serve as a measure of mean Detected Fire Intensity (DFI) observed by the satellite. The DFI from the forest fire pixels is the lowest one, ranging from 35 MW/pixel in June to 96 MW/pixel in August. For shrubs and cultivated LC types, the DFI range is 72–101 MW/pixel and 70–107 MW/pixel, respectively.

The satellite-derived FRP are likely to estimate a specific portion of the actual fire-emitted FRP (see Section 2.1). Data in Table 2 show that shrubs LC is characterized by the largest amount of total energy released by fires for the summer season. In order to assess how the portion of satellite-detected fire energy depends on the vegetation type during the fire season, the DFI values of forest and cultivated LC are divided by the DFI obtained for shrubs in each month (Table 3). It can be seen that the ratio between the DFI parameters of forest and shrubs is 0.48 in June, then increases gradually to 0.95 in August and becomes 0.89 in September. At the same time, the ratio between the DFI values for cultivated–shrubs LC is close to 1.0, ranging from 0.96 in June to 1.06 in August–September. This result comes from the different ability of satellite observations (FRP product) to estimate actual emitted energy from wildfires at different LC types, and this also depends on the biophysical status of the vegetation along the fire season.

Table 3. Ratios of the monthly accumulated DFI of forest and cultivated LC fires to the DFI of shrubs fires for the period of 2007–2018.

Ratio between DFI for Different LC Types	June	July	August	September
DFI Forest/DFI Shrubs	0.48	0.73	0.95	0.89
DFI Cultivated/DFI Shrubs	0.96	0.98	1.06	1.06

3.3.1. Box Plots Analyses

Boxplots of the distribution of fire radiative energy (MW) according to the FRP-Pixel product and biophysical indexes for forest and shrubs LCs are shown in Figures 6 and 7.

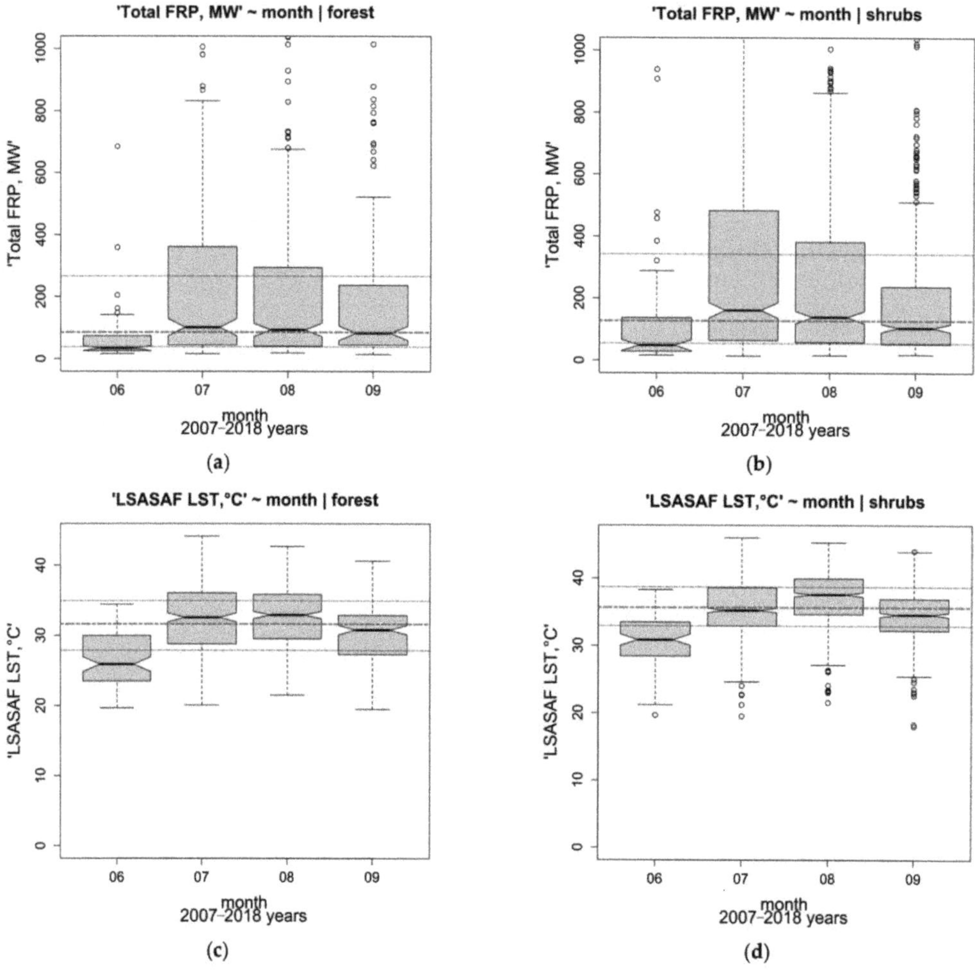

Figure 6. Boxplots of monthly mean values (2007–2018) for the region of Bulgaria during the summer season (June–September): from the LSASAF FRP-Pixel (MW) for (**a**) forest and (**b**) shrubs; from the LSASAF LST 0900 UTC °C for (**c**) forest and (**d**) shrubs. The boxes represent the interquartile range (from the 25th to 75th percentile, first and third red line), whiskers cover 99.3% of the data, and the middle (red) lines represent the mean values.

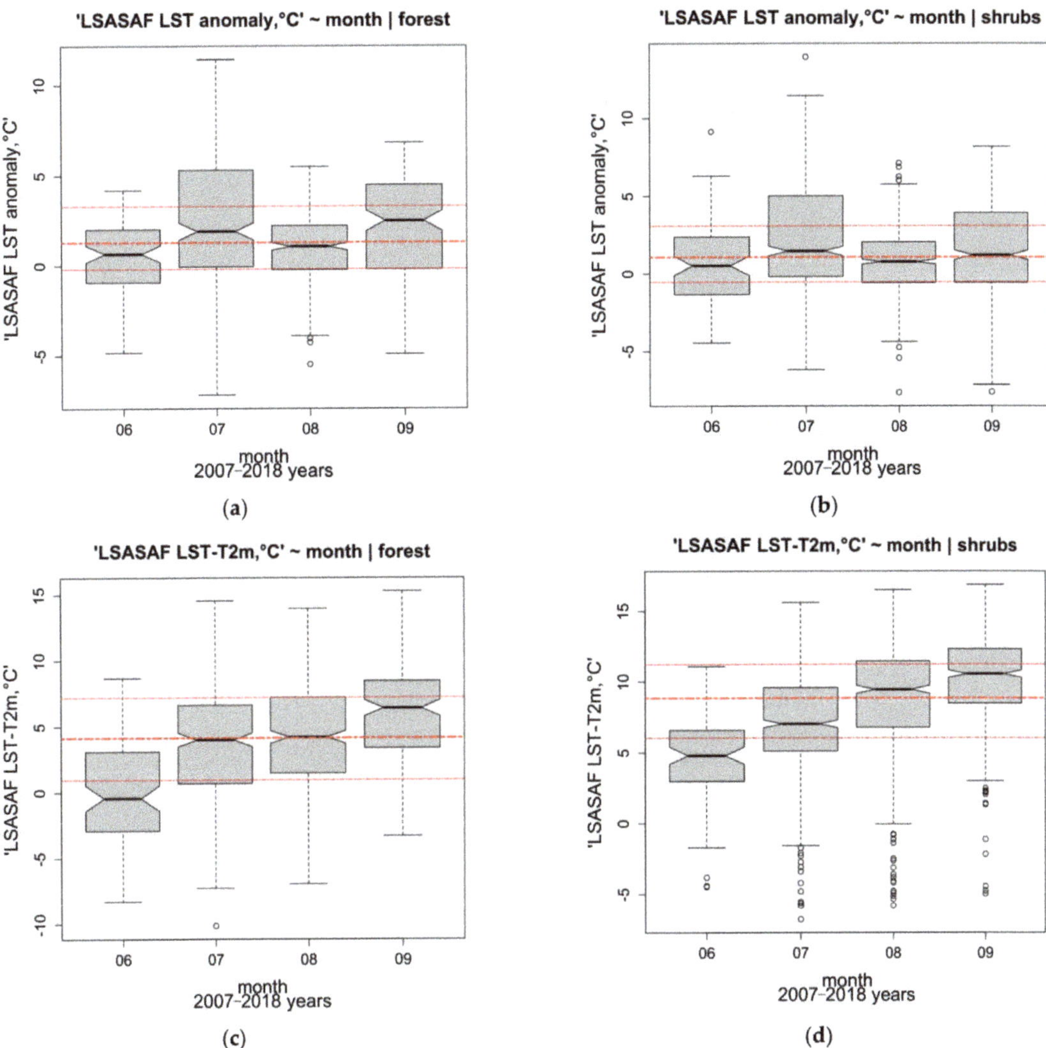

Figure 7. Boxplots of monthly mean values (2007–2018) for the region of Bulgaria during the summer season (June–September) from the LSASAF LST anomaly for (**a**) forest and (**b**) shrubs, and the (LST-T2m) difference for (**c**) forest and (**d**) shrubs. The boxes represent the interquartile range (from 25th to 75th percentile, first and third red line), whiskers cover 99.3% of the data, and the middle (red) lines represent the mean values.

The plots in Figure 6 show different aggregation of data by months. Through a visual comparison of the corresponding notches, it can be seen that there is sufficient reason to accept the statistical hypothesis that the medians are significantly different. The red lines show the median and first and fourth quartile averages of the same statistics for all individual boxplots presented and are used to visually assess the degree of difference depending on the month.

The boxplots in Figure 6a,b demonstrate the variability of the monthly mean values of FRP (MW) for forest and shrubs LC during the summer. The statistical analyses confirm that for July, the ratio between the median of total FRP box plots for shrubs and forest fires is higher than the corresponding ratio for the DFI parameter in Table 3: shrub fires were detected with about 50% higher released FRP (MW), while the DFI parameter was only 35% larger than this for forest fires. This effect seems to be due to higher energy emitted by some forest fires in July.

On the contrary, for June, the medians of total FRP are not much different for these two LC types (Figure 6a,b), but the DFI parameter is twice higher for shrubs than the corresponding DFI value obtained for forest fires (Table 3). This suggests the occurrence of more forest fires under the canopy of dry dead forest fuel with much less energy measured by the satellite and that a certain amount of radiant energy may have been intercepted (scattered and absorbed) by the forest canopy.

The summer course of the FRP median corresponds well to the LST statistical behavior for the forest LC (Figure 6a,c). In June, optimal SMA is present in the deep forest root zone that leads to fully open stomata conductance and unlimited evapotranspiration, leading to decreasing canopy temperature and minimum LST. For shrubs (Figure 6b,d), there is a disagreement: the LST median maximum appears in August, but the maximum of the total FRP median is in July. The reason could be that in shrub lands, the SMA easily reaches the wilting point, in which the stomata conductance strongly decreases in reaction to the increasing water stress conditions through the transpiration regulation mechanism. These lead to an increasing shrub canopy LST in August.

All fires in the period of July–September occur in days associated with positive values of LST anomalies (averaged in the area of 3×3 pixels) around the same location (Figure 7a,b). All 25% of the fire detections in the days associated with negative anomalies of LST occur in June in forest and shrubs LCs. This result suggests that in June, some fires may mostly occur in dead fuel where the canopy temperature is below the historical average for today's day-of-year.

The trend of increasing (LST-T2m) from June to September (Figure 7c,d) corresponds to the gradual decrease of SMA to vegetation cover on the one hand, and to the increasing LST on the other. This is consistent with the result on the disagreement in the behavior of LST versus FRP in August for shrubs LC (Figure 6b,d).

3.3.2. Quantile Regression

Results from the quantile regression analysis of the relationship between log FRP (MW) and biophysical parameters (LST, LST anomalies, (LST-T2m), and SMA anomalies) are presented on Figure 8a–d.

These graphs can quite comprehensively characterize the dependences between each pair of quantities. QR examples of the "all LC" type are here presented. Figure 8a shows at which LST values the maximum FRP (MW) might be expected and the corresponding probability level the FRP might reach the maximum. Accordingly, the QR in Figure 8a shows how the log(FRP, MW) maximum depends on the predictor LST, with a confidence of 5%, 25%, and other levels of significance. The QR lines indicate that the high values of the log(FRP, MW) > 5 can be expected to occur at LST = 35 °C with a 25% probability level; extreme values of log(FRP, MW) > 7 are possible with 5% reliability at LST above 40 °C, and with 10% reliability above 45 °C.

On the plot of Figure 8b, the probability level of FRP maxima at a specific temperature difference (LST-T2m) is illustrated. Very high values of log(FRP, MW) > 6 can be expected to occur at (LST-T2m) = 5 °C with 10% reliability; extreme values of log(FRP, MW) > 7 are possible at a 5% significance level for (LST-T2m) above 12 °C. This illustrates how the log(FRP, MW) maximum depends on the predictor (LST-T2m), with confidence levels of 5%, 10%, and other levels of significance.

The QR between FRP and LST anomalies (Figure 8c) shows that very high values of log(FRP, MW) > 6 can be expected to occur at negative LST anomalies, with a probability of less than 10%. Extreme values of log(FRP, MW) > 7 are not possible even at a 5% significance level, with negative LST anomalies. In parallel, higher FRP energy occurs at negative SMAI anomalies (Figure 8d). The decreasing trend of energy released, along with the decreasing values of negative or low positive anomalies, i.e., decreasing drought (agricultural and ecological), at corresponding probability levels are shown. Extreme values of log(FRP, MW) > 7 are possible at 10% significance with negative SMAI anomalies.

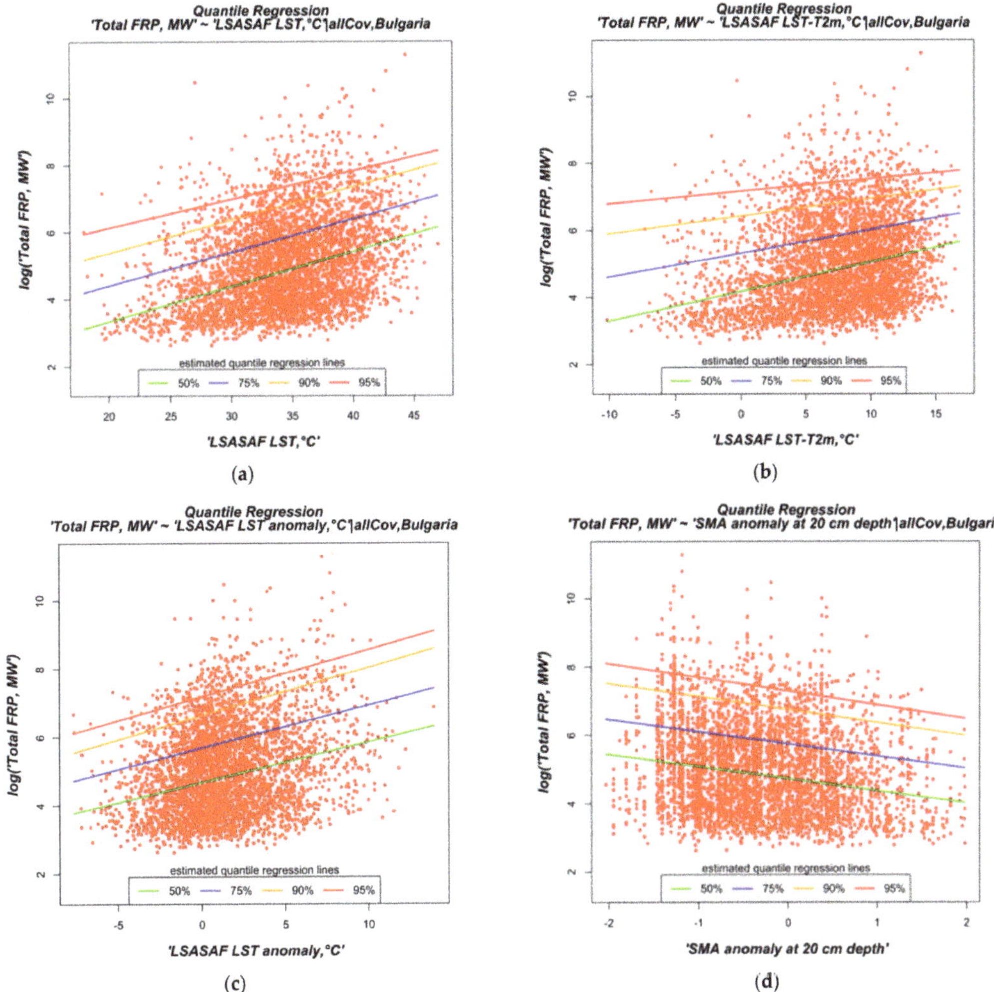

Figure 8. Quantile regression of the total energy released by biomass burning (all land cover types) in Bulgaria (June–September 2007–2018): log (Total FRP-Pixel, MW) vs. (**a**) LSASAF LST 0900 UTC; (**b**) (LSASAF LST-T2m) temperature difference; (**c**) LSASAF LST anomaly (2007–2018); (**d**) SMAI anomaly at 20 cm soil depth. Regression lines for 50%, 75%, 90%, and 95% quantile are shown.

Such QR graphs have been developed for forest, shrubs, and cultivated vegetation types and confirm similar relations for the fire activity over the whole country. The results obtained for some smaller samples over specific regions of Bulgaria show an intersection of the probability lines that is an indication of insufficient data.

3.3.3. Correlation Analyses

The results of the least-squares method applied for the quantitative evaluation of the relation between biogeophysical indexes (LST, LST anomaly, (LST−T2m) temperature difference, including the SMAI anomaly at 20, 50, 100 cm depth) and the severity of wildfire events are presented in Table 4. Monthly means LST at around 0900 UTC, which is 1200 local time (selected as a less cloudy period), are considered. For all land cover types (all biomes included in a sample), exponential regression lines fit the relations at high R^2 values and high significance levels (Figure 9).

Table 4. Coefficients of determination R^2 of the regression between the energy of biomass burning, FRP (MW), and the monthly mean values of biophysical indexes (July–August, 2007–2018). Only days with fire detections are considered.

Fire Characteristics vs. Biophysical Indexes	Month	Forest	Shrubs	Cultivated	All LC Types
Total FRP (MW) vs. LST 0900 UTC	July	0.772	0.593	0.683	0.773
	August	0.407	0.481	0.564	0.540
	September	0.436	0.432	0.609	0.671
Total FRP (MW) vs. LST anomaly 0900 UTC	July	0.779	0.554	0.660	0.668
	August	0.739	0.501	0.598	0.627
	September	0.611	0.584	0.678	0.682
Total FRP(MW) vs. SMA 100 soil depth	July	0.874	0.569	0.669	0.695
	August	0.457	0.494	0.599	0.569
Total FRP (MW) vs. SMA 50 soil depth	July	0.884	0.523	0.713	0.690
	August	0.512	0.444	0.524	0.546
Total FRP (MW) vs. SMA 20 soil depth	July	0.715	0.423	0.665	0.607

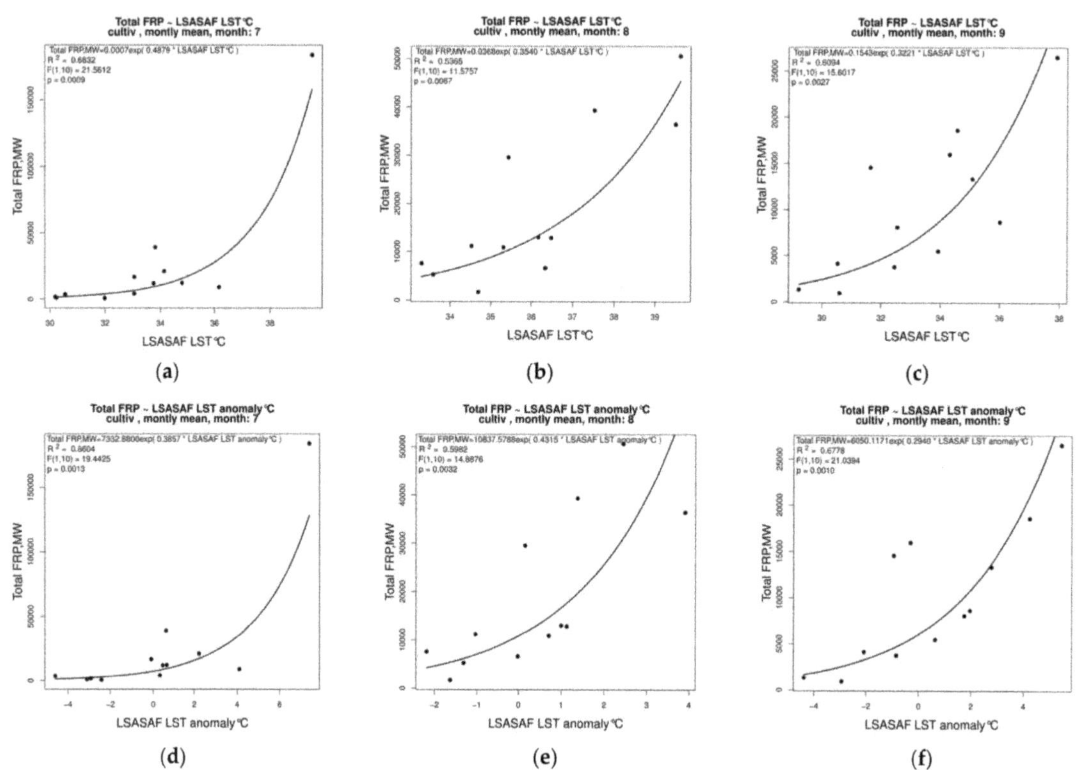

Figure 9. Regression models of the total energy released by biomass burning in Bulgaria according to the FRP-Pixel product, MW vs. LSASAF LST 0900 UTC for: (**a**) July; (**b**) August; (**c**) September, and vs. LSASAF LST anomaly (2007–2018) for: (**d**) July; (**e**) August; (**f**) September. Examples for cultivated land cover are presented.

The SMA index is especially efficient as a measure of fuel dryness, taken at 50 cm or 100 cm soil depth, for all cover types in the period of July–August. With the progressive establishment of dry conditions during August and September, the correlation decreases when the SMA at 20 cm and 50 cm becomes fully exhausted, and the SMAI is equivalent to a constant value of zero.

Exponential regression models better fit the FRP (MW)–LST relations from Table 4. Examples for cultivated LC are shown in Figure 9a–c, indicating high significance levels ($p < 0.007$) of the dependences for July, August, and September. Similar exponential models for the regression of FRP (MW) versus the LST anomalies, R^2 at about 0.6 ($p < 0.003$), are presented in (Figure 9d–f). Results confirm a statistically significant high correlation between biophysical drivers (LST, LST anomalies) and the wildfire severity.

All these results suggest that the biophysical index LST and the related parameters LST anomalies and (LST-T2m) are sensitive to the dynamics of the occurrence and severity of vegetation fires in all studied LC types.

3.3.4. Spatial Pattern and Trends

To characterize fire activity, plots of monthly sums of FRP detections are considered. Figure 10 shows a color-coded map of accumulated fire pixels in the period of June–September 2004–2019. The fire numbers are categorized into the following classes: 1–30 (low, yellow color), 30–90 (moderate, orange color), 90–150 (high, red color), and >150 (extreme, dark red). Thus, the spatial distribution of the spots with higher fire activity is revealed. The "+" sign on the map indicates that the trend passed the Mann–Kendall significance test at the 5% level; thus, regions with a statistically significant positive trend in fire activity over the studied period are localized.

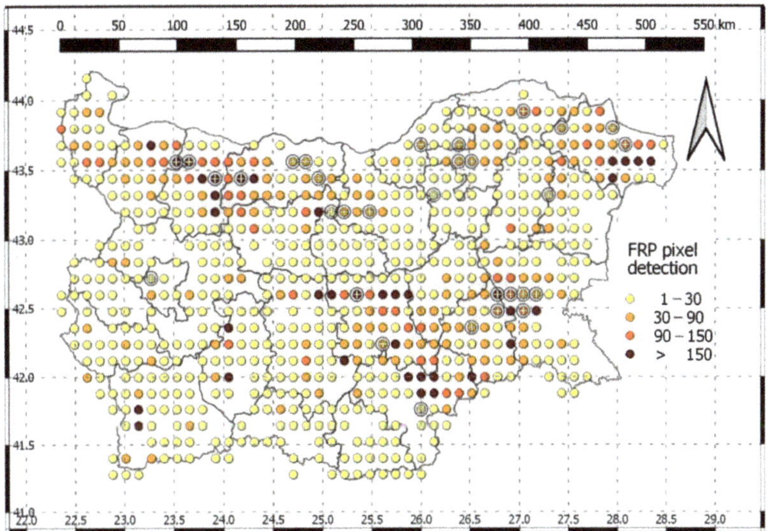

Figure 10. Map of "hot spots" of fire activity over Bulgaria, SE Europe based on long-term satellite observations (June–September, 2004–2019) using the LSASAF FRP-Pixel product. The fire numbers are the sum of the fire pixels (4 × 5 km MSG resolution over Bulgaria) detected and resampled in a 10 × 10 km plot. The "+" sign indicates locations where the trend passed the Mann–Kendall significance test at the 5% level.

The long-term trends of fire activity (number of detections) are different when it is delineated into different land cover types (superposition of Figures 1b and 10). These considerations show that there is no positive trend of fire activity in regions with forest land cover types. There are limited hot spots for fires, located in the northern and southeastern parts of Bulgaria, indicating an increasing trend in fire activity that is mostly scattered in the shrublands. Positive trends are also seen in limited areas of cultivated land cover types, but in this case, it is rather a result of human activities than a climate trend.

4. Discussion

The results of our study confirm the driving role of land surface temperature and the related biophysical parameters in the spatial–temporal evolution of fire activity in Eastern Mediterranean as well as in forcing the vulnerability of forest, shrubs, and cultivated land cover types to wildfire. Based on the 2007–2018 summer time series, the trend analyses (Section 3.2.1) show the synchronized behavior of annual LST values with the released FRP energy (MW) from biomass burning (respectively to the number of fire pixels). The analysis of satellite retrievals from geostationary satellite data show a statistically significant declining trend for the wildfire characteristics, consistent with the course of LST for all considered vegetation types (Figure 3).

The box-plot statistics (Figure 6a,c) show that the median of the FRP (MW) energy released from forest fires in the summer course corresponds well to the behavior of the LST median, which confirms the relationship between the biophysical index LST and fire activity. All fires in the period of July–September occurred in days associated with positive values of LST anomalies, while the other 25% of the fire detections in days associated with negative anomalies of LST occurred in June at forest and shrubs LCs (Figure 7a,b).

A high correlation between the monthly mean values of the biophysical indexes (LST, LST anomalies, LST-T2m, and SMA) and FRP-detected fire pixels was obtained. Exponential regression models fit the relations for all LC types at high significance levels (Section 3.3.3). This result is a step forward from the findings of Song [21], who stressed on the spatial–temporal information provided by LST anomalies (2001–2019 time series over Australia) on fire activity, confirming a significant trend before the fire events but not establishing a linear relationship with fires.

The values of LST, LST anomalies, and LST-T2m, for which the maximum FRP (MW) values might be expected at the corresponding probability level, were statistically estimated (Section 3.3.2) as being able to contribute to fire preparedness activities. For example, extreme values of log(FRP, MW) > 7 were possible at a 5% significance level for (LST-T2m) above 12 °C. A previous study reported that forest fires in August over Bulgaria might occur under land surface drought conditions defined by middy (MSG LST-T2m difference) >12 °C prior to the fire [62].

These dependences obtained for the Eastern Mediterranean confirm and extend the reported results in the literature that land surface state can significantly contribute to the enhancement of weather impacting on the fire environment such as drought (long-term and/or short-term), terrain, and fuel conditions [62,75,76]. Plants constitute the main ignition material in the landscape, and their moisture content plays an important role because it may serve to retard ignition or mitigate the propagation of a fire [77–80]. For this reason, the fuel moisture content is a common component of fire danger and related fire regime assessments; e.g., in [81,82]. Skin temperature is an important disclosure of fuel moisture content, and the results reported confirm the role of LST as a biophysical index of fire activity. The physical mechanism of LST–SMA relations and related fire occurrence covers a range of coupled biophysical processes: as a moist soil surface dries out, more of the incoming solar energy is reflected, and a larger fraction of the absorbed energy is used to heat the air and soil [83]. The heat flow into the soil increases at first, then decreases as the soil becomes very dry [84]. This results in increasing land surface temperatures, which also influences the rate of drying and evapotranspiration. As a result, high LST is a cause and product of dry periods; in other words, drought begets drought and extreme fire activity.

Land surface temperature explains 80% of the variance in air temperature, and vegetation density also plays an important role in explaining the air temperature variance [85]. Since only data from days with detected fires in the summer months were considered, most of the results of the statistical analyses are related to conditions of limited SMA for the vegetation cover.

In such conditions, the (LST-T2m) parameter is a measure of the evapotranspiration rate, and relations are specified by atmosphere/land weather and climate conditions as well as by physiological and structural features of the LC [86,87] Hence, the trend of increasing (LST-T2m) from June to September (see Figure 7c,d) is related to gradually decreasing evapotranspiration during the summer months due to the gradual decrease of SMA to vegetation cover on the one hand, and/or the increasing LST on the other.

All these results suggest that the biophysical index LST and the related parameters of LST anomalies and (LST-T2m) are sensitive to the dynamics of vegetation fire occurrence and severity. The dependences are valid for forest, shrubs, and cultivated LCs at high significant levels and indicate that satellite IR retrievals of radiative temperature is a reliable source of information for vegetation dryness and fire occurrence. Some existing mismatches are due to highly inhomogeneous land cover from one side and the complicated nature of their biogeophysical relations. LST is directly linked to precipitation, cloudiness, and solar irradiance, while SMAI is influenced by the cumulative effect of these meteorological parameters and with functional links to vegetation. In the case of when the period of moisture depletion is long, the LST-SMA relation is weak because the LST continues to increase; on the other hand, after full SM depletion, the "SVAT_bg" model does not assess any further SM changes. In terms of the model simulations, full SM depletion denotes an interruption of liquid and water vapor flow because SM has reached its constant minimum value (at specific soil/climate), which is set to be the maximum hygroscopic value [69].

In addition, there are some limitations in the remote sensing approach for assessing the relation of fire activity to the distribution of its biophysical drivers. Such cases may be due to different abilities of the satellite observations and the derived FRP product to estimate actual emitted energy from wildfire in different LC types, and this also depends on the biophysical status of the vegetation during the fire season. This ability of FRP is significantly different as compared to wildfires in the forest and shrubs LCs in June (reported here, Table 2 and Figure 6a,b) due to the specific physical properties of the forest canopy, which leads to larger amounts of radiant energy to be intercepted. This in turn makes some of the low-energy forest fires impossible to be detected by the satellite. In August and September, with the decrease in SMA and the fraction of vegetation cover, the canopy biophysical properties changed so that a larger amount of energy released by forest fires could be evaluated by satellite measurements, and most of the forest fires could be detected as well.

Although landscape fires in Mediterranean countries are mainly caused by human activities related to agricultural practices, this study confirms that fire regime components (occurrence and severity) exhibit characteristic spatial and temporal patterns that reflect differences in the relative importance of various environmental drivers. In this regard, an identification of the regions most vulnerable to biomass burning is of importance for some prevention activities. Different land covers have widely differing flammability, which depends on species composition, stand age and density, microclimate, and soil conditions [31,88]. For this reason, a detailed discrimination of the land cover types is needed in studying short-term weather and land surface state influences on the fire ignition and spread. In our study, the LC typology is simplified to consider three types: forests, scrubland, and cultivated; this is because relationships between monthly mean values of the physical indexes and fire activity are explored over long-term 12-year periods. Moreover, the forest LC is considered to be spatially stable throughout all 12 years while more land cover changes may occur in the category scrubland during the observation period, and more detailed classification would be difficult to apply. Our qualitative comparative analysis of the ESA CCI LC maps for 2006–2015 does not show any significant changes in the land cover over Bulgaria for the test period.

Recent concerns about the potential increases of forest fires under climate change underline the importance of fire–climate feedbacks [89–91]. In this context, an important result of our study based on a dataset of satellite FRP detections from the last 16 years is that no positive trend in fire activity for forest land cover type has been observed over Bulgaria.

There are significant increasing trends in fire activity for single spots in the low land areas of shrubs LC types in the northern and southeastern part of Bulgaria. A reasonable explanation for these results is that the forests are a more stable environment over a long-term period, which also confirms the validity of our approach for the simplification of the LC variety into three main types for the purposes of the current climate study.

5. Conclusions

This study quantified the relationship between the spatial–temporal variability of land surface temperature (LST) and fire activity on a short-term climatic scale over the Eastern Mediterranean (Bulgaria), accounting for physical properties such as land cover and soil moisture that, combined with LST, provide a valuable metric for surface state. Based on satellite observations from the geostationary Meteosat LSASAF-FRP-Pixel product (summer season 2007–2018), the distribution of fire activity on a monthly/annual basis in relation to the biophysical forcing effect of LST (assessed by IR MSG satellite measurements) was observed in land cover types of different fire vulnerability (forest, shrubland, cultivated).

A synchronized annual behavior between LST and FRP with a declining trend (2007–2018) in wildfire characteristics was seen. Exponential models fit the relationships of LST monthly means, LST anomalies, and LST-T2, as a first proxy of sensible heat exchange with atmosphere, as well as SMA with FRP fire characteristics (MW). All fires in the period of July–September occurred in days associated with positive LST anomalies.

The values of LST, LST anomalies, and LST-T2m, for which the maximum FRP energy (MW) might be expected at corresponding probability levels, were estimated as being capable of contributing to fire preparedness activities. For example, monthly mean extremes of log(FRP, MW) > 7 could be observed with a 5% reliability at an LST monthly mean above 40 °C and (LST-T2m) above 12 °C, and with a 10% reliability above 45 °C. The reported biophysical forcing effects of LST on vegetation fires provided more understanding of the relationships between drought and wildfire; more specifically, of how drought is related to fire danger outputs. Since forest fires have the potential to affect regional climate through changes in the energy budget [91–93], this knowledge is especially important for the Mediterranean region, where land–atmosphere coupling has become one of the important aspects of global environmental change.

To advance the added value of this study by using LST as a biophysical index of drought for fire management, further evidence in support of this approach is observed: First, the use of related FRP products from the MODIS sensor to validate the usefulness of this approach for short-term applications; Second, the evaluation and adaptation of the LST applications in the operational mode for the analyses of real fire situations in the scope of early warnings of fire risk assessment, thus contributing as the final step in the fire management practice.

Supplementary Materials: The following supporting information can be downloaded at: https://www.mdpi.com/article/10.3390/rs14071747/s1, Figure S1: (**a**) FRPdetections & LST.shrubs. (**b**) FRPdetections & LST.cultiv; Figure S2: (**a**) July and September 2007–2018 Shrubs LC. (**b**) August 2007–2018 Shrubs LC; Figure S3: (**a**) FRP-Pixel_detections.SMA_anomaly_at_50cm_depth.forest.montly.7; (**b**) FRP-Pixel_detections.SMA_anomaly_at_50cm_depth.forest.montly.8; (**c**) FRP-Pixel_detections. SMA_anomaly_at_50cm_depth.shrubs.montly.7.

Author Contributions: Conceptualization, J.S.S.; methodology, J.S.S. and C.G.G.; software, C.G.G. and P.N.N.; formal analysis, C.G.G. and P.N.N.; investigation, J.S.S.; writing—original draft preparation, J.S.S.; writing—editing, J.S.S. and C.G.G.; project administration, J.S.S.; funding acquisition, J.S.S. All authors have read and agreed to the published version of the manuscript.

Funding: This research was partially funded by EUMETSAT LSASAF Project, CDOP-3, 2017–2022.

Data Availability Statement: Data is available at EUMETSAT Product Navigator. Available online: https://navigator.eumetsat.int/start (accessed on 24 February 2022).

Acknowledgments: Data processing and preparation of illustration material in Sections 2.5 and 3.1 were performed by Andrey Kulishev within the frame of the LSASAF project implementation. LSASAF is acknowledged for providing the requested LST and FRP data. Data in Section 2.3 have been kindly provided by the State Forest Agency of Bulgaria. Thanks are due to the anonymous reviewers for the valuable suggestions and comments.

Conflicts of Interest: The authors declare no conflict of interest. The funder had no role in the design of the study; in the analyses, or interpretation of data; in the writing of the manuscript, or in the decision to publish the results.

References

1. Bradstock, R.A. A biogeographic model of fire regimes in Australia: Current and future implications. *Global Ecol. Biogeogr.* **2010**, *19*, 145–158. [CrossRef]
2. Krawchuk, M.A.; Moritz, M.A. Constraints on global fire activity vary across a resource gradient. *Ecology* **2011**, *92*, 121–132. [CrossRef] [PubMed]
3. Vlassova, L.; Pérez-Cabello, F.; Mimbrero, M.R.; Llovería, R.M.; García-Martín, A. Analysis of the Relationship between Land Surface Temperature and Wildfire Severity in a Series of Landsat Images. *Remote Sens.* **2014**, *6*, 6136–6162. [CrossRef]
4. Liu, Z.; Ballantyne, A.P.; Cooper, L.A. Biophysical feedback of global forest fires on surface temperature. *Nat. Commun.* **2019**, *10*, 214. [CrossRef] [PubMed]
5. Bowman, D.M.; Balch, J.; Artaxo, P.; Bond, W.J.; Cochrane, M.A.; D'Antonio, C.M.; DeFries, R.; Johnston, F.H.; Keeley, J.E.; Krawchuk, M.A.; et al. The human dimension of fire regimes on Earth. *J. Biogeogr.* **2011**, *38*, 2223–2236. [CrossRef]
6. Hessl, A.E. Pathways for climate change effects on fire: Models, data, and uncertainties. *Prog. Phys. Geogr.* **2011**, *35*, 393–407. [CrossRef]
7. Jensen, D.; Reager, J.T.; Zajic, B.; Rousseau, N.; Rodell, M.; Hinkley, E. The sensitivity of US wildfire occurrence to pre-season soil moisture conditions across ecosystems. *Environ. Res. Lett.* **2018**, *13*, 014021. [CrossRef]
8. Chuvieco, E.; Aguado, I.; Dimitrakopoulos, A.P. Conversion of fuel moisture content values to ignition potential for integrated fire danger assessment. *Can. J. For. Res.* **2004**, *34*, 2284–2293. [CrossRef]
9. Bartsch, A.; Balzter, H.; George, C. The influence of regional surface soil moisture anomalies on forest fires in Siberia observed from satellites. *Environ. Res. Lett.* **2009**, *4*, 045021. [CrossRef]
10. Krueger, E.S.; Ochsner, T.E.; Engle, D.M.; Carlson, J.D.; Twidwell, D.L.; Fuhlendorf, S.D. Soil Moisture Affects Growing-Season Wildfire Size in the Southern Great Plains. *Soil Sci. Soc. Am. J.* **2015**, *79*, 1567–1576. [CrossRef]
11. Krueger, E.S.; Ochsner, T.E.; Quiring, S.M.; Engle, D.M.; Carlson, J.D.; Twidwell, D.; Fuhlendorf, S.D. Measured soil moisture is a better predictor of large growing-season wildfires than the Keetch-Byram drought index. *Soil Sci. Soc. Am. J.* **2017**, *81*, 490–502. [CrossRef]
12. Seneviratne, S.I.; Corti, T.; Davin, E.L.; Hirschi, M.; Jaeger, E.B.; Lehner, I.; Orlowsky, B.; Teuling, A.J. Investigating soil moisture–climate interactions in a changing climate: A review. *Earth Sci. Rev.* **2010**, *99*, 125–161. [CrossRef]
13. Burapapol, K.; Nagasawa, R. Mapping soil moisture as an indicator of wildfire risk using landsat 8 images in Sri Lanna National Park, Northern Thailand. *J. Agric. Sci.* **2016**, *8*, 107. [CrossRef]
14. Kustas, W.; Anderson, M. Advances in thermal infrared remote sensing for land surface modeling. *Agric. For. Meteorol.* **2009**, *149*, 2071–2081. [CrossRef]
15. Monson, R.; Baldocchi, D. *Terrestrial Biosphere-Atmosphere Fluxes*; Cambridge University Press: Cambridge, MA, USA, 2014; p. 487.
16. Lambin, E.; Goyvaerts, K.; Petit, C. Remotely-sensed indicators of burning efficiency of savannah and forest fires. *Int.J. Remote Sens.* **2003**, *24*, 3105–3118. [CrossRef]
17. Veraverbeke, S.; Verstraeten, W.W.; Lhermitte, S.; van de Kerchove, R.; Goossens, R. Assessment of post-fire changes in land surface temperature and surface albedo, and their relation with fire–burn severity using multitemporal MODIS imagery. *Int. J. Wildland Fire* **2012**, *21*, 243–256. [CrossRef]
18. Maffei, C.; Alfieri, S.M.; Menenti, M. Relating Spatiotemporal Patterns of Forest Fires Burned Area and Duration to Diurnal Land Surface Temperature Anomalies. *Remote Sens.* **2018**, *10*, 1777. [CrossRef]
19. Beringer, J.; Hutley, L.; Tapper, N.; Coutts, A.; Kerley, A.; O'grady, A. Fire impacts on surface heat, moisture and carbon fluxes from a tropical savanna in Northern Australia. *Int. J. Wildland Fire* **2003**, *12*, 333–340. [CrossRef]
20. Yang, H. Land Surface Temperature Anomalies and Fire Occurrence. The Relationship between Land Surface Temperature Anomalies and Fire Occurrence in Cariboo Region in 2017. 2021. Available online: https://storymaps.arcgis.com/stories/06a6 acff8a544fb187b2cb4ce262e614 (accessed on 23 February 2022).
21. Song, W. Relating Forest Fire Occurrences to Diurnal Land Surface Temperature Anomalies in Victoria, Australia from 2001 to 2019. 2021. Available online: https://open.library.ubc.ca/collections/researchdata/items/1.0396657 (accessed on 24 February 2022).
22. Nakayama, M.; Maki, M.; Elvidge, C.D.; Liew, S.C. Contextual algorithm adapted for NOAA-AVHRR fire detection in Indonesia. *Int. J. Remote Sens.* **1999**, *20*, 3415–3421. [CrossRef]
23. Justice, C.O.; Giglio, L.; Korontzi, S.; Owens, J.; Morisette, J.T.; Roy, D.; Descloitres, J.; Alleaume, S.; Petitcolin, F.; Kaufman, Y. The MODIS fire products. *Remote Sens. Environ.* **2002**, *83*, 244–262. [CrossRef]

24. Giglio, L.; Descloitres, J.; Justice, C.O.; Kaufman, Y.J. An enhanced contextual fire detection algorithm for MODIS. *Remote Sens. Environ.* **2003**, *87*, 273–282. [CrossRef]

25. Giglio, L.; Csiszar, I.; Restás, Á.; Morisette, J.T.; Schroeder, W.; Morton, D.; Justice, C.O. Active fire detection and characterization with the advanced spaceborne thermal emission and reflection radiometer (ASTER). *Remote Sens. Environ.* **2008**, *112*, 3055–3063. [CrossRef]

26. Schroeder, W.; Oliva, P.; Giglio, L.; Quayle, B.; Lorenz, E.; Morelli, F. Active fire detection using Landsat-8/OLI data. *Remote Sens. Environ.* **2016**, *185*, 210–220. [CrossRef]

27. Ardakani, A.S.; Valadan Zoej, M.J.; Mohammadzadeh, A.; Mansourian, A. Spatial and temporal analysis of fires detected by MODIS data in northern Iran from 2001 to 2008. *IEEE J. Sel. Top. Appl. Earth Obs. Remote Sens.* **2011**, *4*, 216–225. [CrossRef]

28. Palumbo, I.; Grégoire, J.; Simonetti, D.; Punga, M. Spatio-temporal distribution of fire activity in protected areas of Sub-Saharan Africa derived from MODIS data. *Procedia Environ. Sci.* **2011**, *7*, 26–31. [CrossRef]

29. Molinario, G.; Davies, D.K.; Schroeder, W.; Justice, C.O. Characterizing the spatio-temporal fire regime in Ethiopia using the MODIS-active fire product: A replicable methodology for country-level fire reporting. *Afr. Geogr. Rev.* **2014**, *33*, 99–123. [CrossRef]

30. Chuvieco, E. Global characterization of fire activity: Toward defining fire regimes from Earth observation data. Glob. *Chang. Biol.* **2008**, *14*, 1488–1502. [CrossRef]

31. Wei, X.; Wang, G.; Chen, T.; Fiifi, D.; Hagan, T.; Ullah, W. A Spatio-Temporal Analysis of Active Fires over China during 2003–2016. *Remote Sens.* **2020**, *12*, 1787. [CrossRef]

32. Prins, D.M.; Menzel, W.P. Geostationary satellite detection of biomass burning in South America. *Int. J. Remote Sens.* **1992**, *13*, 2783–2799. [CrossRef]

33. Prins, E.M.; Feltz, J.M.; Menzel, W.P.; Ward, D.E. An overview of GOES-8 diurnal fire and smoke results for SCAR-B and 1995 fire season in South America. *J. Geophys. Res. Atmos.* **1998**, *103*, 31821–31835. [CrossRef]

34. Roberts, G.; Wooster, M.J.; Perry, G.L.W.; Drake, N.; Rebelo, L.-M.; Dipotso, F. Retrieval of biomass combustion rates and totals from fire radiative power observations: Application to southern Africa using geostationary SEVIRI imagery. *J. Geophys. Res.* **2005**, *110*, D21111. [CrossRef]

35. Roberts, G.; Wooster, M.J. Fire detection and fire characterization over Africa using Meteosat SEVIRI. *IEEE Trans. Geosci. Remote Sens.* **2008**, *46*, 1200–1218. [CrossRef]

36. Wooster, M.J.; Roberts, G.; Freeborn, P.H.; Xu, W.; Govaerts, Y.; Beeby, R.; He, J.; Lattanzio, A.; Mullen, R. Meteosat SEVIRI Fire Radiative Power (FRP) Products from the Land Surface Analysis Satellite Applications Facility (LSA SAF): Part 1-Algorithms, Product Contents & Analysis. *Atmos. Chem. Phys.* **2015**, *15*, 15831–15907.

37. Govaerts, Y.; Wooster, M.; Roberts, G.; Freeborn, P.; Xu, W.; He, J.; Lattanzio, A. Algorithm Theoretical Basis Document for MSG SEVIRI Fire Radiative Power (FRP) Characterisation. 2015. Available online: https://nextcloud.lsasvcs.ipma.pt/s/Wc7xRm3soPwjj56 (accessed on 24 February 2022).

38. Xu, W.; Wooster, M.J.; He, J.; Zhang, T. Improvements in high-temporal resolution active fire detec-tion and FRP retrieval over the Americas using GOES-16 ABI with the geostationary fire thermal anomaly (FTA) algorithm. *Sci. Remote Sens.* **2021**, *3*, 100016. [CrossRef]

39. Xu, W.; Wooster, M.J.; Kaneko, T.; He, J.P.; Zhang, T.R.; Fisher, D. Major advances in geosttionary fire radiative power (FRP) retrieval over Asia and Australia stemming from use of Himarawi-8 AHI. *Remote Sens. Environ.* **2017**, *193*, 138–149. [CrossRef]

40. Wooster, M.J.; Roberts, G.J.; Giglio, L.; Roy, D.P.; Freeborn, P.H.; Boschetti, L.; Justice, C.; Ichoku, C.; Schroeder, W.; Davies, D.; et al. Satellite remote sensing of active fires: History and current status, applications and future requirements. *Remote Sens. Environ.* **2021**, *267*, 112694. [CrossRef]

41. Kaufman, Y.J.; Justice, C.O.; Flynn, L.P.; Kendall, J.D.; Prins, E.M.; Giglio, L.; Ward, D.E.; Menzel, W.P.; Setzer, A.W. Potential global fire monitoring from EOS-MODIS. *J. Geophys. Res.* **1998**, *103*, 32215–32238. [CrossRef]

42. Wooster, M.J.; Zhukov, B.; Oertel, D. Fire radiative energy for quantitative study of biomass burning: Derivation from the BIRD experimental satellite and comparison to MODIS fire products. *Remote Sens. Environ.* **2003**, *86*, 83–107. [CrossRef]

43. Ichoku, C.; Giglio, L.; Wooster, M.J.; Remer, L.A. Global characterization of biomass-burning patterns using satellite measurements of Fire Radiative Energy. *Remote Sens. Environ.* **2008**, *112*, 2950–2962. [CrossRef]

44. Amraoui, M.; Liberato, M.; Calado, T.; Dacamara, C.; Coelho, L.; Trigo, R.; Gouveia, C. Fire activity over Mediterranean Europe based on information from Meteosat-8. *For. Ecol. Manag.* **2013**, *294*, 62–75. [CrossRef]

45. Gao, X.; Giorgi, F. Increased aridity in the Mediterranean region under greenhouse gas forcing estimated from high-resolution simulations with a regional climate model. *Glob. Planet. Change* **2008**, *62*, 195–209. [CrossRef]

46. Seneviratne, S.I.; Lüthi, D.; Litschi, M.; Schär, C. Land–atmosphere coupling and climate change in Europe. *Nature* **2006**, *443*, 205–209. [CrossRef] [PubMed]

47. Stéfanon, M.; Drobinski, P.; D'Andrea, F.; Lebeaupin-Brossier, C.; Bastin, C. Soil moisture-temperature feedbacks at meso-scale during summer heat waves over Western Europe. *Clim. Dyn.* **2013**, *42*, 1309–1324. [CrossRef]

48. Sifakis, N.I.; Iossifidis, C.; Kontoes, C.; Keramitsoglou, I. Wildfire Detection and Tracking over Greece Using MSG-SEVIRI Satellite Data. *Remote Sens.* **2011**, *3*, 524–538. [CrossRef]

49. Di Biase, V.; Laneve, G. Geostationary Sensor Based Forest Fire Detection and Monitoring: An Improved Version of the SFIDE Algorithm. *Remote Sens.* **2018**, *10*, 741. [CrossRef]

50. Gudmundsson, L.; Rego, F.C.; Rocha, M.; Seneviratne, S.I. Predicting above normal wildfire activity in southern Europe as a function of meteorological drought. *Environ. Res. Lett.* **2014**, *9*, 084008. [CrossRef]

51. Trenberth, K.E.; Dai, A.; van der Schrier, G.; Jones, P.D.; Barichivich, J.; Briffa, K.R.; Shef-field, J. Global warming and changes in drought. *Nat. Clim. Chang.* **2014**, *4*, 17–22. [CrossRef]

52. Aguado, I.; Chuvieco, E.; Borén, R.; Nieto, H. Estimation of dead fuel moisture content from meteorological data in Mediterranean areas. Applications in fire danger assessment. *Int. J. Wildland Fire* **2007**, *16*, 390–397. [CrossRef]

53. Westerling, A.L.; Brown, T.J.; Gershunov, A.; Cayan, D.R.; Dettinger, M.D. Climate and wildfire in the western United States. *Bull. Am. Meteorol. Soc.* **2003**, *84*, 595–604. [CrossRef]

54. Littell, J.S.; McKenzie, D.; Peterson, D.L.; Westerling, A.L. Climate and wildfire area burned in western US ecoprovinces, 1916–2003. *Ecol. Appl.* **2009**, *19*, 1003–1021. [CrossRef]

55. Riley, K.L.; Abatzoglou, J.T.; Grenfell, I.C.; Klene, A.E.; Heinsch, F.A. The relationship of large fire occurrence with drought and fire danger indices in the western USA, 1984–2008: The role of temporal scale. *Int. J. Wildland Fire* **2013**, *22*, 894–909. [CrossRef]

56. McEvoy, J.; Hobbins, M.; Brown, T.J.; VanderMolen, K.; Wall, T.; Huntington, J.L.; Svoboda, M. Establishing Relationships between Drought Indices and Wildfire Danger Outputs: A Test Case for the California-Nevada Drought Early Warning System. *Climate* **2019**, *7*, 52. [CrossRef]

57. Brown, J.F.; Wardlow, B.D.; Tadesse, T.; Hayes, M.J.; Reed, B.C. The Vegetation Drought Response Index (VegDRI): A new integrated approach for monitoring drought stress in vegetation. *GISci. Remote Sens.* **2008**, *45*, 16–46. [CrossRef]

58. Sohrabi, M.M.; Ryu, J.H.; Abatzoglou, J.T.; Tracy, J. Development of soil moisture drought index to characterize droughts. *J. Hydrol. Eng.* **2015**, *20*, 04015025. [CrossRef]

59. Carrão, H.; Russo, S.; Sepulcre-Canto, G.; Barbosa, P. An empirical standardized soil moisture index for agricultural drought assessment from remotely sensed data. *Int. J. Appl. Earth Obs.* **2016**, *48*, 74–84. [CrossRef]

60. Kotroni, V.; Cartalis, C.; Michaelides, S.; Stoyanova, J.; Tymvios, F.; Bezes, A.; Christoudias, T.; Dafis, S.; Giannakopoulos, C.; Giannaros, T.M.; et al. DISARM Early Warning System for Wildfires in the Eastern Mediterranean. *Sustainability* **2020**, *12*, 6670. [CrossRef]

61. Stoyanova, J.S.; Georgiev, C.G. Drought and vegetation fires detection using MSG geostationary satellites. In Proceedings of the 2010 EUMETSAT Meteorological Satellite Conference, Córdoba, Spain, 20–24 September 2010.

62. Stoyanova, J.S.; Georgiev, C.G.; Barroso, C. MSG land surface temperature product as a biogeophysical diagnostic parameter of terrestrial water status. In Proceedings of the 2012 EUMETSAT Meteorological Satellite Conference, Sopot, Poland, 7–10 September 2012.

63. Anderson, M.C.; Norman, J.M.; Mecikalski, J.R.; Otkin, J.A.; Kustas, W.P. A climatological study of evapotranspiration and moisture stress across the continental United States based on thermal remote sensing: 2. Surface moisture climatology. *J. Geophys. Res. Atmos.* **2007**, *112*, D10117. [CrossRef]

64. Vicente-Serrano, S.M.; Beguería, S.; López-Moreno, J.I. A multiscalar drought index sensitive to global warming: The standardized precipitation evapotranspiration index. *J. Clim.* **2010**, *23*, 1696–1718. [CrossRef]

65. Hobbins, M.T.; Wood, A.W.; McEvoy, D.J.; Huntington, J.L.; Morton, C.; Anderson, M.C.; Hain, C.R. The Evaporative Demand Drought Index: Part I—Linking drought evolution to variations in evaporative demand. *J. Hydrometeorol.* **2016**, *17*, 1745–1761. [CrossRef]

66. Stoyanova, J.; Georgiev, C.; Neytchev, P.; Kulishev, A. Spatial-Temporal Variability of Land Surface Dry Anomalies in Climatic Aspect: Biogeophysical Insight by Meteosat Observations and SVAT Modeling. *Atmosphere* **2019**, *10*, 636. [CrossRef]

67. Trigo, I.F.; Dacamara, C.C.; Viterbo, P.; Roujean, J.L.; Olesen, F.; Barroso, C.; Camacho-de-Coca, F.; Carrer, D.; Freitas, S.C.; Garcia-Haro, J.; et al. The satellite application facility for land surface analysis. *Int. J. Remote Sens.* **2011**, *32*, 2725–2744. [CrossRef]

68. Trigo, I.F.; Peres, L.F.; DaCamara, C.C.; Freitas, S.C. Thermal Land Surface Emissivity retrieved from SEVIRI/Meteosat. *IEEE Trans. Geosci. Remote Sens.* **2008**, *46*, 307–315. [CrossRef]

69. Stoyanova, J.S.; Georgiev, C.G. SVAT modelling in support to flood risk assessment in Bulgaria. *Atmos. Res.* **2013**, *123*, 384–399. [CrossRef]

70. Stoyanova, J.S.; Georgiev, C.G. Operational drought detection and monitoring over Eastern Mediterranean by using MSG data. In Proceedings of the 2013 EUMETSAT Meteorological Satellite Conference/19th American Meteorological Society AMS Satellite Meteorology, Oceanography, and Climatology Conference, Vienna, Austria, 16–20 September 2013; EUMETSAT: Darmstadt, Germany, 2013.

71. ESA-CCI Land Cover Map product. Available online: https://www.esa-landcover-cci.org/?q=node/164 (accessed on 24 February 2022).

72. Stoyanova, J.; Plamen, N.; Georgiev, C. Characterizing Fire Activity in Eastern Mediterranean Europe by Surface Temperature and Soil Moisture Variability. In Proceedings of the AGU Fall Meeting Abstracts, Virtual, 1–17 December 2020. [CrossRef]

73. Stoyanova, J.; Georgiev, C.; Neytchev, P.; Kulishev, A. Synoptic and climatic aspects of fire activity and emission effects, Part I. In Proceedings of the 7th SALGEE Virtual Workshop "Drought & Vegetation Monitoring: Energy–Water Cycle", Virtual, 24–26 November 2021; Available online: https://training.eumetsat.int/mod/folder/view.php?id=14866 (accessed on 24 February 2022).

74. R Core Team. *R: A Language and Environment for Statistical Computing*; R Foundation for Statistical Computing: Vienna, Austria, 2017.

75. Stoyanova, J.S.; Georgiev, C.G.; Neytchev, P.N.; Vladimirov, E.V. Combined assessment of terrestrial drought and atmospheric conditions through a composite index for fire risk forecast. In Proceedings of the 2016 EUMETSAT Meteorological Satellite Conference, Darmstadt, Germany, 26–30 September 2016.
76. COMET®Program. Critical Fire Weather Patterns. University Corporation for Atmospheric Research, Boulder, CO 80307-3000. 2019. Available online: https://www.meted.ucar.edu/training_module.php?id=1599#.Xa29R2ZS9aQ (accessed on 23 February 2022).
77. Dimitrakopoulos, A.P.; Mitsopoulos, I.D.; Gatoulas, K. Assessing ignition probability and moisture of extinction in a Mediterranean grass fuel. *Int. J. Wildland Fire* **2010**, *19*, 29–34. [CrossRef]
78. Viegas, D.; Piñol, J.; Viegas, M.; Ogaya, R. Estimating live fine fuels moisture content using meteorologically-based indices. *Int. J. Wildland Fire* **2001**, *10*, 223–240. [CrossRef]
79. Xiao, J.; Zhuang, Q. Drought effects on large fire activity in Canadian and Alaskan forests. *Environ. Res. Lett.* **2007**, *2*, 044003. [CrossRef]
80. Chuvieco, E.; González, I.; Verdú, F.; Aguado, I.; Yebra, M. Prediction of fire occurrence from live fuel moisture content measurements in a Mediterranean ecosystem. *Int. J. Wildland Fire* **2009**, *18*, 430–441. [CrossRef]
81. Pausas, J.G.; Fernández-Muño, S. Fire regime changes in the Western Mediterranean Basin: From fuel-limited to drought-driven fire regime. *Clim. Change* **2012**, *110*, 215–226. [CrossRef]
82. Chandler, C.; Cheney, P.; Thomas, P.; Trabaud, L.; Williams, D. Fire in Forestry. In *Forest Fire Behavior and Effects*; John Wiley and Sons: New York, NY, USA, 1983; Volume 1, p. 450.
83. Miller, D.H. *Energy at the Surface of the Earth: An Introduction to the Energetics of Ecosystems*; Chapter VII Radiant Energy Absorbed by Ecosystems; Academic Press: New York, NY, USA, 1981; pp. 114–135. ISBN 0-12-497152-0.
84. Rose, C.W. *Agricultural Physics*; Pergamon Press: Oxford, UK, 1966; p. 230.
85. Park, S.; Feddema, J.J.; Egbert, S.L. MODIS land surface temperature composite data and their relationships with climatic water budget factors in the central Great Plains. *Int. J. Remote Sens.* **2005**, *26*, 1127–1144. [CrossRef]
86. Prigent, C.; Aires, F.; Rossow, W.B. Land surface skin temperatures from a combined analysis of microwave and infrared satellite observations for an all-weather evaluation of the differences between air and skin temperatures. *J. Geophys. Res.* **2003**, *108*, 4310. [CrossRef]
87. Guangmeng, G.; Mei, Z. Using MODIS Land Surface Temperature to evaluate forest fire risk of Northeast China. *IEEE Geoscince Remote Sens. Lett.* **2004**, *1*, 98–100. [CrossRef]
88. Lavorel, S.; Flanningan, M.D.; Lambin, E.F. Vulnerability of land systems to fire: Interactions among humans, climate, the atmosphere, and ecosystems. *Mitig. Adapt Strat. Glob. Change* **2007**, *12*, 33–53. [CrossRef]
89. Bonan, G.B. Forests and climate change: Forcings, feedbacks, and the climate benefits of forests. *Science* **2008**, *320*, 1444–1449. [CrossRef] [PubMed]
90. Seidl, R.; Thom, D.; Kautz, M.; Martin-Benito, D.; Peltoniemi, M.; Vacchiano, G.; Wild, J.; Ascoli, D.; Petr, M.; Honkaniemi, J. Forest disturbances under climate change. *Nat. Clim. Change* **2017**, *7*, 395–402. [CrossRef]
91. Randerson, J.T.; Liu, H.; Flanner, M.G.; Chambers, S.D.; Jin, Y.; Hess, P.G.; Pfister, G.; Mack, M.C.; Treseder, K.K.; Welp, L.R. The impact of boreal forest fire on climate warming. *Science* **2006**, *314*, 1130–1132. [CrossRef] [PubMed]
92. Dintwe, K.; Okin, G.S.; Xue, Y. Fire-induced albedo change and surface radiative forcing in sub-Saharan Africa savanna ecosystems: Implications for the energy balance. *J. Geophys. Res. Atmos.* **2017**, *122*, 6186–6201. [CrossRef]
93. O'Halloran, T.L.; Law, B.E.; Goulden, M.L.; Wang, Z.; Barr, J.G.; Schaaf, C.; Brown, M.; Fuentes, J.D.; Göckede, M.; Black, A. Radiative forcing of natural forest disturbances. *Glob. Change Biol.* **2012**, *18*, 555–565. [CrossRef]

 *remote sensing*

Article

Multilabel Image Classification with Deep Transfer Learning for Decision Support on Wildfire Response

Minsoo Park [1], Dai Quoc Tran [1], Seungsoo Lee [2] and Seunghee Park [1,3,*]

[1] School of Civil, Architectural Engineering & Landscape Architecture, Sungkyunkwan University, Suwon 16419, Korea; pms5343@skku.edu (M.P.); daitran@skku.edu (D.Q.T.)
[2] Department of Convergence Engineering for Future City, Sungkyunkwan University, Suwon 16419, Korea; skklss@skku.edu
[3] Technical Research Center, Smart Inside Co., Ltd., Suwon 16419, Korea
[*] Correspondence: shparkpc@skku.edu; Tel.: +82-31-290-7525

Abstract: Given the explosive growth of information technology and the development of computer vision with convolutional neural networks, wildfire field data information systems are adopting automation and intelligence. However, some limitations remain in acquiring insights from data, such as the risk of overfitting caused by insufficient datasets. Moreover, most previous studies have only focused on detecting fires or smoke, whereas detecting persons and other objects of interest is equally crucial for wildfire response strategies. Therefore, this study developed a multilabel classification (MLC) model, which applies transfer learning and data augmentation and outputs multiple pieces of information on the same object or image. VGG-16, ResNet-50, and DenseNet-121 were used as pretrained models for transfer learning. The models were trained using the dataset constructed in this study and were compared based on various performance metrics. Moreover, the use of control variable methods revealed that transfer learning and data augmentation can perform better when used in the proposed MLC model. The resulting visualization is a heatmap processed from gradient-weighted class activation mapping that shows the reliability of predictions and the position of each class. The MLC model can address the limitations of existing forest fire identification algorithms, which mostly focuses on binary classification. This study can guide future research on implementing deep learning-based field image analysis and decision support systems in wildfire response work.

Keywords: wildfire response; multilabel classification; data augmentation; decision support systems; transfer learning

Citation: Park, M.; Tran, D.Q.; Lee, S.; Park, S. Multilabel Image Classification with Deep Transfer Learning for Decision Support on Wildfire Response. *Remote Sens.* **2021**, *13*, 3985. https://doi.org/10.3390/rs13193985

Academic Editor: Elena Marcos

Received: 8 September 2021
Accepted: 4 October 2021
Published: 5 October 2021

Publisher's Note: MDPI stays neutral with regard to jurisdictional claims in published maps and institutional affiliations.

1. Introduction

Wildfires have become increasingly intense and frequent worldwide in recent years [1]. A wildfire not only destroys infrastructure in fire-hit areas and causes casualties to firefighters and civilians but also causes fatal damage to the environment, releasing large amounts of carbon dioxide [2]. To minimize such damage, decision makers from the responsible agencies aim to detect fires as quickly as possible and to extinguish them quickly and safely [3]. Wildfire response is a continuous decision-making process based on a variety of information that is constantly shared in a spatiotemporal range, from the moment a disaster occurs to when the situation is resolved [4]. Efficient and rapid decision making in urgent disaster situations requires the analysis of decision-support information based on data from various sources [5].

Video and image data are key factors for early detection and real-time monitoring to prevent fires from spreading to uncontrollable levels [6]. Over the past few decades, the use of convolutional neural networks (CNNs) in image analysis and intelligent video surveillance has proven to be faster and more effective than other sensing technologies in minimizing forest fire damage [7]. Nevertheless, several problems must be addressed in forest fire detection and response when using CNNs.

The first problem is that most of the research on wildfires using deep-learning-based computer vision is mainly limited to binary classifications, such as the classification of wildfire and non-wildfire images [8]. In other words, these models are focused on detecting forest fires but ignore other meaningful information, such as information on the surrounding site. Even though a wide range of regions can be filmed via unmanned aerial vehicles (UAVs) or surveillance cameras, information provisions for decision makers are limited in this single-label classification model environment because only one type of result can be obtained from one instance. Unlike single-label classification or multiclass classification, where the classification scheme is mutually exclusive, multilabel classification (MLC) does not have a specified number of labels per image (instance), and the classes are non-exclusive. Therefore, the model can be trained by embedding more information in a single instance [9].

In the context of wildfire response, information is shared to establish a common understanding of wildfire responders regarding the disaster situations they encounter [10]. Information on the disaster site is a vital data source that must be shared to enable timely and appropriate responses. Therefore, information concerning human lives and property at the site of occurrence must be considered to ensure effective and optimized response decisions by decision makers [11].

Another important problem is that the performance of the learning model can be degraded by overfitting owing to insufficient data [12]. The lack of large-scale image data benchmarks remains a common obstacle in training deep neural networks [13]. However, transfer learning and data augmentation can significantly enhance the predictive performance of binary classification models to overcome image data limitations. In particular, transfer learning (with fine-tuning of pretrained models) improves accuracy compared to scanarios when the parameters of the model are initialized from scratch (i.e., without applying transfer learning) [14].

The main purpose of this study was to develop a decision support system for wildfire responders through the early detection and on-site monitoring of wildfire events. A decision support system should perform two functions: (1) process incoming data and (2) provide relevant information [15]. The received data are limited to image data from an optical camera, and the results of deep learning can be analyzed. Therefore, we propose a transfer learning approach for the MLC model to address the following challenges:

1. Does the proposed CNN-based multilabel image classification model for wildfire response decision support show a convincing performance?
2. Are transfer learning and data augmentation methods, which are used to overcome data scarcity, effective in increasing the performance of the proposed MLC model?
3. Images taken from drones are usually collected at a high resolution. However, the CNN-based result is output as a low-resolution image (224 × 224). How can the gap between these two resolutions be addressed?
4. How can the models be used to support forest fire response decision making?

In this study, it is significant that MLC was used to provide multiple pieces of information within the image frame, away from the binary or multi-class classifications mainly covered in previous studies. The reason for using this multi-information framework is to share various pieces of information at disaster sites with disaster responders in near-real time. In the model configuration, we tried to lower the error rate as much as possible by using data augmentation, transfer learning, by adding similar data, and cross validation. In order to minimize the resolution gap between the CNN input model and the actual captured image, a method of dividing and evaluating the image was attempted.

The backbone network of the MLC was constructed using VGG16 [16], ResNet50 [17], and DenseNet121 [18], which are mainly used in CNN-based binary classification. These models were retrained on a dataset built by researchers and were validated using 10-fold cross-validation. The size of the dataset used in the training model was increased by data augmentation to overcome the limitations caused by a lack of data. Finally, the model with the best performance among the three models was selected using the evaluation metric, and the result was visualized as a class activation map (CAM).

The remainder of this paper is organized as follows: Section 2 briefly summarizes previous studies on wildfire detection and response using image data and decision support systems. Section 3 presents the multilabel image classification, transfer learning model, and evaluation methods. The results of relevant experiments are analyzed and discussed in Section 4. Finally, the conclusions are presented in Section 5.

2. Related Work

Effective disaster management relies on the participation and communication of people from geographically dispersed organizations; therefore, information management is critical to disaster response tasks [19]. Because forest fires can cause widespread damage depending on the direction and speed of the fire, strategic plans are required to ensure prioritization and resource allocation to protect nearby homes and to evacuate people. In the past, limitations in data collection techniques constrained these decision-making processes, making them dependent on the subjective experience of the decision-maker [20]. Recent advances in information technology have led to a sharp increase in the amount of information available for decision making. Nevertheless, human capability in information processing is limited, and it is problematic to process information acquired at the scene of a forest fire timely and reliably. To solve this problem, a forest fire decision-support checklist for the information system was developed [21], and machine-learning-based research has steadily increased in the field of forest fire response and management since the 2000s [11]. Analyzing wildfire sites with artificial intelligence can substantially reduce the response time, decrease firefighting costs, and help minimize potential damage and loss of life [5].

Traditionally, wildfires have mainly been detected by human observations from fire towers or detection cameras, which are difficult to use owing to observer errors and time–space limitations [21]. Research on image-based automated detection that can monitor wildfires in real-time or near-real-time according to the data acquisition environment using satellites and ground detection cameras has been steadily increasing over the past decade [22]. Satellites have different characteristics depending on their orbit, which can be either a solar synchronous orbit or a geostationary orbit. Data from solar synchronous orbit satellites have a high spatial resolution but a low time resolution, which limits their applicability in cases of forest fires. Conversely, geostationary orbit satellites have a high temporal resolution but a low spatial resolution. According to previous studies, geostationary orbit satellites can continuously provide a wide and constant field-of-view over the same surface area; however, many countries do not have satellites owing to budget constraints, atmospheric interference, and low spatial resolution [23]. Therefore, satellites are not suitable for the early detection of small-scale wildfires [24]. On the other hand, small UAVs or surveillance cameras incur much lower operating costs than other technologies [25], offer high maneuverability, flexible perspectives, and resolution and have been recognized for their high potential in detecting wildfires early and for providing field information [26].

Previous studies combined image data and artificial intelligence methods to improve the accuracy of forest fire detection or to minimize the factors that cause errors. Damage detection studies often face the problem of data imbalances [27], which previously relied only on images downloaded from the Web and social media platforms [28,29]. Online image databases, such as the Corsican Fire Database, have been used for binary classification as a useful test set for comparing computer vision algorithms [30] but are still not available in MLC. Recent studies have demonstrated its effectiveness using data augmentation or transfer learning for the generalization of the performance of CNN models [31] and have shown its potential in object detection or MLC fields.

Because neural networks cannot be generalized to untrained situations, the importance of the dataset has been steadily emphasized to improve the performance of the model.

During model verification, the smoke color and texture are too similar to other natural phenomena such as fog, clouds, and water vapor, and because it is difficult to detect smoke during the night, algorithms relying on smoke detection generally cause problems such as high false alarm rates [31,32]. The current study was conducted by including the objects that could not be differentiated in the dataset.

3. Materials and Methods

3.1. Data Augmentation

Data augmentation is the task of artificially enlarging the training dataset using modified data or synthesizing the training dataset from a few datasets before training the CNN model, which lowers the test error rate and significantly improves the robustness of the model to avoid overfitting. The most popular and proven effective current practices for data augmentation are affine transformation, including the rotation and reflection of the original image and color modification, including brightness transformation [33]. In this study, the image dataset was pre-processed in terms of reflection, rotation, and brightness, which are commonly used data augmentation techniques in previous studies to increase the richness of the training datasets.

3.2. Transfer Learning

Transfer learning is another approach to prevent overfitting [34]. It is a machine learning method that uses the weights of the pretrained models as weights for the initial or intermediate layers of the new objective model. In computer vision, transfer learning refers mainly to the use of pretrained models. This method is widely used to handle tasks that lack data availability [35]. There are two representative approaches for applying a pretrained model, called a fixed feature extractor and fine-tuning. The fixed feature extractor is a method of learning only the fully connected layer in a pretrained model and fixing the weights of the remaining layers. It is mainly applied when the amount of data is small, but the training data used for pretraining are similar to the training data of the target model. This approach is uncommon for the deep learning of damage detection areas, such as wildfire monitoring images, because of the dissimilarity between ImageNet and the given wildfire images.

On the other hand, fine-tuning not only replaces the fully connected layers of the pretrained model with a new one that outputs the desired number of classes to re-train from the given dataset but also fine-tunes all or part of the parameters in the pretrained convolutional layers and pooling layers by backpropagation. It is used when the amount of data is sufficient, even if the training data are not similar. This is shown in Figure 1.

The pretrained CNN model from ImageNet [36], which contains 1.4 million images with 1000 classes, is used for transfer learning. However, as there are no labels similar to flame or smoke or other on-site images to assist disaster response in the ImageNet label, fine-tuning is introduced.

3.3. Multilabel Classification Loss

Cross-entropy is defined as the calculation of the difference between the two probability distributions p and q, i.e., error calculations. Cross entropy is used as a loss function in machine learning. However, our framework uses binary cross entropy (BCE), which has commonly been used in the loss function for multilabel classification. The CNN model performs training by adjusting the model parameters such that probabilistic prediction is as similar to ground-truth probabilities as possible through the BCE. In other words, the probability of the output and the target similarly adjusts the model parameters. The BCE loss is defined by the following equation:

$$\mathcal{L}_{BCE} = -\frac{1}{N}\sum_{i=1}^{N}[p(y_i)logq(y_i) + \{1 - p(y_i)\}log\{1 - q(y_i)\}], \tag{1}$$

where N denotes the total count of images, $p(y_i)$ denotes the probability of class y_i in the target, and $q(y_i)$ denotes the predicted probability of class y_i.

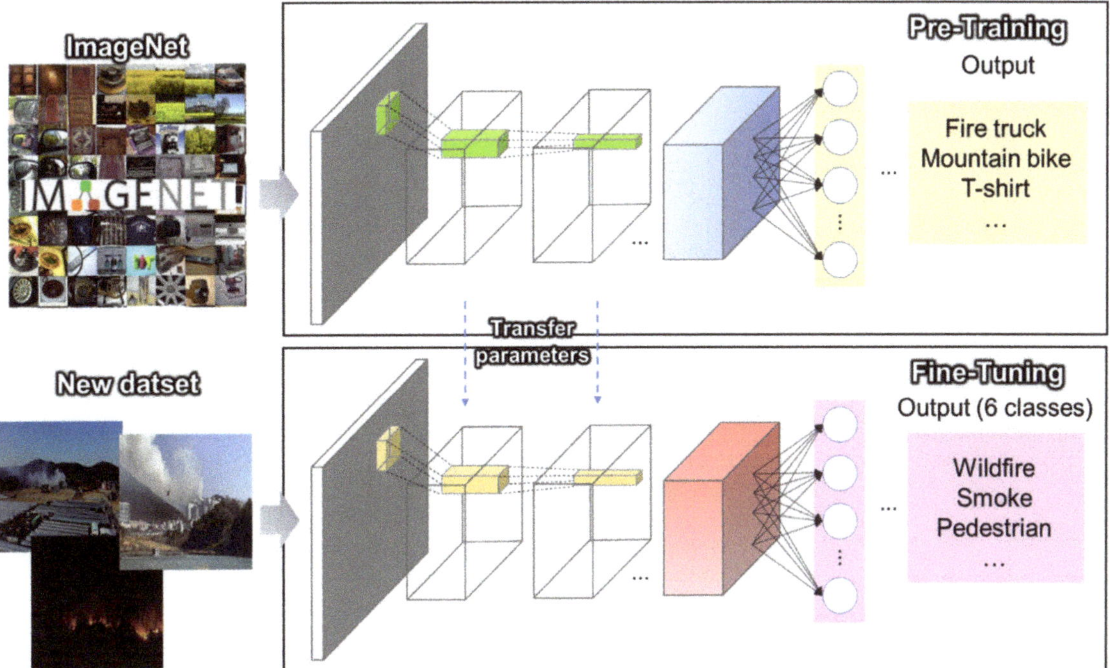

Figure 1. Transfer learning architecture with fine-tuning using the pretrained CNN model initialized with the weights trained from ImageNet.

3.4. Proposed Network

The MLC model used in this study consists of a backbone network pretrained on ImageNet and fully connected layers. In multilabel classification, the training set consists of instances associated with the label set, and the model analyzes the training instances with a known label set to predict the label set of unknown instances. Figure 2 shows an example of a framework for an MLC-based proposed model with DenseNet121 as the backbone. The fully connected layer included dropout [37] and batch normalization [34]. The order of dropout, batch normalization, and rectified linear units (ReLU) were constructed based on the methodology of Ioffe [34] and Li [38].

Six classes were printed out, and the model was configured to achieve the following goals required for disaster response during the event of a forest fire: (a) check whether a forest fire has occurred, (b) detect smoke for the early detection of fires, (c) detect the burning area for extinguishing, and (d) detect the areas where human or property damage may occur.

In this study, we selected each of the three pretrained models mentioned above as a backbone network. An MLC model was constructed to provide information that can be supported for wildfire response from CCTV or UAV images. Finally, we compared the performance of each model.

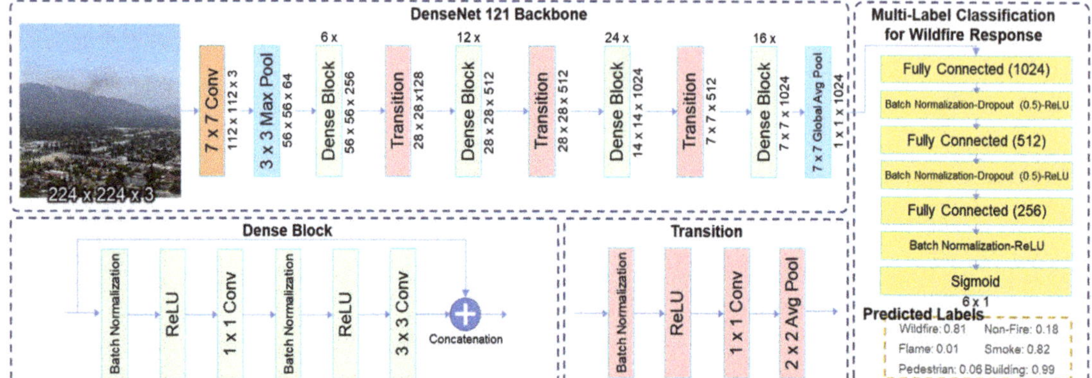

Figure 2. Proposed approach pipeline: framework of multilabel classification with transfer learning from DenseNet-121 as a backbone network. The probabilities generated by the sigmoid function were independently output at the end of the neural network classifier.

3.5. Performance Metrics

Instances in single-label classification can only be classified correctly or incorrectly, and these results are mutually exclusive. However, the classification schemes in multilabel classification are mutually non-exclusive: in some cases, the predicted results from the classification model may only partially match the elements of the real label assigned to the instance. Thus, methods for evaluating multilabel models require evaluation metrics specific to multilabel learning [39]. Generally, there are two main groups of evaluation metrics in the recent literature: example-based metrics and label-based metrics [40]. Label-based measurements return macro/micro averages across all labels after the performance of the training system, for each label is calculated individually, whereas example-based measurements return mean values throughout the test set based on differences in the actual and predicted label sets for all instances. To evaluate the performance of each model and to verify the effectiveness of transfer learning and data augmentation, this study used macro/micro average precision (PC/PO), macro/micro average recall (RC/RO), and macro/micro average F1-score (F1C/F1O). The abbreviations for evaluation metrics are based on the notations of Zhu [41] and Yan [9]. The metrics are defined as follows:

$$PC = \frac{1}{q} \sum_{\lambda=1}^{q} \frac{TP_\lambda^q}{TP_\lambda^q + FP_\lambda^q} \tag{2}$$

$$RC = \frac{1}{q} \sum_{i}^{q} \frac{TP_\lambda^q}{TP_\lambda^q + FN_\lambda^q} \tag{3}$$

$$PO = \frac{\sum_{\lambda=1}^{q} TP_\lambda}{\sum_{\lambda=1}^{q} (TP_\lambda + FP_\lambda)} \tag{4}$$

$$RO = \frac{\sum_{\lambda=1}^{q} TP_\lambda}{\sum_{\lambda=1}^{q} (TP_\lambda + FN_\lambda)} \tag{5}$$

$$F1C = \frac{2 * PC * RC}{PC + RC} \tag{6}$$

$$F1O = \frac{2 * PO * RO}{PO + RO} \tag{7}$$

In the above equations, *TP*, *FP*, and *FN* denote true positives, false positives, and false negatives, respectively, as evaluated by the classifier.

Macro averages are used to evaluate the classification model on the average of all of the labels. In contrast, the micro average is weighted by the number of instances of each label, which makes it a more effective evaluation metric on datasets with class imbalance problems. The $F1$ score is a harmonic average that considers both precision and recall. Therefore, the $F1$ score is generally considered a more important metric for comparing the models. In addition, the datasets for MLC generally suffer from data imbalance, and thus, micro-average-based metrics are considered important.

In addition, this study used Hamming loss (*HL*) and mean average precision (*mAP*), which are represented by example-based matrices. These metrics are defined as follows:

$$Hamming\ Loss = \frac{1}{|D|} \sum_{i=1}^{|D|} \frac{|Y_i \Delta Z_i|}{|L|} \tag{8}$$

$$mAP = \frac{1}{|L|} \sum_{i=1}^{|L|} AP_i \tag{9}$$

In the above equations, $|D|$ is the number of samples, $|L|$ is the number of labels, and AP_i is the average map of label i. Hamming loss is the ratio of a single misclassified label to the total number of labels (considering both the cases when incorrect labels are predicted and when associated labels are not predicted) and is one of the best-known multilabel evaluation methods [42]. The mean average precision was the mean value of the average precision for each class.

3.6. Class Activation Mapping

In the CNN model, the convolutional units of various layers act as object detectors. However, the use of fully connected layers causes a loss in the localizing features of these objects. Class activation mapping (CAM) [43] is used as a CNN model translation method and is a popular tool for researchers to generate attention heatmaps. A feature of CAM is that the network can include the approximate location information of the object even though the network has been trained to solve a classification task [8]. To calculate the CAM value, the fully connected layer is modified with the global average pooling layer (GAP). Subsequently, a fully connected layer connected to each class is attached and fine-tuned. However, it has a limitation in that it must use a GAP layer. When replacing the fully connected layer with GAP, the fine-tuning of the rear part is required again. However, CAM can only be extracted for the last convolutional layer.

Gradient-weighted class activation mapping (Grad-CAM) [44] solves this problem using a gradient. Specifically, it uses the gradient information coming into the last convolutional layer to take into account the importance of each neuron to the target label. In this study, Grad-CAM was used to emphasize the prediction values determined by the classification model and to visualize the location of the prediction target.

4. Results

This section presents the learning process and test results of the MLC model to support wildfire responses. Experiments were conducted in a CentOS (Community Enterprise Operating System) Linux release 8.2.2004 environment with Nvidia Tesla V100 GPU, 32 GB memory, and models were built and trained using PyTorch [45], a deep learning open-source framework.

4.1. Dataset

The dataset used to train and test the deep learning model contained daytime and nighttime wildfire images captured by surveillance cameras or drone cameras downloaded from the Web and cropped images of a controlled fire in the forest captured by a drone by the researchers. This study also included a day–night image matching (DNIM) dataset [46], which was used to reduce the effects of day and night lighting changes, and Korean tourist spot (KTS) [47] datasets generated for deep learning research, which comprise images linked by forest labels containing important wooden cultural properties in the forest. Additionally, wildfire-like images and 154 cloud and 100 sun images were also included as datasets because they have similar properties with early wildfire smoke and flames as the color or shape and are often detected erroneously. As such, they were included in the training dataset to prevent predictable errors in the verification stage and to train the robust model against wildfire-like images.

The collected images were resized or cropped to 224×224 pixels to consider whether the model is applicable to high-definition images. The datasets included 3,800 images. Figure 3 shows samples of the images.

Figure 3. Resized sample of annotated images for MLC: (**a**) downloaded from the Web; (**b**) KTS dataset; (**c**) DNIM dataset; (**d**) dataset for error protection purposes; (**e**) dataset of a controlled fire captured by researchers.

All instances were annotated according to the following classes: "Wildfire", "Non-Fire", "Flame", "Smoke", "Building", and "Pedestrian" (each class was abbreviated as "W", "N", "F", "S", "B", and "P", respectively). Table 1 lists the number of images for each designated label set before data augmentation. It consists of 2165 images downloaded from the Web, 1000 images from the KTS dataset, 101 images from the DNIM dataset, 254 images for error protection purposes, and 280 cropped images captured by the researchers. To ensure the annotation quality and accuracy, all of the annotated images were checked twice by different authors.

Table 1. Number of image datasets for each annotated multilabel instance.

Label	WS	WSF	WSP	WSBP	WSB	WSFB	WSFP	WSFBP	N	NB	NP	NBP
Original	585	419	176	103	82	84	87	67	1567	331	210	89
After data pre-processing (augmentation and partition)												
Train	1464	996	432	240	234	150	216	138	3726	786	462	276
Test	341	253	104	63	43	59	51	44	946	200	133	43

4.2. Data Partition

The dataset used for the experiment was divided into train, validation, and test sets. The test dataset included 2280 images from the entire dataset. The remaining 1520 images were pre-processed by data augmentation techniques, such as rotation, horizontal flip, and brightness, which are typically used in CNN image classification studies to secure sufficient data for learning, as shown in Table 2. Table 1 also lists the number of images for each designated label set after augmentation. Overall, the non-fire label group was the most common, and as the number of multi-labels increased, the number of the label group decreased. In particular, the number of label groups in which pedestrians and houses exist in the wildfire site, which is difficult to obtain, has the lowest number.

Table 2. Number of images generated by each data augmentation method.

Augmentation Method	Original Data	Brightness	Flip	Rotation	Total
Images	1520	3040	1520	3040	9120

Since drones are generally not perpendicular to the horizon or are inverted when photographing wildfires, the rotation was not set up as extreme, such as at 90° or 180°, but was instead set up between 10° and 350°, considering the lateral tilt of the drone. In addition, if the brightness of the image is too high or too low, the boundary line of the objective target becomes unclear, and the object becomes ambiguous. Therefore, data enhancement was performed between the maximum brightness $l = 1.2$ and the minimum brightness $l = 0.8$. After data augmentation, the training and test datasets were divided in a ratio of 4:1. In the model learning phase, 912 randomly sampled instances from the training dataset were evenly divided into 10 groups for evaluation using the cross-validation strategy.

The total number of classes of the prepared data was checked, and the distribution is shown in Figure 4. Due to the nature of wildfire response, most of the early detection was performed by smoke, so the number of smoke classes was higher than the number of flame classes. In addition, the wildfire classes and non-fire classes also had an imbalance, and the building and pedestrian classes also had relatively few classes. Since there was an imbalance between the labeling classification table in Table 1 and the overall class distribution in Figure 4, the micro average-based metric evaluation index should be checked.

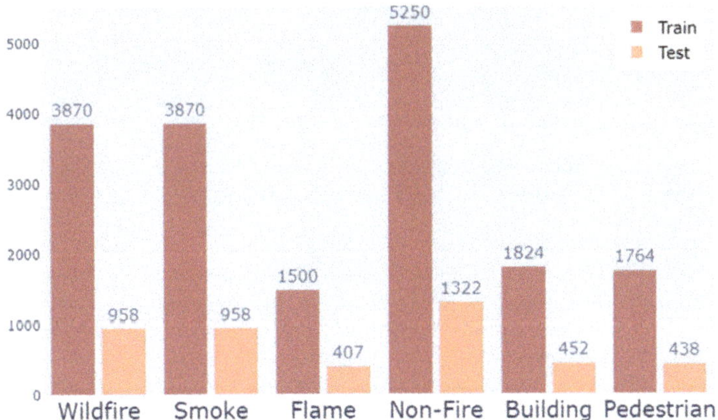

Figure 4. Histogram of class distribution on multi-label classification datasets after data pre-processing.

4.3. Performance Analysis

This study compared the models with different backbones and verified the efficiency of transfer learning and data augmentation. The model was constructed using training and validation sets that had been partitioned by a 10-fold cross-validation strategy, and the final performance was measured according to each performance metric from the test dataset. The initialized learnable parameters (i.e., hyperparameters) for the CNN-based MLC architectures are listed in Table 3.

Table 3. Parameters for each CNN architecture.

	VGG-16	**ResNet-50**	**DenseNet-121**
Mini batch size	48	57	48
Iteration	171	144	171
Number of training epoch	100	100	100
Learning rate	0.001	0.001	0.001
Optimizer	Adam	Adam	Adam

The models were trained using binary cross-entropy as a loss function with the selected parameters. Each model was trained using a 10-fold cross-validation strategy, and the results were calculated 10 times. The training process using the validation scheme of each model with the selected hyperparameter combination is illustrated in Figure 5. In the case of VGG-16, the training loss fell gently and started at a very high validation loss value, while the training loss of ResNet-50 and DenseNet-121 fell sharply to about epoch 10 at the initial stage and then remained close to zero. However, it was found that DensNet-121 remained lower in terms of the validation learning curve. In the final epoch, epoch 100, it was shown that both the training loss and validation loss recorded the lowest values in DensNet-121.

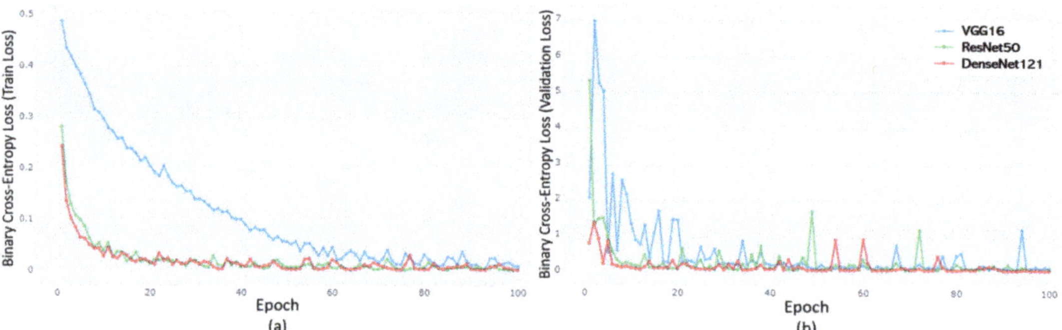

Figure 5. Learning curve over epochs (until 100). (**a**) Training learning curve. Final loss: VGG-16 (0.00789); ResNet-50 (0.00137); DenseNet-121 (0.00048). (**b**) Validation learning curve. Final loss VGG-16 (0.06790); ResNet-50 (0.03352); DenseNet-121 (0.00318).

The models were evaluated using the label-based performance metrics, which are shown in Figure 6 as a box plot. All of the proposed models showed good multilabel classification ability using images of forest and wildfire sites for disaster response, with high scores (above 0.9) for most of the evaluation metrics. Among the proposed models, DenseNet-121 not only showed a significantly higher score for all of the evaluation metrics (distribution of the highest box and median value) but the interquartile ranges of each metric result were also typically smaller (i.e., with fewer distributed results) than in other models. Thus, the model maintained consistently high performance over several tests. Table 4 presents the results of the evaluation measurements with the mean and standard deviation.

However, an evaluation that only uses label-based measurements cannot highlight the dependencies between classes. Therefore, Table 4 presents example-based scores that consider all of the classes simultaneously and thus are considered more suitable for multilabel problems. The mAP score for the best-performing model (DenseNet-121) was 0.9629, whereas the mAP score for HL was 0.009.

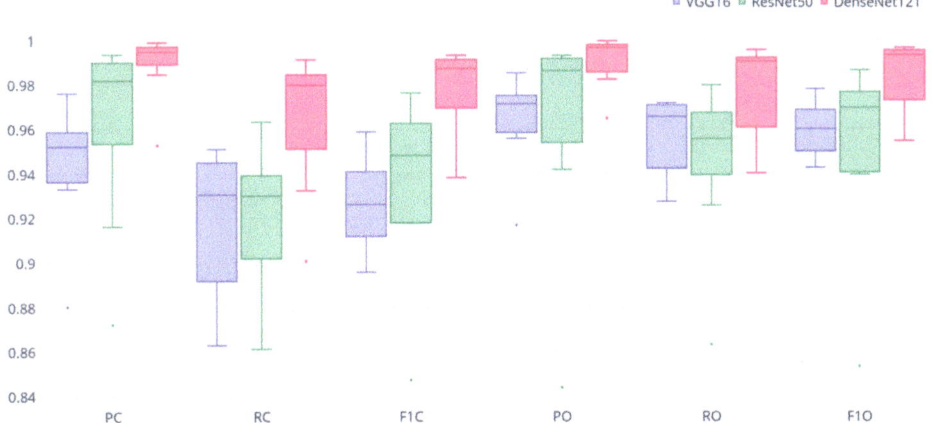

Figure 6. Boxplot of the 10-fold cross-validation results from performance metrics for MLC model with each backbone network.

Table 4. Compared performance scores of each backbone network (mean ± standard deviation).

	VGG-16	**ResNet-50**	**DenseNet-121**
PC	0.9435 ± 0.0308	0.9640 ± 0.0399	0.9899 ± 0.0138
RC	0.9177 ± 0.0338	0.9221 ± 0.0304	0.9661 ± 0.0293
F1C	0.9265 ± 0.0212	0.9368 ± 0.0371	0.9769 ± 0.0215
PO	0.9635 ± 0.0225	0.9655 ± 0.0462	0.9914 ± 0.0110
RO	0.9560 ± 0.0178	0.9485 ± 0.0329	0.9783 ± 0.0214
F1O	0.9595 ± 0.0123	0.9555 ± 0.0390	0.9847 ± 0.0159
mAP	0.8811 ± 0.0312	0.9056 ± 0.0529	0.9629 ± 0.0327
HL	0.0025 ± 0.0008	0.0017 ± 0.0009	0.0009 ± 0.0009

In addition, the per-class score of the area under the receiver operating characteristic curve (ROC-AUC) values of our proposed models were calculated to determine the performance for each class in the image dataset. The ROC curve is a graph showing the performance of the classification model at all possible classification thresholds, unlike the recall and precision values that change as the threshold is adjusted. AUC is a numerical value calculated from the area under the ROC curve and represents the measure of separability. Therefore, the ROC-AUC is a performance metric that is more robust than other performance indicators. AUC values range from 0 to 1, where AUC = 0.5 indicates that the model performed a random guess, and thus, the prediction was the entirely unacceptable. The best performance is when AUC = 1, indicating that all of the instances are properly classified. Table 5 presents the results with mean and standard deviation values.

Table 5. ROC-AUC scores per class for each model (mean ± standard deviation).

	Wildfire	**Smoke**	**Flame**	**Non-Fire**	**Building**	**Pedestrian**
VGG-16	98.70 ± 00.48	98.72 ± 00.47	98.71 ± 00.46	95.68 ± 02.65	91.66 ± 00.53	87.54 ± 02.61
ResNet-50	98.37 ± 01.25	98.36 ± 01.25	98.35 ± 01.25	97.60 ± 01.70	92.55 ± 03.64	92.92 ± 03.67
DenseNet-121	99.01 ± 01.38	99.00 ± 01.40	99.01 ± 01.40	98.29 ± 01.95	96.65 ± 02.75	96.44 ± 03.65

The dataset for the MLC model includes pictures of fires or non-fires (because the results are mutually exclusive). Therefore, the results of two classes—"Wildfire" and "Non-fire"—are calculated in almost the same way. The results of the classes "Wildfire" and "Smoke" were also calculated similarly, as flames are inevitably accompanied by smoke, although this smoke may be invisible because some fires are small or obscured by forests. The accuracy of the pedestrian and building labels was low in all of the models, which can be attributed to the relatively small number of labels assigned to the instances. It was confirmed that the ROC-AUC scores in all of the classes were generally high in the transfer learning algorithm using DenseNet-121 as a network.

Finally, to confirm the effect of transfer learning and data augmentation on the training model, we removed one data limit overcoming strategies each time using the control variable method and obtained the F1-score and the HL value. This method was implemented for DenseNet-121, which showed the highest performance. The training learning curve over epochs is illustrated in Figure 7. In the training stage, there was a significant difference in the slope of the learning cover curve when no data limit overcoming strategies were used and when one or more strategy was used; a steep learning curve was demonstrated with the strategies; a shallow learning curve was demonstrated without the strategies. When all of the strategies were used, the curve was formed the most rapidly, and the lowest final loss was calculated. This means other models require more practice or attempts before a performance begins to improve until the same level is reached. The curve produced in the case of using all of the strategies was formed the most rapidly, and the lowest final loss score was also calculated. There was no significant difference in the data augmentation and transfer learning effects when looking at the gradient slope or the final loss, but it was shown that the roughness of the curve was further reduced when data augmentation was used. In other words, learning was more stable.

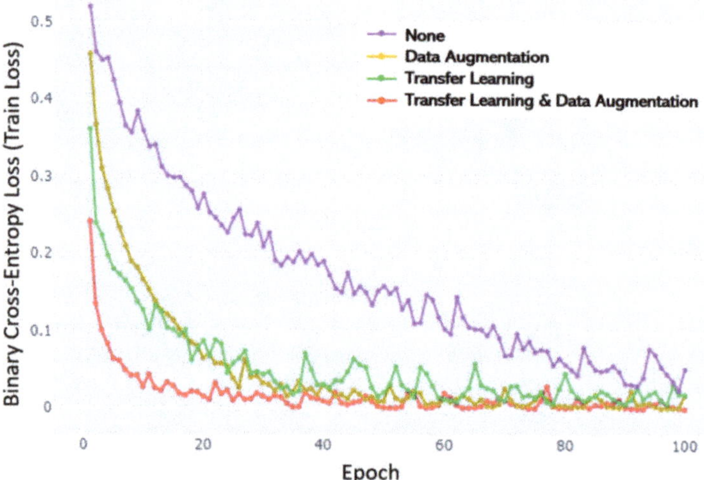

Figure 7. Learning curve over epochs (until 100) trained by the control variable method. Final loss. Transfer learning and data augmentation (0.00048); data augmentation (0.01807); transfer learning (0.01909); none (0.05216).

Additionally, the test results determined by the evaluation metrics are listed in Table 6. The results of this experiment show that transfer learning can significantly improve multilabel classification performance. With the exception of the transfer learning strategy, the macro average F1-score decreased by 0.0745, the micro average F1-score decreased by 0.0466, and the HL increased by 0.0286. In the case where only augmentation was used, the macro average F1-score decreased by 0.1159, the micro average F1-score decreased by 0.0701, and the HL increased by 0.0412.

The performance of transfer learning was further reduced when trained only with the datasets with no data augmentation. Hence, the quantitative number of the datasets required for learning in MLC has a significant impact on the performance of the model.

Table 6. F1 scores and Hamming loss values for the models trained by the control variable method.

Strategies to Overcome Data Limitations	F1C	F1O	HL
Transfer Learning and Data Augmentation	0.9769	0.9847	0.0093
Transfer Learning	0.8610	0.9146	0.0505
Data Augmentation	0.9024	0.9381	0.0379
None	0.7951	0.8634	0.0845

4.4. Visualization

To perform localization, a bounding box is drawn by the thresholding method, which retains over 20% of the grad-CAM result. It also provides a confidence score indicating the extent to which the model's predictions are true, considering a threshold value of 0.5. Figures 8 and 9 show an example of the results obtained with DenseNet-121 as a backbone.

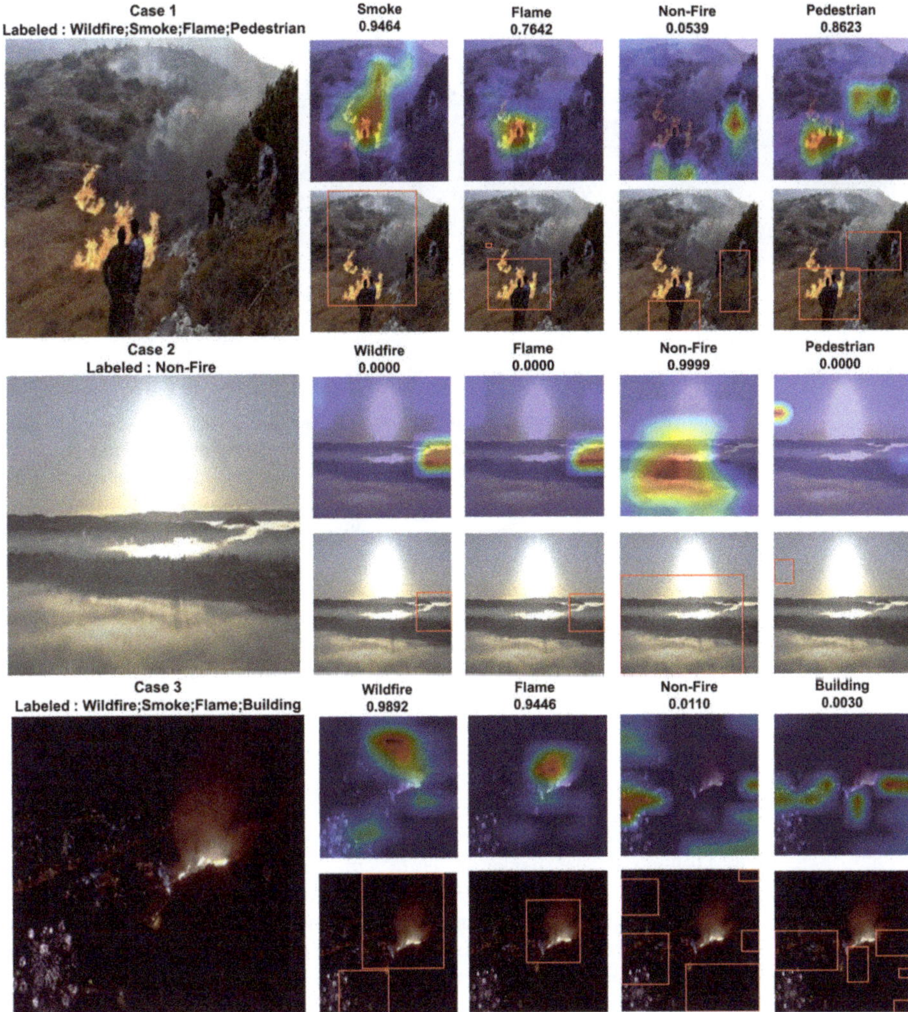

Figure 8. Original image (first column) and the heatmap and bounding box (columns 2–5) for a well-classified example (**Case 1**); an example to review errors for objects with similar colors or shapes (**Case 2**); and an example of an error for a particular class (**Case 3**). Case 3 has an error for the building class. The number above each picture is the predictive confidence score.

Figure 9. Example of a heatmap and bounding box result for the class that has the possibility of an error. Confidence score (**a**): Wildfire (0.8117); Smoke (0.8158); Flame (0.0096); Non-Fire (0.1860); Building (0.0006); Pedestrian (0.9952). Confidence score (**b**): Wildfire (0.8117); Smoke (0.8158); Flame (0.0096); Non-Fire (0.1860); Building (0.0006); Pedestrian (0.9952).

As shown in Figure 8, the sum of the confidence scores between the two classes is almost 100% because the wildfire and non-fire classes are mutually exclusive. Among the test datasets, a Case 1 image was selected as a sample of labeled wildfire with pedestrians, a Case 2 image was selected as a sample with a confusing object, such as sun or fog, that can be treated as a wildfire object for verification. Finally, a sample image with a night fire was selected to evaluate the model in nighttime conditions in Case 3.

In Case 1, the model predicted smoke, flames, non-fire, and a pedestrian with confidence scores of 0.9464, 0.7642, 0.0539, and 0.8623, respectively. The heatmap and bounding box were separated from each other to express the location. Conversely, for non-fire that is not assigned to an instance, a heatmap map without fire or smoke was displayed in the bush area.

Case 2 used an image taken at sunrise in a foggy mountainous area. All of the classes except for the non-fire class showed a score of 0.000, and it was possible to examine whether the model worked correctly to classify the sun and fog, which are frequently used for evaluating errors in wildfire detection.

Case 3 was an image of a wildfire that occurred near a downtown area at nighttime, and fireworks were displayed nearby, which may have some effect on detection. For the wildfire class, a heatmap was formed even in an area unrelated to the wildfire (lower left), which was judged to have detected the smoke generated by the firecrackers on the image. However, although the lighting in the dark at night was considered in the model training process, the heatmap represented the residential area, but the reliability of the building class was still very low (0.0030).

These results were similarly expressed in other test datasets, indicating that the accuracy of the classification was only improved if the CNN model was observable with the naked eye because only clear targets were labeled during the dataset preparation process.

Using an example, Figure 9 shows that the classification model is robust to a small object or noise in a photograph. As shown in Figure 9a, firefighters were dispatched to extinguish the fire in the forest, and nearby hikers were caught on camera. Although the human shape in Figure 9a looks very small, it is detectable with a 0.9952 confidence score, and an approximate location of the object was determined. Figure 9b shows a house with lights on and a nearby forest fire. Despite the similarity of the lamp to the fire image, the heatmap result did not recognize this part as a fire. Thus, the model looked at the appropriate part when identifying each class.

Finally, Figure 10 shows the influence of transfer learning and data augmentation. As discussed in the previous subsection, four cases were classified using the control variable method. For each case, the heatmap and bounding box of a specific class (smoke and person in Figure 10) were visualized, and the probability values were calculated. In the case of not using the data-shortage overcoming strategy, it was found that the heatmap represented the wrong place, and the class that was difficult to distinguish could not be detected at all. (If the value is less than 0.5, it is not treated as a detected value.) Therefore, the accuracy difference is significant if data supplementation strategies are not used for wildfire images, where data are inevitably lacking, as shown in Figures 9 and 10. When a data supplementation method was used, the heatmap distribution was somewhat reasonable, and the confidence value for the class that was to be detected was significantly increased. When both strategies were used, the heatmap distribution was the cleanest, and the positive predictive probability was the highest.

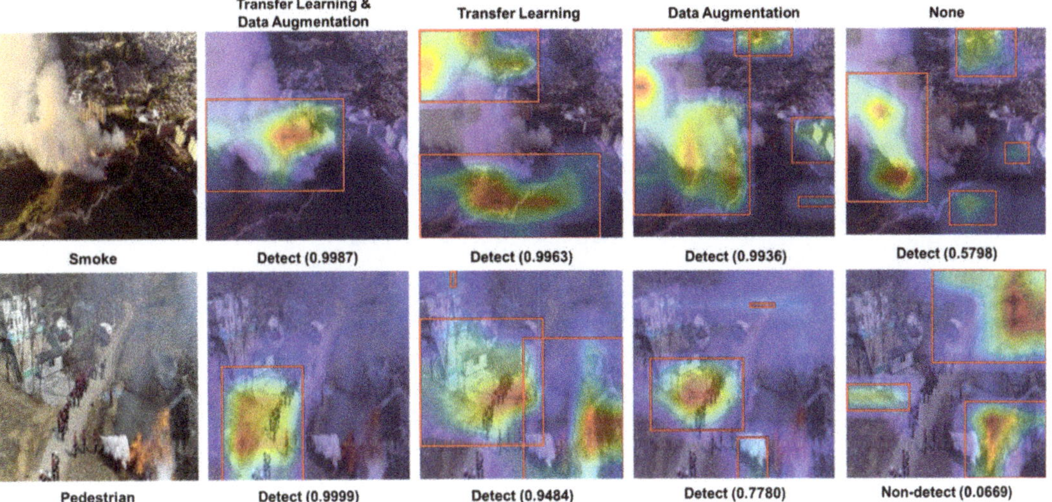

Figure 10. Sample of the Grad-CAM results depending on the model trained by the control variable method.

4.5. Application

The proposed model was applied to an image captured by the researchers using a DJI Phantom 4 Pro RTK drone, which had an image size of 1280×720 high-definition (HD) units. The filming site was composed of a virtual wildfire environment similar to that of a fire created by lighting a drum around the forest.

Although the captured HD image can be resized to the input size of the proposed model, downscaling a high-resolution image may result in the loss of information that is useful for classification, and the model may not operate smoothly [48]. Thus, the images were divided into 28 equal parts of 224×224, and the model was evaluated for the divided pictures. When the pictures are divided without overlapping parts, there is a possibility of a blind spot where the object to be found is cut off. Thus, the images are divided such that there are overlapping parts. The predicted classification values of each part of the picture were merged into the entire image and were visualized. The results of applying the proposed model to the drone shooting screen are shown in Figure 11. The confidence value was over 50%, and the label corresponding to the photograph part was predicted. In Figure 11, a forest fire was detected based on smoke in the central part of the whole picture. However, there was also an error (9.02%) in the building area, which was important for preserving residential and cultural assets. This error can be explained as follows: the large object was still cut off in the cropped image despite the application of the overlapping method. The white dotted circle was drawn to highlight the area with people, and the model correctly predicted that there were people in the area at 99.34% and 77.03%.

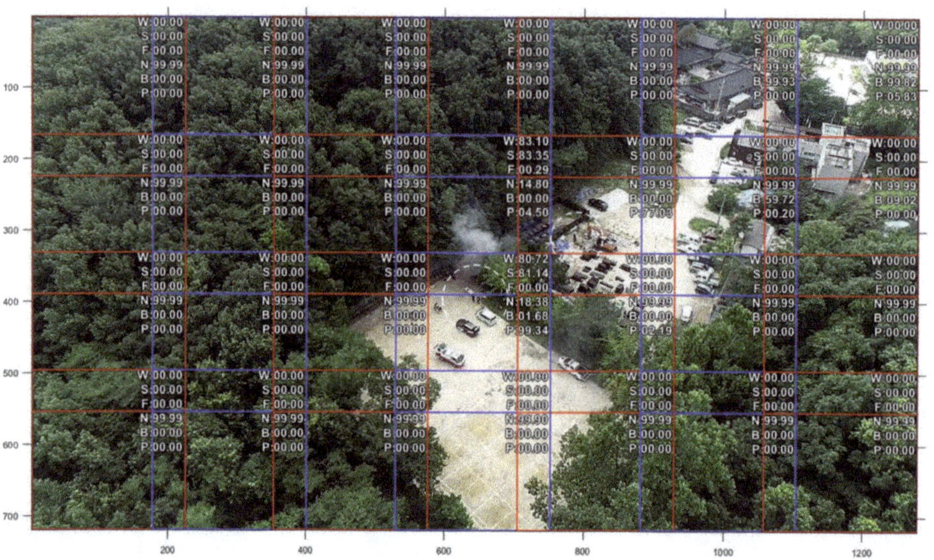

Figure 11. Sample of model application with the confidence score for each class. The blue and red boxes represent 224 × 224 images.

5. Discussion

In this study, transfer learning and data augmentation were combined to improve the capabilities of the model. Three different pretrained models were used to handle data limitations, data augmentation was performed, and each model was evaluated using label-based evaluative metrics and example-based evaluative metrics. In conclusion, DenseNet-121 surpassed VGG-16 and ResNet-50 in the proposed MLC model. This is confirmed by the results of the evaluation metrics. With the advancement in camera technology, the image resolution increases, but training a CNN to handle large images is particularly difficult. The problems are the cost and learning time caused by excessive computational load in the initial layer. Because of the discrepancy between the image size in these models and the image size taken from the imaging device, we split the high-resolution image into smaller parts and processed them separately. The method proposed in this study loses less data and is expected to better classify small objects compared to scenarios when the original image is reduced in size and only a single image is processed. The proposed framework can be converted into other applications of image-based decision-making systems for disaster response fields to extract redundant information from one object.

Some previous studies used public data for binary classification problems (fire and non-fire). However, a dataset with multiple labels or classes changes according to the requirements of the system, and it is difficult to use the datasets from previous studies. The classified labels were defined to solve the need for a response from the image sources collected at the site. Fire is often accompanied by smoke, which is released faster than flames. The flame of a forest fire is barely visible from a distance. However, the smoke columns caused by fires are usually visible on camera. Therefore, early smoke detection is an effective way to prevent potential fire disasters [49]. Based on the detection of flames, field responders can be informed as to where the flames need to be extinguished. Decision makers for wildfire response, who receive information on the life and property at the fire site, use this basic information to decide on the evacuation route by considering the spot of fire occurrence to preferentially protect the area where there is a possibility of severe damage and to establish a line of defense. Instructions for prioritizing such tasks and for efficiently allocating limited support resources must be provided. Therefore, label categories can be defined for wildfire response and building large image benchmarks for disaster response.

This is a basic study that provides multilabel information of target areas from cameras by applying CNN to wildfire response. Considering that multilabeling was performed manually by the researchers, the distinction between instances was vague in some of the images collected from external data sources. Instances that were too small to distinguish were not labeled to avoid overfitting. In addition, in the case of an instance that cannot be easily distinguished with the naked eye, it was not possible to easily add a class to be classified because of a label classification error.

Therefore, future research should aim to construct a formal annotated data benchmark for wildfire response in deep learning systems to enable the use of field information for supporting disaster decision-makers from the perspective of the wildfire detection algorithm. For example, the state of the wildfire may be understood from the fire shape. Crown fires are the most intense and dangerous wildfires, and surface fires cause relatively little damage. It is also important to identify forest species in disaster areas using videos. If the forests of the target area are coniferous, fires may spread to a large area. To provide this additional information, it is important to ensure communication between photographers, labeling workers, and deep learning model developers. From the perspective of wildfire response, future studies should also aim to develop an integrated wildfire-response decision-support system that can provide decision makers with various insights. Location can be retrieved from the global positioning system (GPS) of drones filming in disaster areas, and this can be combined with data on weather conditions that greatly affect wildfire disasters, such as wind direction, wind speed, and drying rate at the target site. In addition, when combined with a geographic information system (GIS), it is possible to determine the slope of the target area because a steep slope is difficult to control during a wildfire.

6. Conclusions

To the best of our knowledge, previous computer vision-based frameworks for managing fires have only used binary classification. However, in disaster response scenarios, decision makers must prioritize extinguishing operations by considering the range of flames, major surrounding structures such as residential facilities or cultural assets, and residents at the site. Various types of information on the scene of a wildfire can be obtained and analyzed using the photographs from an imaging device. However, annotation work is limited because of a lack of training datasets and the fact that previous wildfire detection research has only focused on binary classification. To solve these problems, we proposed a basic MLC-based framework to support wildfire responses.

The proposed model was verified through well-known evaluation indicators from the dataset selected by the researchers, and DenseNet-121, the most effective of the three representative models, was selected as the final model. Then, we visualized the result through grad-cam, and proposed a method to divide and evaluate each image to prevent data omission when applied to FHD or higher photos according to recently developed camera technology.

Author Contributions: Conceptualization, M.P.; methodology, M.P. and D.Q.T.; data curation, S.L.; writing—original draft preparation, M.P. and S.P.; writing—review and editing, M.P. and S.P.; visualization, M.P. and D.Q.T.; project administration, M.P.; funding acquisition, S.P. All authors have read and agreed to the published version of the manuscript.

Funding: This research was supported by a grant (2019-MOIS31-011) from the Fundamental Technology Development Program for Extreme Disaster Response funded by the Ministry of Interior and Safety (MOIS, Korea).

Data Availability Statement: Data is contained within the article and reference.

Conflicts of Interest: The authors declare no conflict of interest.

References

1. Goss, M.; Swain, D.L.; Abatzoglou, J.T.; Sarhadi, A.; Kolden, C.A.; Williams, A.P.; Diffenbaugh, N.S. Climate change is increasing the likelihood of extreme autumn wildfire conditions across California. *Environ. Res. Lett.* **2020**, *15*, 094016. [CrossRef]
2. Guggenheim, D. *An Inconvenient Truth*; Hollywood Paramount Home Entertainment: Hollywood, CA, USA, 2006.
3. Roldán-Gómez, J.J.; González-Gironda, E.; Barrientos, A. A survey on robotic technologies for forest firefighting: Applying drone swarms to improve firefighters' efficiency and safety. *Appl. Sci.* **2021**, *11*, 363. [CrossRef]
4. Chaudhuri, N.; Bose, I. Exploring the role of deep neural networks for post-disaster decision support. *Decis. Support Syst.* **2020**, *130*, 113234. [CrossRef]
5. Muhammad, K.; Ahmad, J.; Baik, S.W. Early fire detection using convolutional neural networks during surveillance for effective disaster management. *Neurocomputing* **2017**, *288*, 30–42. [CrossRef]
6. Akhloufi, M.A.; Castro, N.A.; Couturier, A. UAVs for wildland fires. In Proceedings of the SPIE 10643, Autonomous Systems: Sensors, Vehicles, Security, and the Internet of Everything. International Society for Optics and Photonics, Orlando, FL, USA, 15–19 April 2018; Volume 10643.
7. Wang, Y.; Dang, L.; Ren, J. Forest fire image recognition based on convolutional neural network. *J. Algorithms Comput. Technol.* **2019**, *13*. [CrossRef]
8. Gong, T.; Liu, B.; Chu, Q.; Yu, N. Using multi-label classification to improve object detection. *Neurocomputing* **2019**, *370*, 174–185. [CrossRef]
9. Yan, Z.; Liu, W.; Wen, S.; Yang, Y. Multi-label image classification by feature attention network. *IEEE Access* **2019**, *7*, 98005–98013. [CrossRef]
10. Hanashima, M.; Sato, R.; Usuda, Y. The standardized disaster-information products for disaster management: Concept and formulation. *J. Disaster Res.* **2017**, *12*, 1015–1027. [CrossRef]
11. Kwak, J.; Bhang, K.; Kim, M.-I. Developing a decision making support information checklist based on analyses of two large-scale forest fire cases. *Crisis Emerg. Manag. Theory Praxis* **2020**, *16*, 21–30. [CrossRef]
12. Li, T.; Zhao, E.; Zhang, J.; Hu, C. Detection of wildfire smoke images based on a densely dilated convolutional network. *Electronics* **2019**, *8*, 1131. [CrossRef]
13. Namozov, A.; Cho, Y.I. An efficient deep learning algorithm for fire and smoke detection with limited data. *Adv. Electr. Comput. Eng.* **2018**, *18*, 121–129. [CrossRef]
14. Taylor, M.E.; Stone, P. Transfer learning for reinforcement learning domains: A survey. *J. Mach. Learn. Res.* **2009**, *10*, 1633–1685.
15. Wybo, J.-L. FMIS: A decision support system for forest fire prevention and fighting. *IEEE Trans. Eng. Manag.* **1998**, *45*, 127–131. [CrossRef]
16. Simonyan, K.; Zisserman, A. Very deep convolutional networks for large-scale image recognition. *arXiv* **2014**, arXiv:1409.1556.
17. He, K.; Zhang, X.; Ren, S.; Sun, J. Deep Residual Learning for Image Recognition. In Proceedings of the 2016 IEEE Conference on Computer Vision and Pattern Recognition (CVPR), Las Vegas, NV, USA, 27–30 June 2016; pp. 770–778.
18. Huang, G.; Liu, Z.; Van der Maaten, L.; Weinberger, K.Q. Densely Connected Convolutional Networks. In Proceedings of the IEEE Conference on Pattern Recognition and Computer Vision 2017, Honolulu, HI, USA, 21–26 July 2017; pp. 4700–4708.
19. Santiago, J.S.S.; Manuela Jr, W.S.; Tan, M.L.L.; Sañez, S.K.; Tong, A.Z.U. Of timelines and timeliness: Lessons from typhoon haiyan in early disaster response. *Disasters* **2016**, *16*, 644–667. [CrossRef]
20. Jung, D.; Tuan, V.T.; Tran, D.Q.; Park, M.; Park, S. Conceptual framework of an intelligent decision support system for smart city disaster management. *Appl. Sci.* **2020**, *10*, 666. [CrossRef]
21. Jain, P.; Coogan, S.C.P.; Subramanian, S.G.; Crowley, M.; Taylor, S.; Flannigan, M.D. A review of machine learning applications in wildfire science and management. *Environ. Rev.* **2020**, *28*, 478–505. [CrossRef]
22. Barmpoutis, P.; Papaioannou, P.; Dimitropoulos, K.; Grammalidis, N. A review on early forest fire detection systems using optical remote sensing. *Sensors* **2020**, *20*, 6442. [CrossRef]
23. Aslan, Y. A Framework for the Use of Wireless Sensor Networks in the Forest Fire Detection and Monitoring. Master's Thesis, Department of Computer Engineering, The Institute of Engineering and Science Bilkent University, Ankara, Turkey, 2010.
24. Go, B.-C. IOT technology for forest fire disaster monitoring. *Broadcast. Media Mag.* **2015**, *20*, 91–98.

25. Christensen, B.R. Use of UAV or remotely piloted aircraft and forward-looking infrared in forest, rural and wildland fire management: Evaluation using simple economic analysis. *N. Z. J. For. Sci.* **2015**, *45*, 16. [CrossRef]

26. Chi, R.; Lu, Z.M.; Ji, Q.G. Real-time multi-feature based fire flame detection in video. *IET Image Process.* **2016**, *11*, 31–37. [CrossRef]

27. Park, M.; Tran, D.Q.; Jung, D.; Park, S. Wildfire-detection method using DenseNet and CycleGAN data augmentation-based remote camera imagery. *Remote Sens.* **2020**, *12*, 3715. [CrossRef]

28. Bedo, M.V.N.; De Oliveira, W.D.; Cazzolato, M.T.; Costa, A.F.; Blanco, G.; Rodrigues, J.F.; Traina, A.J.; Traina, C. Fire detection from social media images by means of instance-based learning. In *Springer International Conference on Enterprise Information Systems*; Springer: Berlin/Heidelberg, Germany, 2015; pp. 23–44.

29. Sharma, J.; Granmo, O.; Goodwin, M.; Fidje, J.T. Deep Convolutional Neural Networks for Fire Detection in Images. In Proceedings of the International Conference on Engineering Applications of Neural Networks, EANN2017, Athens, Greece, 25–27 August 2017.

30. Toulouse, T.; Rossi, L.; Campana, A.; Celik, T.; Akhloufi, M.A. Computer vision for wildfire research: An evolving image dataset for processing and analysis. *Fire Saf. J.* **2017**, *92*, 188–194. [CrossRef]

31. Sousa, M.J.; Moutinho, A.; Almeida, M. Wildfire detection using transfer learning on augmented datasets. *Expert Syst. Appl.* **2020**, *142*, 112975. [CrossRef]

32. Muhammad, K.; Ahmad, J.; Mehmood, I.; Rho, S. Convolutional neural networks based fire detection in surveillance videos. *IEEE Access* **2018**, *6*, 18174–18183. [CrossRef]

33. Mikołajczyk, A.; Grochowski, M. Data Augmentation for Improving Deep Learning in Image Classification Problem. In Proceedings of the International Interdisciplinary PhD Workshop (IIPhDW), Swinoujście, Poland, 9–12 May 2018; pp. 117–122.

34. Ioffe, S.; Szegedy, C. Batch Normalization: Accelerating deep network training by reducing internal covariate shift. *arXiv* **2015**, arXiv:1502.03167.

35. Mihalkova, L.; Mooney, R.J. Transfer learning from minimal target data by mapping across relational domains. In *Proceedings of the 21st International Jont Conference on Artifical Intelligence (IJCAI'09)*; Morgan Kaufmann Publishers Inc.: San Francisco, CA, USA, 2009; pp. 1163–1168.

36. Deng, J.; Dong, W.; Socher, R.; Li, L.J.; Li, K.; Fei-Fei, L. Imagenet: A Large-Scale Hierarchical Image Database. In Proceedings of the 2009 IEEE Conference on Computer Vision and Pattern Recognition, Miami, FL, USA, 20–25 June 2009; pp. 248–255.

37. Srivastava, N.; Hinton, G.E.; Krizhevsky, A.; Sutskever, I.; Salakhutdinov, R. Dropout: A simple way to prevent neural networks from overfitting. *J. Mach. Learn. Res.* **2014**, *15*, 1929–1958.

38. Li, X.; Chen, S.; Hu, X.; Yang, J. Understanding the Disharmony between Dropout and Batch Normalization by Variance Shift. In Proceedings of the IEEE Conference on Computer Vision and Pattern Recognition, Long Beach, CA, USA, 15–20 June 2019; pp. 2682–2690.

39. Pereira, R.B.; Plastino, A.; Zadrozny, B.; Merschmann, L.H. Correlation analysis of performance measures for multi-label classification. *Inf. Process. Manag.* **2018**, *54*, 359–369. [CrossRef]

40. Zhang, M.L.; Zhou, Z.H. A review on multi-label learning algorithms. *IEEE Trans. Knowl. Data Eng.* **2014**, *26*, 1819–1837. [CrossRef]

41. Zhu, F.; Li, H.; Ouyang, W.; Yu, N.; Wang, X. Learning Spatial Regularization with Image-Level Supervisions for Multi-Label Image Classification. In Proceedings of the IEEE Conference on Computer Vision and Pattern Recognition, Honolulu, HI, USA, 21–26 July 2017; pp. 5513–5522.

42. Schapire, R.E.; Singer, Y. Boostexter: A boosting-based system for text categorization. *Mach. Learn.* **2000**, *39*, 135–168. [CrossRef]

43. Zhou, B.; Khosla, A.; Lapedriza, A.; Oliva, A.; Torralba, A. Learning Deep Features for Discriminative Localization. In Proceedings of the IEEE Conference on Computer Vision and Pattern Recognition, Las Vegas, NV, USA, 27–30 June 2016; pp. 2921–2929.

44. Selvaraju, R.R.; Cogswell, M.; Das, A.; Vedantam, R.; Parikh, D.; Batra, D. Grad-Cam: Visual Explanations from Deep Networks via Gradient-Based Localization. In Proceedings of the 2017 IEEE International Conference on Computer Vision (ICCV), Venice, Italy, 22–29 October 2017; pp. 618–626.

45. Paszke, A.; Gross, S.; Massa, F.; Lerer, A.; Bradbury, J.; Chanan, G.; Killeen, T.; Lin, Z.; Gimelshein, N.; Antiga, L. PyTorch: An imperative Style, High-Performance Deep Learning Library. In Proceedings of the Advances in Neural Information Processing Systems, Vancouver, BC, Canada, 8–14 December 2019; pp. 8024–8035.

46. Zhou, H.; Sattler, T.; Jacobs, D.W. Evaluating local features for day-night matching. In *Springer European Conference on Computer Vision*; Springer: Berlin/Heidelberg, Germany, 2016; pp. 724–736.

47. Jeong, C.; Jang, S.-E.; Na, S.; Kim, J. Korean tourist spot multi-modal dataset for deep learning applications. *Data* **2019**, *4*, 139. [CrossRef]

48. Sabottke, C.F.; Spieler, B.M. The effect of image resolution on deep learning in radiography. *Radiol. Artif. Intell.* **2020**, *2*, e190015. [CrossRef] [PubMed]

49. Zhou, Z.; Shi, Y.; Gao, Z. Wildfire smoke detection based on local extremal region segmentation and surveillance. *Fire Saf. J.* **2016**, *85*, 50–58. [CrossRef]

Article

A Burned Area Mapping Algorithm for Sentinel-2 Data Based on Approximate Reasoning and Region Growing

Matteo Sali [1], Erika Piaser [1], Mirco Boschetti [1], Pietro Alessandro Brivio [1], Giovanna Sona [2], Gloria Bordogna [1] and Daniela Stroppiana [1,*]

[1] Institute for Electromagnetic Sensing of the Environment (IREA), Consiglio Nazionale delle Ricerche, 20133 Milan, Italy; sali.m@irea.cnr.it (M.S.); piaser.e@irea.cnr.it (E.P.); boschetti.m@irea.cnr.it (M.B.); brivio.pa@irea.cnr.it (P.A.B.); bordogna.g@irea.cnr.it (G.B.)

[2] Department of Civil and Environmental Engineering (DICA), Politecnico di Milano, 20133 Milan, Italy; giovanna.sona@polimi.it

* Correspondence: stroppiana.d@irea.cnr.it; Tel.: +39-02-23699-288

Abstract: Sentinel-2 (S2) multi-spectral instrument (MSI) images are used in an automated approach built on fuzzy set theory and a region growing (RG) algorithm to identify areas affected by fires in Mediterranean regions. S2 spectral bands and their post- and pre-fire date ($\Delta_{post\text{-}pre}$) difference are interpreted as evidence of burn through soft constraints of membership functions defined from statistics of burned/unburned training regions; evidence of burn brought by the S2 spectral bands (partial evidence) is integrated using ordered weighted averaging (OWA) operators that provide synthetic score layers of likelihood of burn (global evidence of burn) that are combined in an RG algorithm. The algorithm is defined over a training site located in Italy, Vesuvius National Park, where membership functions are defined and OWA and RG algorithms are first tested. Over this site, validation is carried out by comparison with reference fire perimeters derived from supervised classification of very high-resolution (VHR) PlanetScope images leading to more than satisfactory results with Dice coefficient > 0.84, commission error < 0.22 and omission error < 0.15. The algorithm is tested for exportability over five sites in Portugal (1), Spain (2) and Greece (2) to evaluate the performance by comparison with fire reference perimeters derived from the Copernicus Emergency Management Service (EMS) database. In these sites, we estimate commission error < 0.15, omission error < 0.1 and Dice coefficient > 0.9 with accuracy in some cases greater than values obtained in the training site. Regression analysis confirmed the satisfactory accuracy levels achieved over all sites. The algorithm proposed offers the advantages of being least dependent on a priori/supervised selection for input bands (by building on the integration of redundant partial burn evidence) and for criteria/threshold to obtain segmentation into burned/unburned areas.

Keywords: Mediterranean ecosystems; convergence of evidence; accuracy assessment

Citation: Sali, M.; Piaser, E.; Boschetti, M.; Brivio, P.A.; Sona, G.; Bordogna, G.; Stroppiana, D. A Burned Area Mapping Algorithm for Sentinel-2 Data Based on Approximate Reasoning and Region Growing. *Remote Sens.* **2021**, *13*, 2214. https://doi.org/10.3390/rs13112214

Academic Editors: Leonor Calvo, Elena Marcos, Susana Suarez-Seoane and Víctor Fernández-García

Received: 19 April 2021
Accepted: 1 June 2021
Published: 5 June 2021

1. Introduction

Wildfires are the largest contributor to global biomass burning (BB) and represent a significant dynamic component of ecosystems, affecting terrestrial and atmosphere systems [1,2]. In vegetated areas of Southern Europe, fire is a major damaging agent and recent years (2017–2018) have witnessed unprecedented fire seasons with countries suffering large forest fires as a consequence of drought and heatwaves. Global warming has been affecting fires with increased frequency and severity, as observed in both real and simulated data [3,4]; this effect is particularly true in Mediterranean ecosystems (object of this work) where, according to models' forecasting, warming and a precipitation deficit will exacerbate fire weather conditions [5].

Fires impact on atmospheric chemistry, with aerosols and greenhouse gas emissions [6], the carbon budgets [7], hydrological cycles, soils and vegetation components of ecosystems [8,9]. In this framework, the extent of the area affected by fires is critical to investigate trends and patterns of fire occurrence and identify drivers of fire occurrence as well as for modeling future fire patterns and fire regimes; this information can therefore support the assessments of fire impacts on both natural and social systems. Several studies can be mentioned on the use of remote sensing (RS) to map burned areas at a regional to global scale [10–12]. Coarse-resolution RS data have been proved to be the most suitable source for depicting fire distribution over large areas and one primary image source for burned area (BA) products is the Moderate Resolution Imaging Spectrometer (MODIS) [13]. To the aim of improving accuracy of global estimates by detecting smaller fires (<50 ha), higher-resolution sensors have been exploited, such as 300 m Medium Resolution Imaging Spectrometer (MERIS) and 100 m Project for On-Board Autonomy–Vegetation (PROBA-V) [10,14,15], although the results were not always more accurate compared to coarser spatial resolution products [16].

At a regional scale and in heterogeneous environments, such as the Mediterranean ecosystems of Southern Europe, coarse-resolution data do not provide enough spatial detail and medium/high- and very high-resolution satellite images are preferable for accurately mapping burned areas. Indeed, small fires could significantly contribute to global effects of fires [15,17]. The greatest challenge in the RS community is the development of a global algorithm for mapping burned areas from decametric satellite images (e.g., Sentinel-2 and Landsat missions) [18–20], although important steps forward are shown by recent works [21,22]. In order to achieve this objective, some issues are yet to be addressed; among them a significant variability, across the ecosystems, of burned area spectral response as a function of pre-fire vegetation conditions and characteristics, fire behavior and intensity as well as time since the fire event. The lower revisiting time of medium spatial resolution satellites reduces the likelihood of observing burned surfaces immediately after the fire when spectral separability is greatest; yet Sentinel-2 (S2) revisiting time, obtained with the combined use of A&B missions, offers unprecedent opportunity with an average revisiting time of about five days [23].

Several algorithms have been proposed for mapping burned areas in the diverse ecosystems of the globe [12] and some key elements can be pointed out as providing the most robust approaches: supervised self-adaptive algorithms that can fit local conditions and contextual, multi-source (combining active fires) and multi-temporal approaches that can reduce the likelihood of commission errors due to spectral confusion with low albedo surfaces and highlight changes induced by fire on the surface [16,24,25].

The algorithm proposed here exploits the abovementioned key elements to deliver a semi-automatic robust and self-adaptive classification algorithm for S2 imagery exploiting pre-fire and post-fire acquisitions to maximize mapping accuracy. The algorithm inherits from the conceptual framework of the multi-criteria soft aggregation approach of burn evidence proposed by Stroppiana et al. (2012) [26] for burned area mapping and also applied for flooding mapping [27].

Major improvements with respect to the previous algorithm characteristics are (i) the use of S2 band reflectance in post-fire images and of their temporal difference between post- and pre-fire acquisitions and (ii) the definition of soft constraints by membership functions of fuzzy sets based on statistics (percentiles) of reflectance as derived from training areas. These improvements build on (i) the exploitation of a greater frequency of acquisition of S2 data (nominal five days when A and B constellations are combined), that allows the implementation of a robust change detection approach and (ii) a more automatic way for defining membership functions based on frequency distribution of training pixels over burned and unburned surfaces [19]. In particular, the algorithm aggregates partial evidence of burn, extracted from the information provided by S2 bands through the membership functions, into a synthetic score of global evidence by means of an ordered weighted averaging (OWA) operator [28]. Each S2 band could potentially and independently be used as a source of evidence of burn for the identification of areas affected by fires, hereafter named 'partial evidence' since it is given by a single input feature. However, the concurrent aggregation of multiple spectral bands can provide a more reliable evaluation of the occurrence of fires by modeling a convergence of evidence provided by redundant information, hereafter named 'global evidence' since it is given by multiple input features. This multi-criteria soft aggregation computes distinct pixel-based global evidence obtained with different OWAs (e.g., ranging from extreme conditions of minimum and maximum operators) that are exploited as input to a region growing (RG) algorithm.

The algorithm is trained over a large fire occurred in the Vesuvius National Park, Italy; this site is characterized by a complex and fragmented forest ecosystem and, in 2017, was affected by fires that generated different degrees of severity, providing a wide range of burn spectral conditions. After the training phase, the algorithm was automatically applied (with no need of repeating training) to five sites located in Southern Europe to assess exportability (i.e., robustness of membership functions, seed and growing layer selection strategy and map accuracy with respect to local characteristics). Copernicus Emergency Management Service (EMS) products (https://emergency.copernicus.eu/, last access 1 May 2021) from major events in the 2017 summer fire season were used as reference data for assessing the accuracy of burned area mapping over these sites.

The major novelty of this work with respect to our previous work was to assess the robustness and exportability of the multi-criteria soft aggregation algorithm developed for post-fire Landsat data to multi-temporal S2 data. Indeed, a specific novel aspect was the full exploitation of the temporal component as information for burned area mapping together with improvement of automatization of the algorithm to reduce the dependence on expert knowledge in the definition of the membership functions.

2. Study Areas and Datasets

In Southern Europe, 2017 was characterized by abnormal droughts and heatwaves [29]. Summer was the second warmest on record, with temperatures over 1.7 °C above the 1981–2010 average; the warmest being 2003 at more than 2.0 °C above average (https://climate.copernicus.eu/node/358, last access 1 May 2021). Extreme weather conditions led to severe fires affecting, in particular, Portugal, Spain, Southern France, Greece and Italy [30]. In this framework, we selected sites for algorithm training and testing that are described in the sections below.

2.1. Training and Exportability Sites

Six sites situated in Mediterranean ecosystems were selected in this work among the regions most affected by forest fires in 2017 (Figure 1). Vesuvius National Park, Italy, was used for algorithm training (development, tuning and thematic accuracy assessment) and the other five sites (Spain, Greece and Portugal) were exploited for testing algorithm exportability. The Corine Land Cover map (CLC2012, https://land.copernicus.eu/pan-european/corine-land-cover, last access 1 May 2021) was used to extract information on major land covers that are summarized in Table 1. Most of the sites are mainly covered by natural vegetation (forest and shrub/grasslands), with the exception of Kalamos and Zakynthos where croplands are predominant, covering approximately 40% and 52%, respectively; in the Vesuvius site, forest and croplands cover similar proportions (38.7% and 36.5%). Table 1 also shows the proportion of area burned within fire polygons of the Copernicus EMS dataset among the land cover classes: in all sites, fires affected mainly forested areas, except for Kalamos, Greece, where fires affected mostly shrub/grassland (~36%).

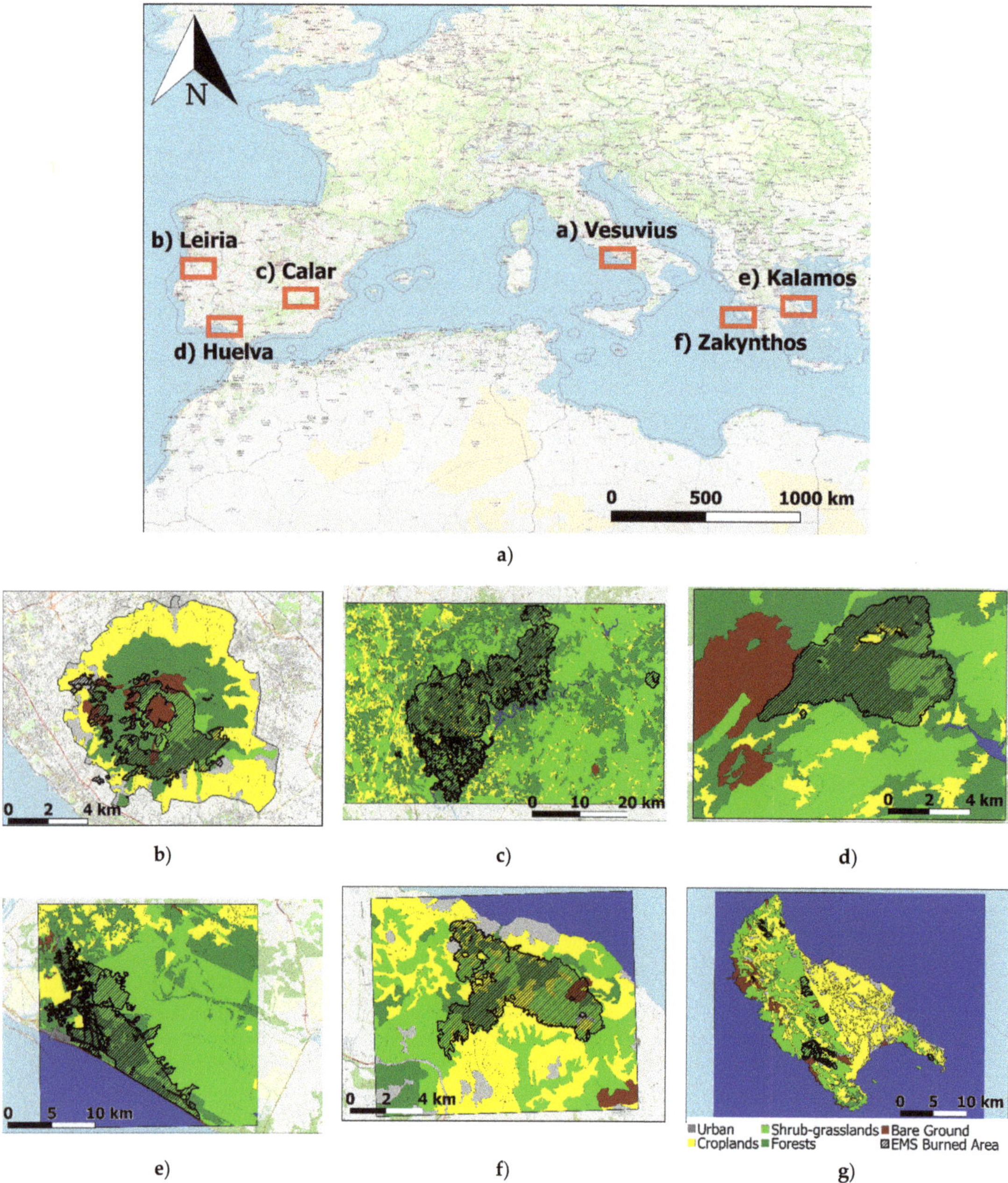

Figure 1. The CLC2012 land cover classes over the six sites located in Southern Europe (**a**): Vesuvius, Italy (**b**), Leiria, Portugal (**c**), Calar, Spain (**d**), Huelva, Spain (**e**), Kalamos, Greece (**f**), Zakynthos, Greece (**g**).

Table 1. CLC2012 land cover classes over the entire site and within fire perimeter identified by EMS. In the last column, the proportions of fire damage levels in the burned area according to EMS fire grading products (CD = Completely Destroyed, HD = Highly Damaged, MD = Moderately Damaged and ND = Negligible to Slightly Damaged), where available.

		CLC2012 Class (%)					EMS Fire Damage			
		Bare	Crops	Forest	Shrub	Urban	CD	HD	MD	ND
Vesuvius	Site	6.6	38.7	36.5	10.4	7.78	0.0	12.8	3.7	2.3
Italy	BA	3.4	4.8	47.2	34.4	10.26	0.0	68.2	19.7	12.0
Leiria	Site	0.2	15.4	43.1	40.2	1.10	0.0	8.9	5.6	2.3
Portugal	BA	-	5.6	66.7	27.5	0.17	0.0	53.0	33.1	13.8
Calar	Site	13.6	8.2	61.2	17.0	-	18.9	16.5	3.6	2.0
Spain	BA	0.3	2.4	76.4	20.9	-	46.2	40.2	8.7	4.9
Huelva	Site	0.8	11.9	39.7	46.6	0.89	0.0	15.4	1.1	0.0
Spain	BA	0.2	0.1	61.2	37.7	0.71	0.0	93.1	6.7	0.2
Zakynthos	Site	5.1	52.5	32.5	3.7	6.17	-	3.2	-	-
Greece	BA	2.3	14.6	81.8	1.3	0.02	-	100	-	-
Kalamos	Site	2.0	40.6	24.2	24.7	8.36	-	20.0	-	-
Greece	BA	3.3	25.1	33.9	36.3	1.37	-	100	-	-

2.2. Sentinel-2 Dataset

The remote sensing data source used for algorithm training and exportability tests was the Sentinel-2 (S2) multispectral instrument (MSI) which measures the Earth's reflected radiance in 13 spectral bands from VIS/NIR to SWIR with a spatial resolution ranging from 10 m to 60 m (https://earth.esa.int/web/sentinel/home, last access 1 May 2021).

Over each site, pre-fire and post-fire S2 images were selected and downloaded by considering the occurrence of the major fire events, the dates of available reference datasets and the most clear sky conditions (Figure 2). Since post-fire S2 images simultaneous with reference data were desirable but hardly feasible, post-fire S2 dates were selected as close as possible to the date of reference fire perimeters (Table 2) to minimize bias in accuracy metrics due to spectral signal changes and further burning occurring after the date of reference perimeters. S2 images were downloaded as Level 1C from Copernicus Open Access Hub (https://scihub.copernicus.eu/, accessed 1 May 2021) since at the time of data processing no Level 2A products were available. S2 images were processed with Sen2r [31] Toolbox developed in R and released under GNU General Public License version 3 (GPL-3) and available on GitHub (https://ranghetti.github.io/sen2r, accessed 1 May 2021): Level-1C products were corrected with Sen2Cor [32] to derive bottom of atmosphere (BOA) reflectance in the VIS–NIR–SWIR wavelengths (S2 bands 2 to 12). In pre-processing steps, pixels with high and medium cloud probability were masked out while low cloud probability and cloud shadow pixels were retained to avoid discarding of burned pixels. In Sen2r, masking of cloudy pixels was done with information from the scene classification map (SCL) [33].

Table 2. Pre- and post-fire S2 dates over the six sites, reference date and source for the EMS products.

Study Site	Pre-Fire S2	Post-Fire S2	EMS Date	EMS Source (https://emergency.copernicus.eu/mapping/list-of-activations-rapid, access 1 May 2021)
Vesuvius—Italy	08/04	22/07	16/07	EMSR213
Leiria—Portugal	04/06	04/07	20/06	EMSR207
Calar—Spain	15/07	04/08	04/08	EMSR216
Huelva—Spain	11/06	01/07	27/06	EMSR209
Zakynthos—Greece	25/07	03/09	18/08	EMSR224
Kalamos—Greece	28/07	17/08	18/08	EMSR224

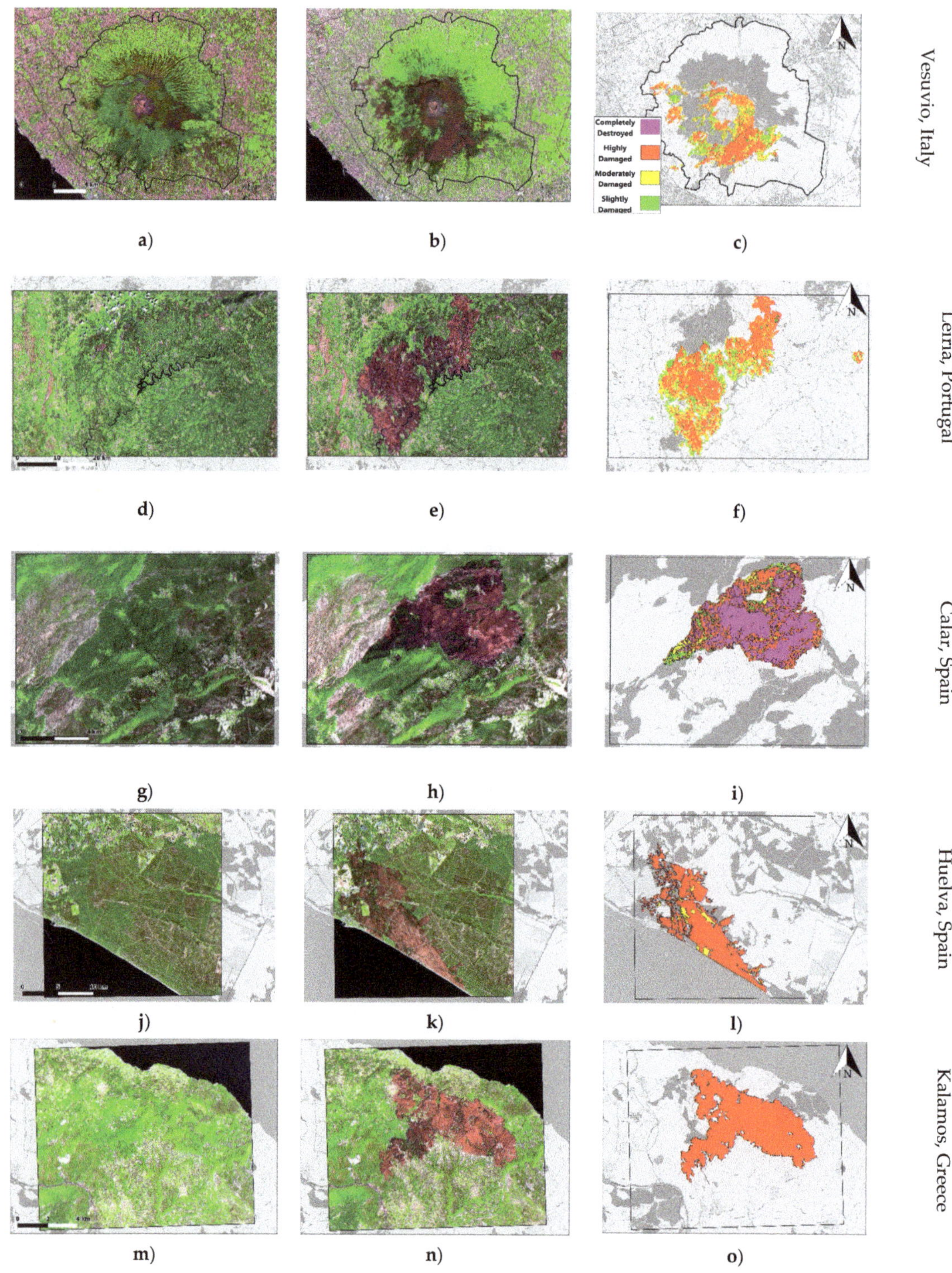

a)　　b)　　c)

d)　　e)　　f)

g)　　h)　　i)

j)　　k)　　l)

m)　　n)　　o)

Figure 2. *Cont.*

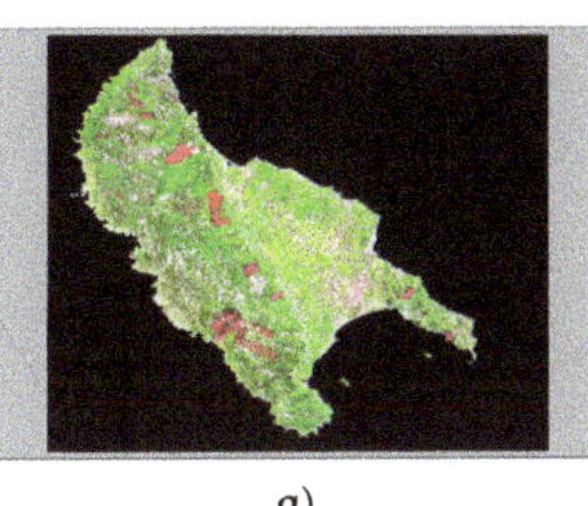
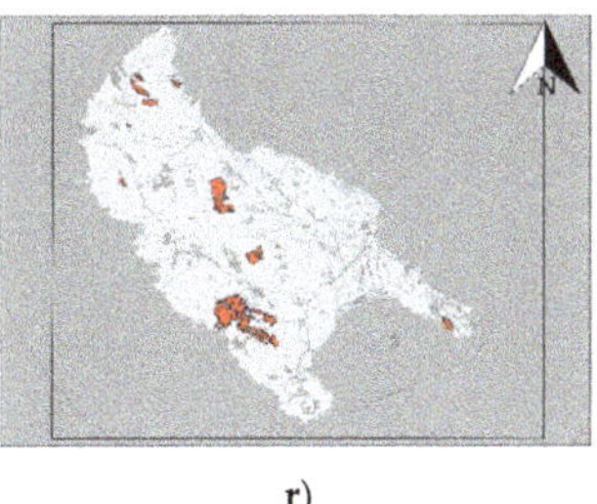

p) q) r)

Zakynthos, Greece

Figure 2. Pre- and post-fire S2 images (first and second column) and EMS fire grading maps (last column) for each site: Vesuvius, Italy, (**a–c**); Leiria, Portugal, (**d–f**); Calar, Spain (**g–i**); Huelva, Spain (**j–l**); Kalamos, Greece (**m–o**); Zakynthos, Greece (**p–r**). S2 images are displayed as RGB false color composites (SWIR2, NIR, red). Notice that for Kalamos and Zakynthos sites, Greece, no fire damage grading maps are available from EMS.

Output pre-processed S2 images for the selected dates were available as bottom of atmosphere (BOA) reflectance values in VIS–NIR–SWIR wavelengths (S2 bands 2 to 12) resampled to a 10 m spatial resolution with a nearest neighbor method. The temporal difference between post- and pre-fire ($\Delta_{post-pre}$) reflectance was computed and, together with post-fire reflectance, band values were used as input to a separability analysis to identify features that were most suited for burned area identification. Hereafter, Δ is always meant as post-pre reflectance difference.

Training data were collected over the Vesuvius site, Italy, by photointerpretation of false color RGB (SWIR–NIR–Red) composites of S2 images (Figure 2): polygons over burned and unburned surfaces were extracted and labeled by considering as 'burned' only areas that were affected by fires between the two S2 dates.

2.3. Reference Fire Perimeters

Burned area maps output from the proposed algorithm were validated by comparison with fire reference perimeters obtained from independent source data. In all sites, major fires occurred during the 2017 summer season, affecting, to a large extent, ecosystems, houses and people, so the Copernicus Emergency Management Service (EMS) was activated. EMS consists of the on-demand and fast provision (hours–days) of geospatial information in support of emergency management activities and derived from processing and analysis of satellite imagery acquired immediately after natural or human-made disasters such as floods, droughts and forest fires. EMS products are delivered as ready-to-print maps and geographic datasets (vector package). Two types of geo-products are delivered: fire delineation (fire perimeter) and fire damage grading (burn severity) derived from very high-resolution multispectral images. Fire damage grading is provided in four classes: "Completely Destroyed", "Highly Damaged", "Moderately Damaged" and "Negligible to Slightly Damaged". Quality checks are performed by the European Commission Joint Research Centre (EC JRC) to assure fast delivery of high-quality products. Further validation activities can be carried out by the EC JRC if triggered by the European Commission and/or suggested by authorized users; further information can be found in the 'Online Manual for Rapid Mapping Products' (https://emergency.copernicus.eu/mapping/ems/online-manual-rapid-mapping-products, last access 1 May 2021).

Table 2 summarizes the reference dates of the EMS delineation and grading map source (where available). EMS delineation maps were used as source datasets for fire reference perimeters for all sites, except Vesuvius where EMS maps depicted the status of the surface on 16 July 2017. Yet, ongoing fires made the time gap between the EMS and S2 post-fire date (22 July 2019) critical for the comparison and for the estimation of reliable accuracy metrics (Appendix A, Figure A1). Hence, to generate a reference dataset suitable for validation, very high-resolution (VHR) PlanetScope images [34] were used.

PlanetScope is a constellation composed of more than 120 optical CubeSat 3U satellites carrying a multi-spectral sensor with four bands: three in the visible wavelengths (b1: 455–515 nm; b2: 500–590 nm; b3: 590–670 nm) and one in the NIR wavelengths (b4: 780–860 nm) (https://earth.esa.int/web/guest/missions/3rd-party-missions/current-missions/planetscope, accessed 1 May 2021). PlanetScope has a swath of about 25 Km and a spatial resolution of 3 m for all bands [35]. Imagery is captured as a continuous strip of single frame images known as 'scenes'.

Pre- and post-fire PlanetScope images (22 April and 22 July 2017) were downloaded and classified with a supervised random forest (RF) algorithm to extract fire perimeters (Figure 3). The RF classifier is a machine learning algorithm that is largely used in remote sensing [36,37]. It builds an ensemble of decision trees (CART) and merges them together to yield a more accurate and stable prediction than any single tree alone [38]. The map generated at high resolution (~3 m) can be considered spatial explicit ground truth information for the assessment of S2 products (namely 10 m resolution).

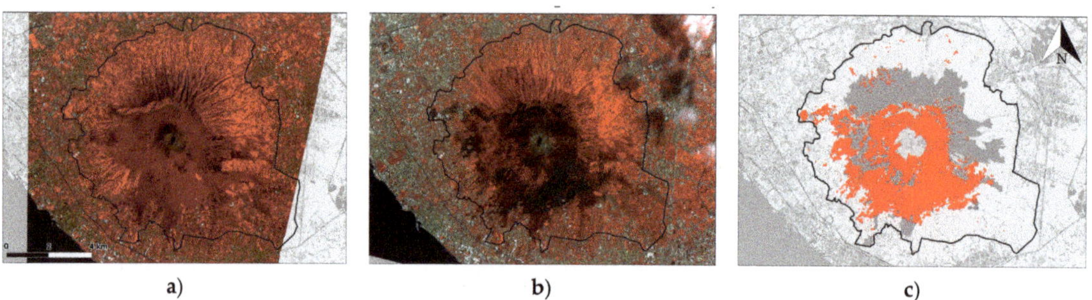

a)   b)   c)

Figure 3. Pre- (**a**) and post-fire (**b**) PlanetScope images: 22 April and 22 July 2017, over the Vesuvius sites. Images are displayed as RGB false color composites (NIR, red, green) and the reference burned area map (**c**) obtained with RF algorithm. Black polygon shows the border of Vesuvius National Park.

3. Methods

The proposed algorithm relies on a multi-criteria approximate reasoning approach that aggregates information brought by multiple features into synthetic global degrees of evidence of burn. Each input feature could be used as source for deriving evidence of burn conditions (partial evidence) but aggregation reinforces evidence by exploiting the convergence of partial evidence from multiple possibly redundant sources and by compensating the inconsistency/conflict of evidence from multiple possibly complementary sources. This step allows strengthening of the likelihood of the presence of burn and reduces confusion between burned areas and surfaces with similar spectral characteristics [26]. Aggregation was carried out with ordered weighted averaging operators (OWAs): a parameterized family of soft-mean-like aggregation operators. Different operators were used to represent attitudes ranging between pessimistic (the maximum extent of the phenomenon to minimize the chance of underestimating: modeling a compensative aggregation to integrate complementarities of multiple criteria) and optimistic (minimize the chance of overestimating: modeling a concurrent aggregation to integrate mutual reinforcement of multiple criteria). Layers of global evidence derived with different OWAs were input to a region growing (RG) algorithm. The approach was applied independently to each pixel of the input RS images as follows:

1. Selection of the input features;
2. Definition of the soft constraints (membership function, MF) for each input feature from training data and application to derive partial evidence of burn (MD);
3. Selection of OWAs, according to their semantic, for the soft integration;
4. Computation of the global degree of evidence of burn for generating seed and growing layers;

5. Implementation of the RG algorithm;
6. Segmentation of the RG output score to derive burned area maps.

The algorithm was applied to vegetated areas, while not burnable (bare soil and urban classes) and agricultural areas were masked out based on the CLC2012 land cover map. Output burned area maps were compared to reference datasets for estimating accuracy metrics. Steps (1) to (6) were applied over the training site (Vesuvius, Italy) for the selection of the best input features, the definition of the membership functions and the customization of the RG algorithm. Exportability was tested over the other sites by applying steps (4) to (6) to assess the algorithm's performance over geographic locations with different environmental conditions compared to those of the training site.

3.1. Separability Analysis

A separability analysis was preliminarily carried out over the Vesuvius training site for selecting the most suitable bands for the discrimination of burned and unburned surfaces. Separability metric M (1) was computed from frequency distributions of training pixels [39].

$$M = \left| \frac{\mu_u - \mu_b}{\sigma_u + \sigma_b} \right| \tag{1}$$

where μ_u, μ_b are mean values and σ_u, σ_b are standard deviation values of the unburned (u) and burned (b) classes in the training data. Spectral bands with M > 1 were selected as input features for the computation of the partial evidence of burn.

3.2. Definition of the Membership Functions

Membership functions (MFs) are soft constraints that can be defined with different approaches according to the available expertise and training data [40]. Here, soft constraints were defined from training data over the Vesuvius site for each input feature identified by the separability analysis. MFs are sigmoid Functions (2) used to convert a pixel's values of the input feature into degrees of membership to the burned class (membership degree, MD), i.e., the partial evidence of burn that is a score in the range [0, 1]. The extremes of the sigmoid-shaped MFs were defined based on percentiles of the unburned and burned histogram distributions, respectively [19].

$$f(x) = \frac{L}{1 + e^{-k(x-x0)}} \tag{2}$$

where L is the upper limit of the function, in this case $L{\rightarrow}1$ to quantify the maximum degree of membership, k is the curve's slope and $x0$ is the inflection point. The two parameters k and $x0$ are estimated by using percentiles.

Based on Roteta et al. (2019) [19], we selected a different shape for the sigmoid functions depending on the spectral characteristics of burned areas in the specific input feature: a z-shaped function or s-shaped function. The upper limit of the function $f(x){\rightarrow}1$ represents the greatest partial evidence of burn defined by a pixel's values of the input features below and above the 50th percentile of the frequency distribution function of the burned training pixels for z-shaped and s-shaped functions, respectively (Figure 4). On the opposite side, $f(x){\rightarrow}0$ (no partial evidence of burn) and it is given by the 10th and 90th percentile of the frequency distribution function of the unburned training pixels for the z-shaped and s-shaped function, respectively. Training pixels for burned and unburned categories, extracted over the Vesuvius training site, were used to estimate k and $x0$ parameters: a z-shaped function was used when fire occurrence led to a decrease in the pixel's value in a given input feature, for example, in the NIR S2 reflectance band. On the contrary, an s-shaped function represents the case when per-pixel input feature value increases over a burned area.

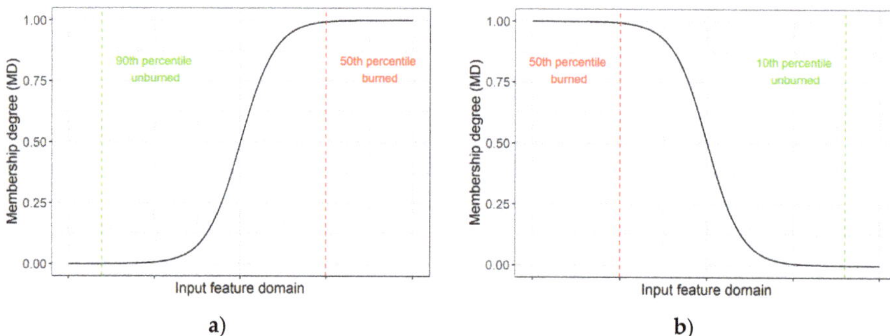

Figure 4. The s-shaped (**a**) and z-shaped (**b**) sigmoid functions chosen as MFs to quantify the partial evidence of burn.

3.3. OWA Operators for Computing Global Evidence

Layers of partial evidence were integrated to derive global scores of burn evidence with ordered weighted averaging (OWA) operators [28]. An OWA of dimension N (OWA: $[0, 1]^N \rightarrow [0, 1]$), where N is the number of input values $[d_1, \ldots , d_N]$ to aggregate, has a weighting vector $W = [w_1, \ldots , w_N]$, with $\sum_{i=1,\ldots,N} w_i = 1$, so that it computes an aggregated output value $a \in [0, 1]$ by applying the following formula [41]:

$$a = OWA([d_1, \ldots ,d_N]) = \sum_{i=1}^{N} w_i * g_i \qquad (3)$$

where g_i is the ith largest value among the input $d_1, \ldots ,d_N$.

Input values $d_1, \ldots ,d_N$, are rearranged from the highest to the smallest; reordering is a key element of OWA operators, meaning that a specific weight w_i is not univocally associated to the specific i^{th} input but rather it is associated with the ith position of the reordered inputs [28]. Since, in our case, OWAs aggregated the MDs expressing partial evidence of burn provided by single features, in each pixel, a different reordering of the MDs may have occurred so different features contributed to determining the aggregated value.

Different operators were tested to represent decision attitudes between pessimistic (the maximum extent of the phenomenon to minimize the chance of underestimating) and optimistic (to minimize the chance of overestimating). For example, a weighting vector W of the OWA operator with the last weight $w_N = 1$ considers only the contribution of the smallest input value, i.e., the minimum partial evidence degree after reordering; hence, it implements an optimistic attitude by computing the minimum total burned area (AND aggregation). Conversely, by setting the first weight $w_1 = 1$, the maximum partial evidence will determine the largest burned area, thus modeling the pessimistic case (OR aggregation). Intermediate cases, in which all or most components of W are not null, model soft integrations.

In this work, we compare the results obtained by applying five different OWA operators with the following weighting vectors:

$W_{AND} = [0, \ldots, 0, 1]$ thus $OWA_{AND}([d_1, \ldots, d_N]) = \min\{d_1, \ldots, d_N\}$

$W_{OR} = [1, \ldots, 0, 0]$ thus $OWA_{OR}([d_1, \ldots, d_N]) = \max\{d_1, \ldots, d_N\}$

$W_{AlmostAND} = [0, \ldots, 0.5, 0.5]$ thus $OWA_{AlmostAND}([d_1, \ldots, d_N]) = \frac{1}{2}\min\{d_1, \ldots, d_N\} + \frac{1}{2}\min\{\{d_1, \ldots, d_N\} - \{\min\{d_1, \ldots, d_N\}\}\}$

$W_{Aaverage} = \left[\frac{1}{N}, \ldots, \frac{1}{N}\right]$ thus $OWA_{Average}([d_1, \ldots, d_N]) = \frac{1}{N} \sum_{j=1}^{N} d_j$

$W_{AlmostOR} = [0.5, 0.5, 0, \ldots, 0]$ thus $OWA_{AlmostOR}([d_1, \ldots, d_N]) = \frac{1}{2}\max\{d_1, \ldots, d_N\} + \frac{1}{2}\max\{\{d_1, \ldots, d_N\} - \{\max\{d_1, \ldots, d_N\}\}\}$

The output of an OWA operator applied to all pixels in an image is a gray-level image whose pixels take values in $[0, 1]$, where each pixel's value is the global evidence of burn:

in Figure 5, this step performs the dimension reduction from N input features (MD scores for each pixel) to 1 (synthetic score). OWA layers are then input to the region growing algorithm.

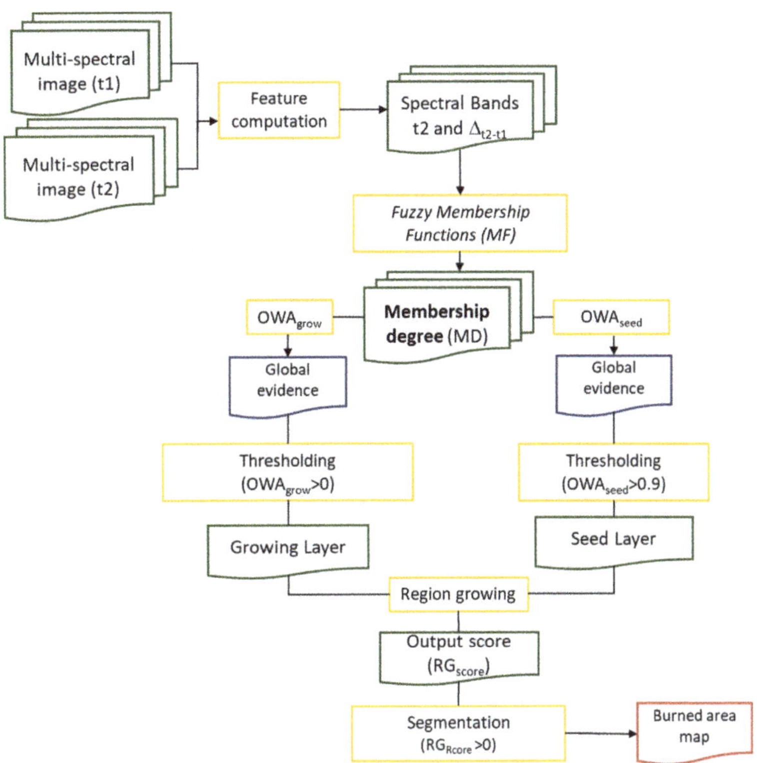

Figure 5. Flowchart of the processing steps from input S2 MSI images to generate BA maps.

3.4. Region Growing

The RG algorithm, implemented in Harris IDL language (https://www.l3harrisgeospatial.com/docs/region_grow.html, last access 1 May 2021), needs as input a seed layer (OWA_{seed}) and a growing layer (OWA_{grow}) and two conditions (thresholds) on OWA_{seed} and OWA_{grow} to identify seed pixels and to delimit maximum growing boundaries. Seed and growing layers are selected among the layers of global evidence generated with distinct OWAs: one concurrent for the seed layer whose segmentation is set to minimize commission errors, and one compensative for the grow layer, whose segmentation is set to minimize omission errors by expanding seeds. Hence, seeds are extracted from the most restrictive OWA_{AND} while growing boundaries are derived from less restrictive OWAs (OWA_{OR}, $OWA_{Average}$, $OWA_{AlmostOR}$). While the requirement on the seed layer is very high, the strategy for identifying candidate boundaries (i.e., the limits for the region growing) can be looser, thus allowing the algorithm to also expand over pixels with low burn signal but connected to more reliable pixels, i.e., the seeds (e.g., partially burned pixels along the perimeter of the burned patches). The RG algorithm is an iterative algorithm that at each iteration expands the seed pixels: starting from initial seeds (pixel with OWA_{AND} above a given threshold), it searches the 8-neighbor connected pixels and it includes in the seed layer only those pixels with $OWA_{grow} > 0$. These expanded pixels update the seeds for the next iteration cycle. The iteration ends when boundaries of maximum growth is reached in the OWA_{grow}

layer. The output layer is a gray-level image ($RG_{score} \in [0, 1]$) whose values can be further segmented to generate the maps of burned/unburned areas (binary maps).

Preliminary analysis carried out over the training site allowed the definition of the implementation conditions of the RG algorithm: thresholds for seed and growing layers as well as a threshold for the segmentation of the RG_{score}. The criterion set for seed selection is $OWA_{AND} > 0.9$; no change in burned area mapping accuracy was observed for different thresholds applied to OWA_{AND} due to its frequency distribution with a bimodal shape centered over extreme values 0 (no evidence of burn) and 1 (full membership to the burned class, greatest evidence of burn) (not shown). For the growing layer, different conditions on the value of OWA_{grow} were analyzed to identify the maximum growing boundary (S.2, S.3), showing that the highest accuracy is achieved when all pixels with not null evidence are retained ($OWA_{grow} > 0$). Finally, in the segmentation step on RG_{score}, the analysis of variable thresholds showed that $RG_{score} > 0$ provides the greatest accuracy (S.2, S.3). As a result of these preliminary analyses, Figure 5 shows the flowchart of the algorithm proposed in this work.

3.5. Validation

The accuracy of the output burned area maps was estimated by comparison with reference fire perimeters (i.e., fire polygons from RF classification of PlanetScope images for the Vesuvius site and Copernicus EMS fire delineation layers). Over the Vesuvius site, accuracy assessment was part of the training phase for the definition of the best implementation criteria of the algorithm. Over the other sites, accuracy assessment of the burned area maps obtained by applying the algorithm in its final form contributed to the evaluation of the exportability. In both cases, validation was carried out by estimating metrics from the confusion matrix (commission error, omission error, Dice coefficient and relative bias) [42]. In order to build the confusion matrix, score map output from the RG algorithm was segmented to extract burned/unburned areas in a binary form ($RG_{score} > 0$, S.3). In the Results section, accuracy metrics are presented and discussed as a function of the OWA_{grow} layer.

4. Results

4.1. Separability and Membership Functions

Results of the separability analysis depict the distance between burned and unburned classes, as observed in the post-fire and $\Delta_{post\text{-}pre}$ S2 reflectance bands for the Vesuvius site (Table 3); $M > 1$ highlights a good separability that is, in this case, achieved by S2 post-fire red–edge (RE2, RE3) and NIR bands and their temporal difference ($\Delta_{post\text{-}pre}$). S2 SWIR2 post-fire reflectance provides very poor separability ($M < 0.1$) that increases when temporal difference (Δ_{SWIR2}) is computed, suggesting that the difference with respect to pre-fire unburned conditions enhances separability; the opposite occurs for the S2 SWIR1 band. Reflectance of burned surfaces is the result of a mixture of bare soil, unburned vegetation and combustion products (ash, charcoal) that are present on the surface after a fire. The combustion of vegetation significantly influences the post-fire spectral signature by generally decreasing reflectance (μ_b), thus enhancing the difference with respect to unburned conditions (μ_u), especially in the NIR wavelengths. For longer wavelength bands (i.e., SWIR2), the spectral reflectance of dry unburned vegetation (green vegetation proportion absorbing radiation due to water content) and burned surfaces could be equally low, thus reducing separability (lower M value). The separability power of SWIR wavebands in the temporal change detection algorithm has also been widely exploited as an indicator of burn severity although the sensitivity of these bands has been found to vary geographically [43]. Red–edge bands show good separability for wavelengths longer than 740 nm (RE2 and RE3), and are certainly of great interest, although these bands are not present on all space-borne sensors and they are mainly selected for vegetation chlorophyll content estimation and monitoring [44].

Table 3. Separability metric M measuring the distance between burned and unburned surface spectral signal in the post-fire and post-pre fire reflectance of the S2 bands. Bold numbers highlight values M > 1 of the selected features.

S2 Band	M Post fFire	M ΔPost Fire-Pre Fire
Green (b3)	0.577	0.027
Red (b4)	0.321	0.454
RE1 (b5)	0.879	0.214
RE2 (b6)	**2.091**	**1.571**
RE3 (b7)	**1.917**	**1.561**
NIR (b8)	**1.812**	**1.530**
SWIR1 (b11)	0.873	0.099
SWIR2 (b12)	0.029	**1.100**

Based on these results, the following seven input layers were selected as features for the implementation of the algorithm (Figure 5): post-fire NIR, post-fire RE2 and RE3 and temporal difference ($\Delta_{\text{post-pre}}$) of the same three bands and additionally of SWIR2. From the same training dataset, statistics for burned and unburned surfaces were computed for the estimation of the k and $x0$ parameters of the membership functions (Table 4). MFs map the input feature's values into the [0, 1] domain where values closer to 1 (0) represent the greatest (lowest) likelihood of being burned. Among the selected features, only Δ_{SWIR2} was properly described by an s-shaped function since, according the training dataset, over burned areas, $\Delta_{\text{SWIR2}} > 0$.

Table 4. Percentiles of the frequency distribution functions extracted from the training pixels of the study area and used for defining MFs.

S2 Band	Burned			Unburned			MF Parameters	
	10%	50%	90%	10%	50%	90%	k	x_0
Post$_{\text{RE2}}$	0.058	0.074	0.102	0.147	0.220	0.286	−125.89	0.111
Post$_{\text{RE3}}$	0.061	0.077	0.112	0.156	0.249	0.339	−115.77	0.116
Post$_{\text{NIR}}$	0.054	0.073	0.115	0.147	0.264	0.370	−123.66	0.109
Δ_{RE2}	−0.126	−0.098	−0.063	−0.021	0.012	0.088	−120.29	−0.06
Δ_{RE3}	−0.158	−0.124	−0.075	−0.026	0.012	0.108	−93.721	−0.075
Δ_{NIR}	−0.180	−0.139	−0.085	−0.034	0.011	0.111	−87.14	−0.086
Δ_{SWIR2}	0.025	0.063	0.114	−0.030	0.0084	0.024	236.98	0.044

4.2. Partial and Global Evidence of Burn

MFs are applied to the input features to derive maps of partial evidence of burn (MD = membership degree score). Figure 6 shows the partial evidence of burn for the training site: greater MD values represent higher likelihood of burn (from blue to yellow in the figure) and the frequency distribution of pixel values varies with the input feature according to its sensitivity in detecting different burned conditions. Areas located in the southernmost regions of the park that were severely affected by fires during summer 2017 are consistently identified by all features with the greatest values (yellow regions). Differences in the partial evidence of burn brought by single features are mainly observed in the northern regions; in fact, each feature is sensitive to different characteristics of the burned surfaces and/or different degrees of burn. In the figure, non-forested areas within the border of the national park are masked out and shown in gray.

Global evidence of burn is computed with OWA operators integrating partial evidence, as shown in Figure 7 for the Vesuvius training site. OWA score ranges within [0, 1], with the greatest values showing pixels with the highest likelihood of being burned according to convergent evidence of burn from the input layers. All OWA maps highlight regions of the Vesuvius site most affected by fires in the southernmost areas, where all input layers agree on identifying higher partial evidence of burn. The 'concurrent–strict' to 'complementary–relaxed' integration conditions implemented by the different OWAs supported the choice of seed and growing layers for the RG algorithm: we selected OWA_{AND} as the seed layer combined with average and OR-like OWAs as the growing layer ($\text{OWA}_{\text{Average}}$, $\text{OWA}_{\text{AlmostOR}}$ and OWA_{OR}). Once seed pixels were selected ($\text{OWA}_{\text{AND}} > 0.9$), the RG algorithm expanded the initial selection in an iterative way over the OWA_{grow} layer in order to also capture pixels with lower values of global evidence (less likely to be burned).

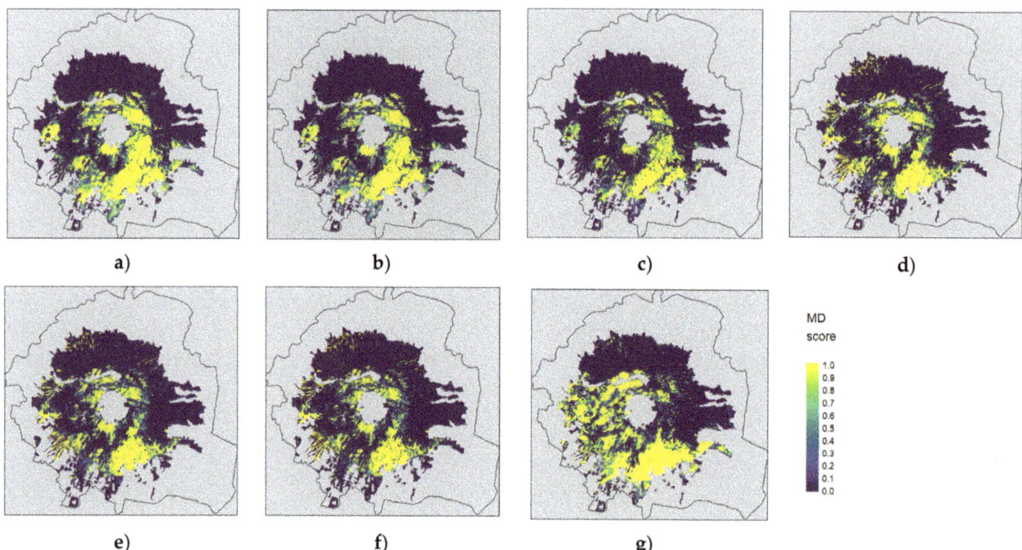

Figure 6. Partial evidence of burn as given by the input features interpreted by sigmoid MFs: Post_{RE2} (**a**), Post_{RE3} (**b**) Post_{NIR} (**c**), Δ_{RE2} (**d**), Δ_{RE3} (**e**), Δ_{NIR} (**f**) and Δ_{SWIR2} (**g**). Color scale shows increasing degree and likelihood of burn (blue to yellow).

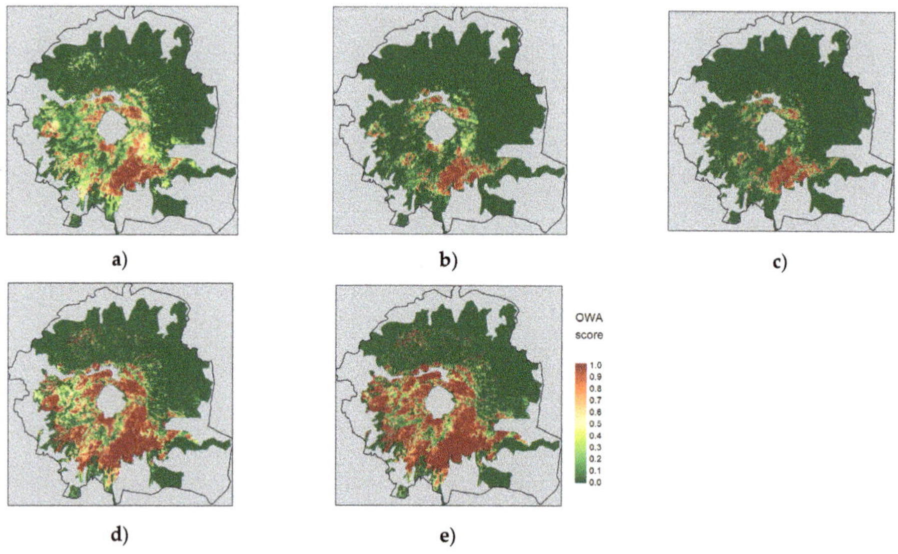

Figure 7. Global evidence of burn shown by OWA score [0, 1] (left column) over the Vesuvius study site: OWA_{AND} (**a**), $\text{OWA}_{\text{almostAND}}$ (**b**), $\text{OWA}_{\text{Average}}$ (**c**), $\text{OWA}_{\text{AlmostOR}}$ (**d**) and OWA_{OR} (**e**); borders of Vesuvius National Park are highlighted in black and masked areas in gray.

4.3. RG Burn Score and Validation

Figure 8 shows the RG_{score} (top row) and agreement (bottom row) maps that depict the spatial distribution of agreement between the two-class algorithm's output and the reference datasets: full agreement over burned (orange) and unburned classes (white) as well as errors of omission (green) and commission (blue).

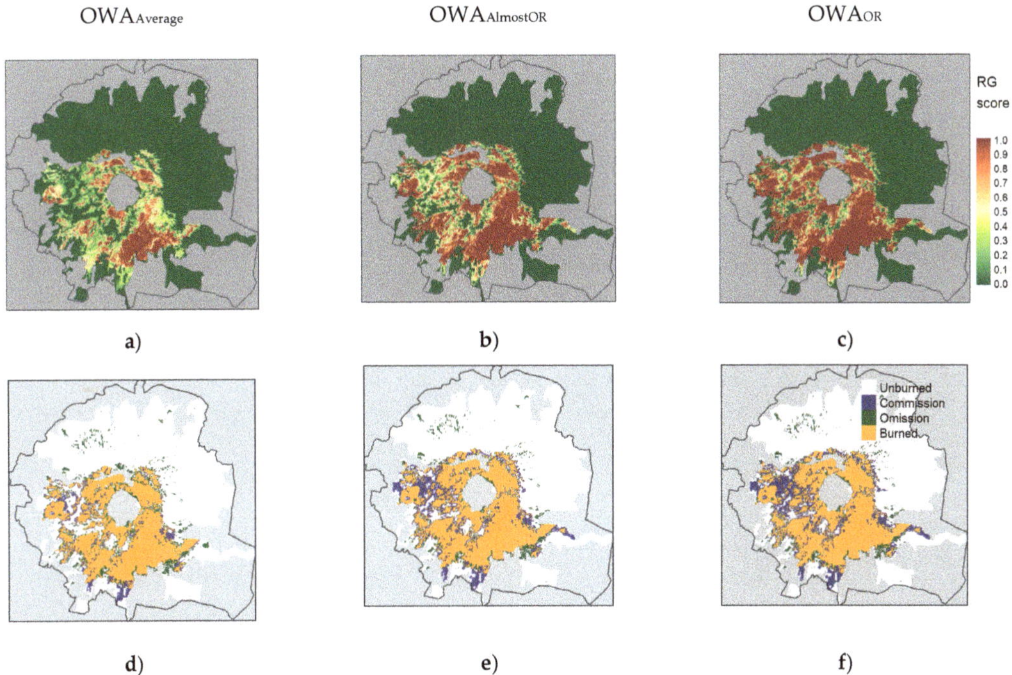

Figure 8. RG_{score} and agreement maps for the Vesuvius training site for growing layers OWA$_{Average}$ (**a,d**), OWA$_{AlmostOR}$ (**b,e**) and OWA$_{OR}$ (**c,f**).

In the agreement maps, omission errors in the northern slopes of the volcano are mainly produced by the lack of seed pixels in the OWA$_{seed}$ layer (Figure 7a), likely due to low-intensity and/or below-canopy fires; these regions are common to all output RG_{score} maps since the omission error descends from the seed layer rather than the growing layer. Commission errors are larger in the maps produced with OWA$_{AlmostOR}$ and OWA$_{OR}$ that implement 'complementary–relaxed' aggregation. Accuracy metrics quantifying the errors depicted in Figure 8 are summarized in Table 5 together with total estimated hectares of area burned (Tot BA).

Table 5. Accuracy metrics (oe = omission error, ce = commission error, dc = Dice coefficient, relB = relative bias) over the Vesuvius site for the three (OWA$_{Average}$, OWA$_{AlmostOR}$ and OWA$_{OR}$) growing layers. The total amount of burned area from the algorithm (Tot BA RG) and the reference (Tot BA REF) are also given.

OWA$_{grow}$	oe	ce	dc	RelB (%)	Tot BA RG (ha)	Tot BA REF (ha)
Average	0.15	0.12	0.87	+1.82	1676.39	
AlmostOR	0.10	0.20	0.85	−5.81	1959.69	1744.07
OR	0.09	0.22	0.84	−7.70	2029.95	

Results show that omission error is below 0.15 while commission error is in the range [0.12–0.22]. Commission is greater than omission for $OWA_{AlmostOR}$ and OWA_{OR} growing layers, while $OWA_{Average}$ leads to the greatest underestimation. Over the training site, total estimated burned area from the RG algorithm ranges between 1676.39 ha and 2029.95 ha, while the reference dataset provides 1744.07 ha of area burned. Saulino et al. (2020) [45] estimated that the area burnt by summer wildfires in 2017 in Vesuvius National Park amounted to 3350.23 ha, although this estimate covers the entire summer season by including fires that occurred later than the S2 post-fire date, 22 July.

4.4. Exportability Results

The algorithm developed over the Vesuvius site was applied to the sites selected for testing and located in Spain (2), Portugal (1) and Greece (2) with the following criteria:

1. Seed layer: OWA_{AND};
2. Seed selection: $OWA_{AND} > 0.9$;
3. Growing layers: $OWA_{Average}$, $OWA_{AlmostOR}$ and OWA_{OR};
4. RG algorithm: $OWA_{grow} > 0$;
5. Burned area mapping: $RG_{score} > 0$.

Figure 9 shows the agreement maps for the five sites and the corresponding accuracy metrics are summarized in Figure 10 and compared to metrics estimated for the training site. Increasing commission errors for OR-like operators (ce > 0.15) are visible in Calar and Huelva sites, Spain. In particular, in the Calar site, the OWA_{OR} growing layer generates a significantly greater commission error (ce > 0.30) by mistakenly classifying as burned a region of woodland–shrubland located in the northeastern part of the site. Additionally, in the Zakynthos site, Greece, commission error for the OR-like OWA operators is greater than $OWA_{Average}$ and mainly located in sparsely vegetated land covers. Commission is greater than omission in all sites except Kalamos, Greece, and Leiria, Portugal. In all sites, the difference in the estimates of the Dice coefficient for the three growing layers is negligible while relative bias shows values significantly below zero (overestimation) for Vesuvius, Calar and Huelva sites and OR-like operators confirm the 'complementary–relaxed' aggregation of these operators. Estimates of the relative bias clearly show that $OWA_{Average}$ provides burned area maps that tend to underestimate the area actually burned; again, the opposite occurs for OWA_{OR}. In terms of relB, the Zakynthos and Leiria sites show the lowest values and no difference among the three growing layers tested. Finally, a negligible difference is observed in accuracy metrics obtained over the Leiria site, Portugal, probably due to the clear burn spectral signal produced by intense and severe fires affecting forest cover; over this site, we obtained the overall greatest Dice coefficient and lowest relative bias.

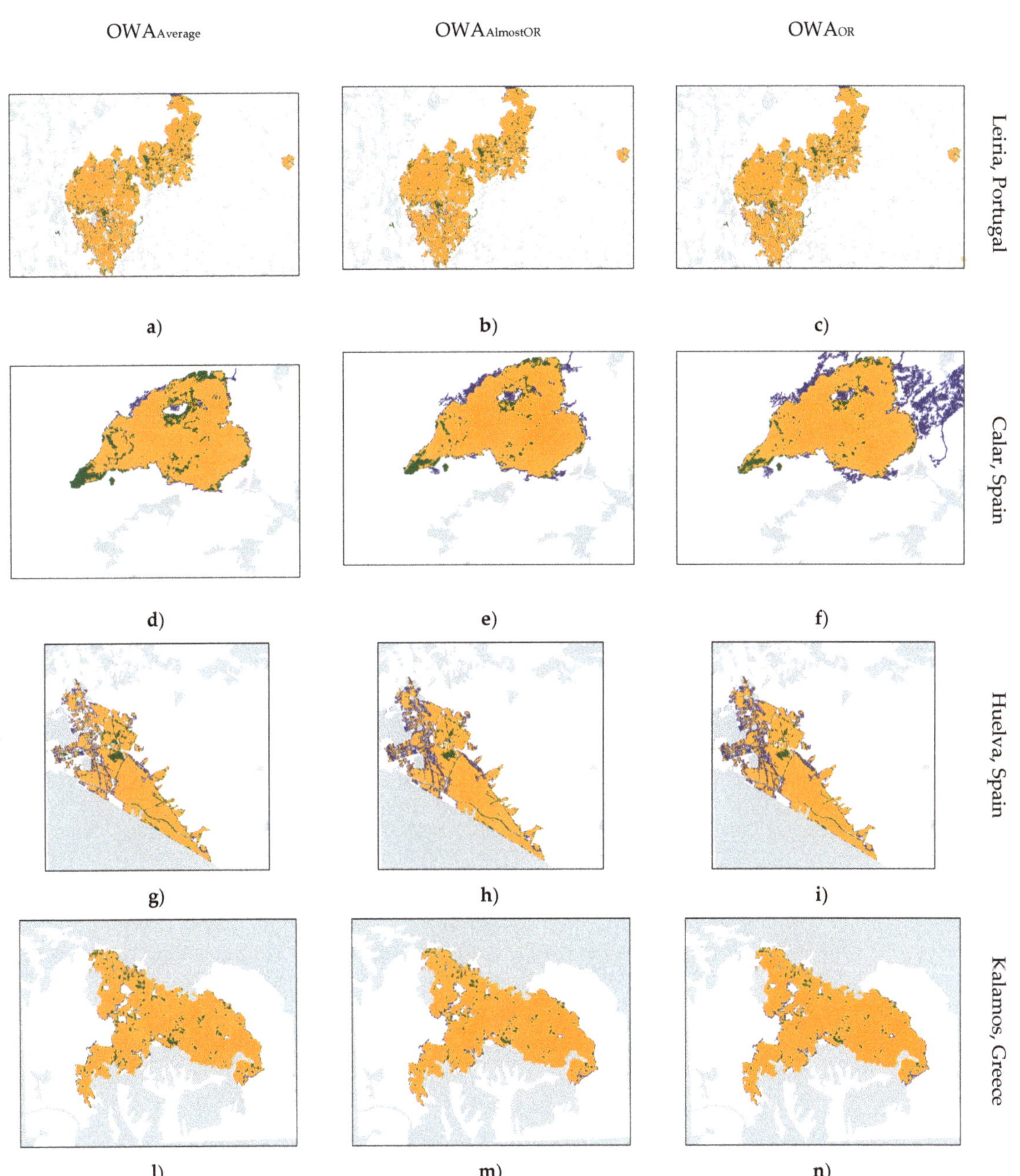

Figure 9. *Cont.*

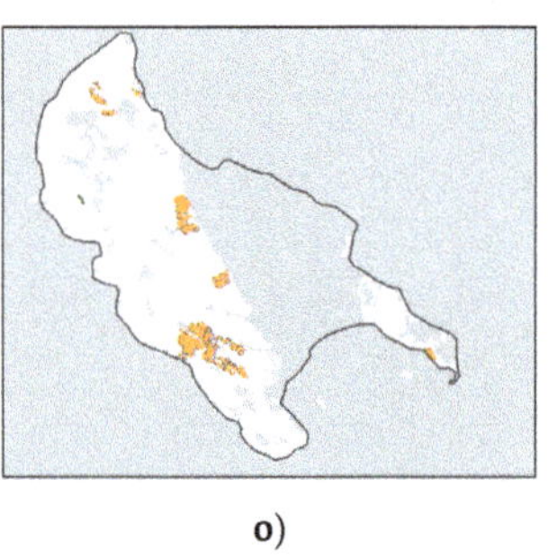 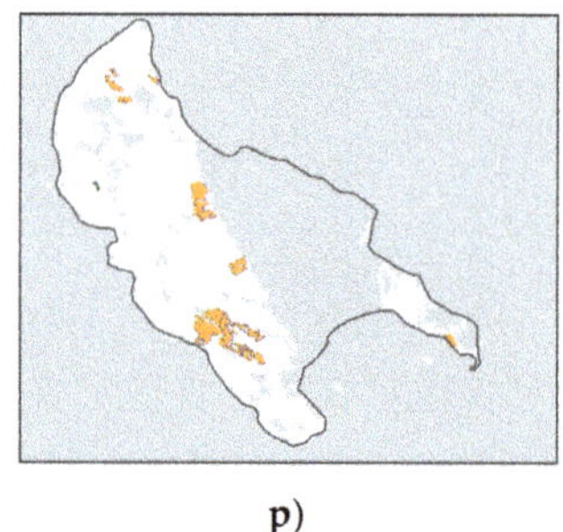 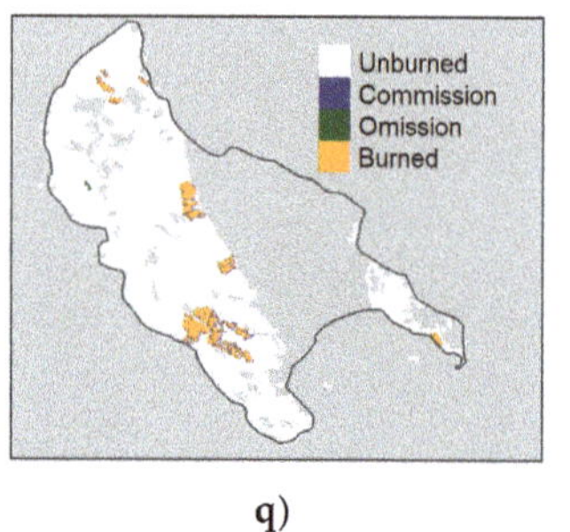

Figure 9. Agreement maps obtained for OWA$_{Average}$ (left column), OWA$_{AlmostOR}$ (middle column) and OWA$_{OR}$ (right column) over exportability sites: Leiria, Portugal (**a–c**), Calar, Spain (**d–f**), Huelva, Spain (**g–i**), Kalamos, Greece (**l–n**) and Zakynthos, Greece (**o–q**). The four classes represent: correctly classified burned pixels (orange), correctly classified unburned pixels (white), pixels mistakenly classified as burned—commission (blue) and pixel mistakenly classified as unburned—omission (green).

A regression analysis was carried over a grid layer spacing of 500 m × 500 m to compare the proportion of grid cells labeled as burned in the RG output and the reference maps. This analysis allows a more robust quantitative evaluation of the agreement between classified and reference data [46] by reducing the effect of error compensation. The results are displayed as regression scatter plots and the agreement was quantified by regression coefficients and error metrics: the coefficient of determination (R2) and the root mean squared error (RMSE) computed from the proportion and the total amount (hectares) of area burned within each grid cell (Figure 11). Slopes of the linear regression are generally very close to 1, showing a more than satisfactory agreement between classified and reference maps; in particular, slope values slightly below 1 can be observed for the Leiria, Calar and Kalamos sites, pointing out an underestimation of the area burned in S2 maps; on the contrary, overestimation occurs for the Zakynthos site (slope > 1.07). Indeed, underestimation error is rather expected when burned area mapping is carried out by coarser-resolution data [46] despite the contribution of RG in reducing commission (Figure 12). These trends appear to be least influenced by the OWA$_{grow}$ layer that is selected in the RG (columns in Figure 11). Only Vesuvius and Huelva show the slope of the regression model changing from negative to positive with OWA$_{grow}$; indeed, commission errors brought by the OR-like OWAs lead to a slope > 1. The R2 values confirm the very good agreement with lower values obtained over the Vesuvius training site and with OR-like operators (R2~0.85). By looking at the RMSE, the best results are obtained by applying different OWAs in the different sites: OWA$_{Average}$ yields the best results in Vesuvius, Huelva and Zakynthos, OWA$_{OR}$ performs best in Kalamos and Leiria, while OWA$_{AlmostOR}$ performs best in Calar. Hence, it is not univocally identified which OWA$_{grow}$ layer performs best across the sites, although OWA$_{OR}$ should be discarded due to the high overestimation errors. By choosing OWA$_{Average}$ or OWA$_{AlmostOR}$, the average grid cell RMSE is below 2 ha. As highlighted in Figure 9, the greatest commission errors occur over the Calar site, Spain, and with OWA$_{OR}$ growing layer; this error is represented in the scatter plot by grid cells along the *y*-axis (BA reference cell proportion = 0).

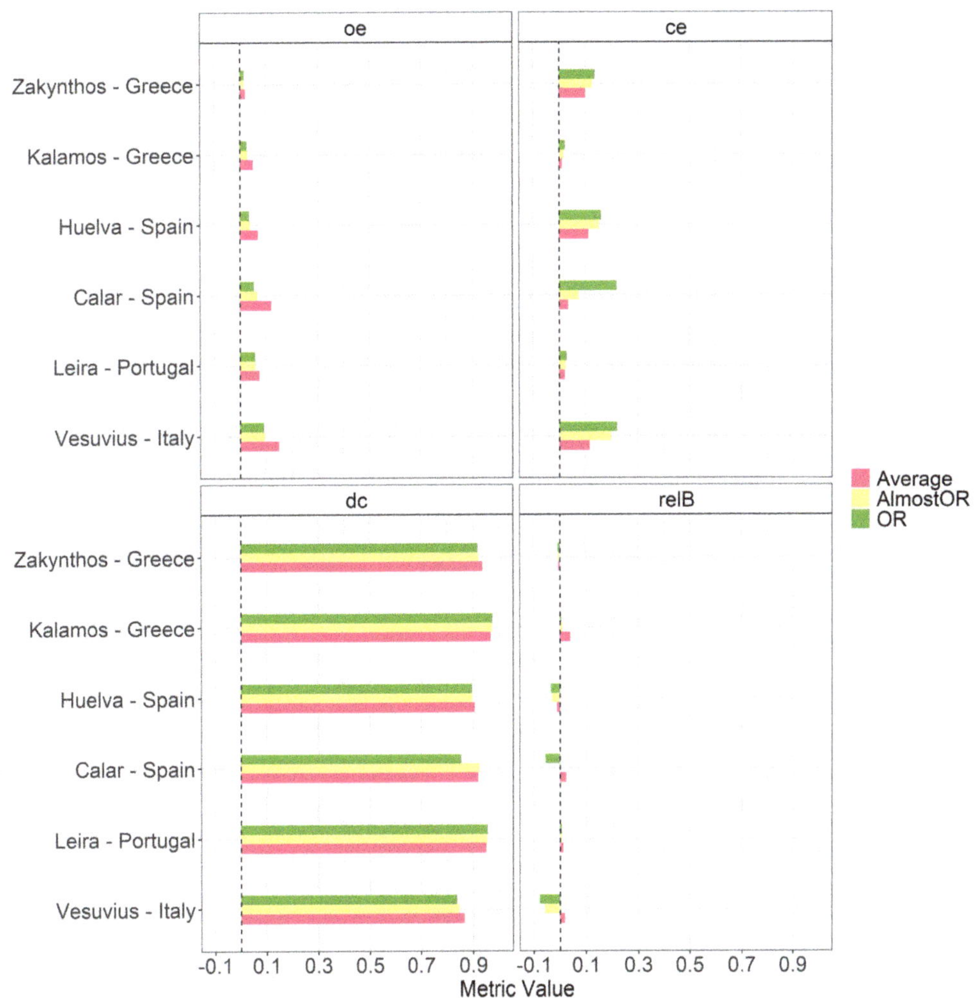

Figure 10. Accuracy metrics (omission error = oe, commission error – ce, Dice coefficient = dc and relative bias = relB) over the exportability and training sites.

By observing, in particular, scatter plots for the Leiria and Huelva sites, a large number of grid cells are fully burned according to the reference (BA reference cell proportion along the *x*-axis = 1) but the burn proportion detected by the RG algorithm is largely variable (BA RG cell proportion along the *y*-axis) with some cases of full omission. An in-depth analysis of these cells by visual inspection of S2 RGB color composite images (RGB = SWIR − NIR − Red) revealed that disagreement is mainly due to unburned islands and/or linear elements, such as roads, that are included as burned in the EMS polygons. Moreover, discrepancy between RG and reference perimeters is also due to differences in the reference date of the pre-fire image. The EMS pre-fire images can date back to previous years while S2 pre-fire images in this study belong to the same year as the fire event (2017); in fact, to limit the influence of changes of surface conditions due to other phenomena and to maximize the burned/unburned separability, we kept a time gap between pre- and post-fire S2 images in the range of 1–2 months. Some examples area given in the (Appendix A, Figure A4).

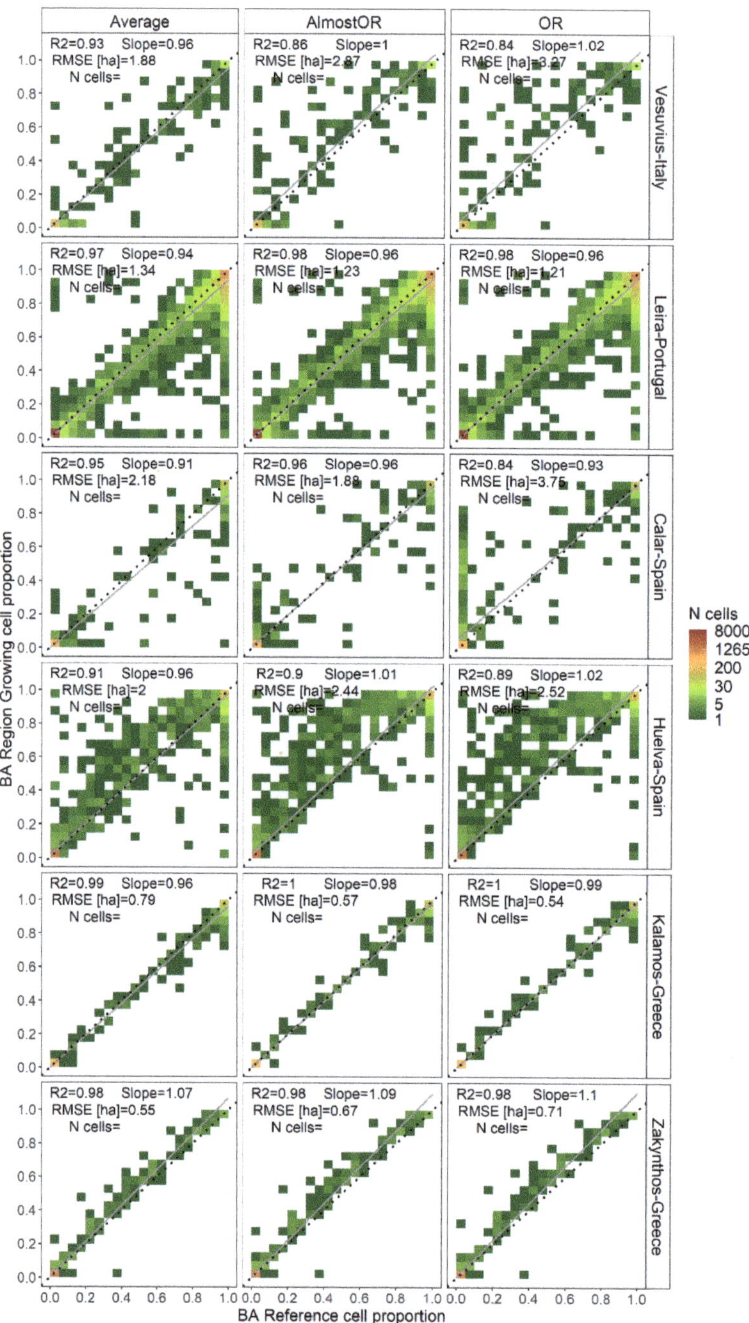

Figure 11. Scatter plot of the proportion of 500 m × 500 m grid cell mapped as burned in the RG output and the reference dataset for each site and OWA$_{grow}$ layer. Scatter plots are displayed as counts of cells for 0.05 step along the x- and *y*-axis to better represent overlapping points and with a logarithmic color scale. The black dotted line is the 1:1 line while the gray continuous line is the linear regression model. The coefficient of determination (R2), slope of the regression linear model (Slope), root mean squared error (RMSE) and total number of cells (N cells) are also shown.

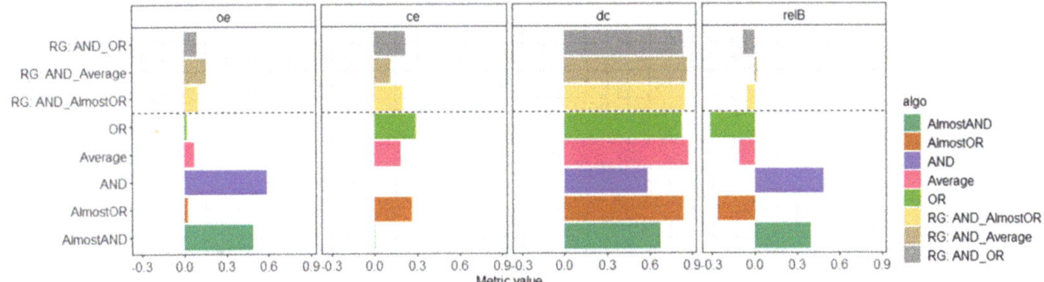

Figure 12. Accuracy metrics (omission error = oe, commission error = ce, Dice coefficient = dc and relative bias = relB) over the training site for burned area maps obtained with the RG algorithm (with three cases of growing layer, OWA$_{Average}$, OWA$_{Almost_OR}$ and OWA$_{OR}$) and from segmentation of the OWA global evidence (from all tested OWAs).

5. Discussion

The algorithm described here relies on S2 spectral bands and their temporal difference (pre-fire to post-fire reflectance change) in mapping burned areas in Mediterranean ecosystems. We chose to rely on spectral bands rather than indices for the more robust relationship between reflectance and surface properties; indeed, spectral indices might provide local good discrimination but their performance can vary in space and time [47]. The approach integrates burn evidence from those S2 bands and their temporal differences that showed the greatest sensitivity in discriminating burned and unburned surfaces. Among the S2 spectral bands, the separability metric M identified red–edge (bands 6 and 7) and NIR (band 8) as the most suitable bands while the short-wave infrared domain showed poor separability (M < 1) with the lowest values for SWIR2 (band 12). In this wavelength domain, separability improves slightly above the threshold value (M = 1.1) only when temporal difference ($\Delta_{post-pre}$) is computed and for the longer wavelength SWIR2 S2 band. Additionally, Δred–edge and NIR reflectance showed high separability (M > 1.5), although lower than post-fire reflectance (M > 1.8), thus, suggesting that, in the case of a lack of temporal series and/or suitable S2 image pairs, single-date images could provide accurate mapping results. This confirms previous findings obtained with Landsat images by Stroppiana et al. (2012) [26].

Seven features ($Post_{RE2}$, $Post_{RE3}$, $Post_{NIR}$, Δ_{RE2}, Δ_{RE3}, Δ_{NIR}, Δ_{SWIR2}) were therefore selected as input for the algorithm that relies on approximate reasoning to model uncertainty on burned areas through the convergence of evidence of burned conditions. In this framework, pixel-based partial burn evidence is the likelihood of observing burned conditions in a given single spectral feature and the global evidence is given by the complementary concurrent aggregation of partial evidence degrees. This way, self-adaptation of the algorithm to slightly similar areas and context is achieved, conferring robustness to the approach. Prior to aggregation, input features are interpreted in terms of burn likelihood by membership functions (MFs); similarly to other works proposed in the literature [19], we chose sigmoid-shaped functions to rescale the domain of each input feature into a common domain [0, 1] and these functions were defined from training sets in a semi-automatic way. This improvement makes the implementation of the burned area mapping algorithm less dependent on supervision and/or expert intervention compared to the previous version where membership functions were defined in a fully expert driven way [26]. Tests carried out on the exportability to new sites (4.4) confirmed that MFs are robust and provide reliable results in similar ecological conditions (Mediterranean ecosystems of Southern Europe). If other new regions have similar characteristics in terms of fire regime and land cover, the algorithm could be applied in its present form since it is robust and self-adaptive, stable and valid. The algorithm could be further automatized with MFs defined from training that are automatically extracted, for example, from active fire points in a hybrid approach able to combine multi-source information to add new evidence for burned area classification [10,16].

Integration of partial evidence to derive global scores was achieved with ordered weighted averaging (OWA) operators; OWA layers were then used as input to a region growing (RG) algorithm, which is largely exploited for image segmentation to balance omission and commission errors [25]. An initial seed layer, in which burned pixels are identified to minimize commission errors (i.e., in this case using an AND-like OWA), expands by adding new neighboring pixels belonging to a grow layer, identified to minimize omission errors (i.e., in this case OR-like OWA). The improved performance of the algorithm achieved with RG is presented in Figure 12 for the Vesuvius training site. First, the implementation of the RG combined with OWA significantly reduces relative bias (relB) that quantifies the difference between overestimation and underestimation. Moreover, RG reduces the higher omission error brought by AND-like OWAs (more restrictive conditions on the convergence of evidence) while it reduces commission of the OR-like operators. The balance between these two types of errors, as quantified by the Dice coefficient (dc), shows that RG brings significant improvement with respect to extreme OWA_{AND} and OWA_{OR} accuracy (dc~0.6).

Hence, two layers of global evidence were selected as input to the RG algorithm for seed selection and growing. Seeds were identified as pixels where $OWA_{AND} > 0.9$ to guarantee the highest reliability relying on the OWA operator able to implement 'concurrent–strict' integration conditions. On the other hand, the grow layer is selected among operators implementing "partially complementary–relaxed" (average) and "complementary–relaxed" (OR-like) aggregation. The output of the RG algorithm (RG_{score}) is a continuous layer in [0, 1] that can be segmented to deliver a binary two-class burned/unburned map to be compared to reference fire perimeters (i.e., validation).

Preliminary analysis over the Vesuvius training site (Appendix A, Figures A2 and A3) showed that all pixels in the OWA_{grow} layer could be retained as potentially burned ($OWA_{grow} > 0$) and all evidence values from the RG algorithm contribute to burned area mapping ($RG_{score} > 0$). Under these conditions, estimated accuracy metrics are commission error < 0.22, omission error < 0.15, Dice coefficient > 0.84.

Over the exportability sites, accuracy metrics fall in the range of values: oe [0.02, 0.15], ce [0.01, 0.22], dc [0.84, 0.97] and relB [−0.077, 0.040]. The range of values is given by the three OWAs used as potential layers for growing boundaries. Accuracy metrics are more than satisfactory for a semi-automatic burned area mapping algorithm covering a wide range of fire and land cover conditions in Mediterranean ecosystems. Where fire severity is greatest, such as in the case of the Leiria site, Portugal, we observed that all three growing layers analyzed provided comparable accuracy.

In the literature, Pulvirenti et al. (2020) [48] proposed an automated algorithm based on S2 spectral indices over forest areas and achieved an average commission error of 6.3% and omission error of 12.7%. Similarly, Smiraglia et al. (2020) [49] obtained commission error = 33% and omission error = 24% by also exploiting S2 spectral indices. Furthermore, Seydi et al. (2021) [50] mapped burned areas with a random forest algorithm (proved to be the algorithm providing greatest accuracy) with ce = 8.7% and oe = 9.2%. Hence, the performance of the algorithm proposed here is consistent with published results.

The regression analysis over 500 m × 500 m grid cells confirmed the high spatial accuracy achieved over all sites and, in particular, over Kalamos and Zakynthos, Greece (R2~1, RMSE < 0.1 ha). Overall accuracy metrics (Figure 10) showed that the algorithm tends to overestimate, with commission errors larger than omissions; the greatest overestimation rate being from the OWA_{OR} (RMSE > 3 ha), as also shown by the agreement maps (Figure 9, third column). In these cases, areas erroneously classified as burned are mainly located in sparsely vegetated land covers. On the contrary, disagreement between reference and S2 classification and resulting in large omission errors occurred in the Leiria and Huelva sites. In these cases, the regression analyses highlighted local omission errors better than the overall accuracy metrics: visual comparison of RGB S2 composite images pointed out that these errors are due to unburned islands and linear patterns (e.g., roads) that are erroneously included in the reference burned polygons. Additionally, differences in pre-fire image date for S2 (our algorithm) and EMS products could lead to biased accuracy metrics; an example is reported in the (Appendix A, Figure A4). Although EMS delineation maps proved to be suitable source of information for validation [18] by delivering reliable fire perimeters in rapid mapping mode, inconsistency might occur locally. Since no detailed information is available on the accuracy of the EMS fire perimeters, we could not further investigate this issue.

From the algorithm point of view, even if a single OWA_{grow} layer was not identified as the best performing across all sites, we can be confident in stating that $OWA_{Average}$ and $OWA_{AlmostOR}$ are the best ones. If the characteristics of a new region are known in advance and comparable to those of one of our test sites, we could select the best OWA case by case.

6. Conclusions

We propose a burned area mapping algorithm that is an improvement over a previous version [26] in several aspects:

1. Customization to S2 imagery for implementing a convergence of evidence approach;
2. Exploitation of additional spectral bands available from the S2 MSI instrument;
3. Automatic interpretation of input features (e.g., post-fire and $\Delta_{post\text{-}pre}$ reflectance) through membership functions (MFs) defined from training statistics (partial evidence of burn);
4. Tests of OWA operators from AND-like (for seed selection) to OR-like (for growing layer) integration criteria;
5. Implementation of OWA global evidence in a region growing (RG) algorithm;
6. Accuracy assessment over a wide range of conditions/locations in Southern Europe for the 2017 summer fire season.

Accuracy over training and exportability sites confirmed that the semi-automatic algorithm is robust and self-adaptive over different land cover and fire regime conditions in Mediterranean landscapes. Overall, accuracy metrics (oe < 15%, ce < 22%, dc > 0.84) are consistent with values from the literature for regional applications, although effort should be made in reducing commission errors. A key issue in the validation activity is the availability of reference fire perimeters comparable in space and time with the burned area maps from classification that could induce biased estimation of accuracy metrics. Finally, future activity will be focused on the exploitation of the output burn evidence from OWA operators and RG (RG_{score}) as an indicator of variable degrees of burn severity.

Author Contributions: Conceptualization, D.S., M.B., G.B. and P.A.B.; methodology, D.S., E.P., M.B. and G.S.; software, D.S. and E.P.; validation, M.S., D.S., M.B. and E.P.; formal analysis, M.S., D.S. and E.P.; data curation, D.S.; writing—original draft preparation, D.S. and M.B.; writing—review and editing, D.S., M.S., P.A.B. and G.S.; supervision, G.B. and P.A.B. All authors have read and agreed to the published version of the manuscript.

Funding: This research received no external funding.

Acknowledgments: We would like to acknowledge Luigi Ranghetti, CNR IREA, for his valuable support in processing S2 data with Sen2r Toolbox. The authors would like to thank the editors and the anonymous reviewers for their suggestions for improving the manuscript.

Conflicts of Interest: The authors declare no conflict of interest.

Appendix A

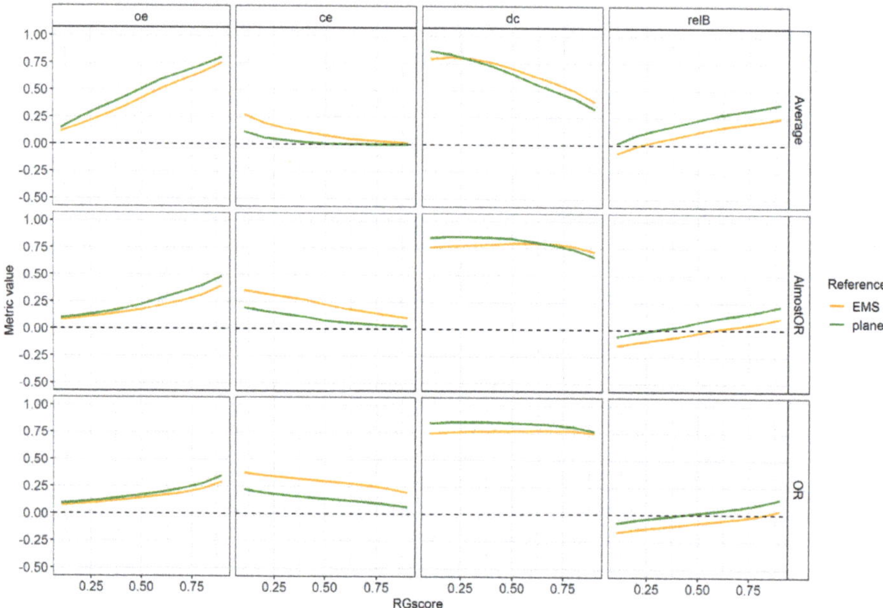

Figure A1. Estimates of the accuracy metrics omission error (oe), commission error (ce), Dice coefficient (dc) and relative bias (relB) over the Vesuvius site estimated by comparison with PlanetScope (green lines) and EMS (orange line) fire reference polygons, horizontal red dashed line shows y = 0.

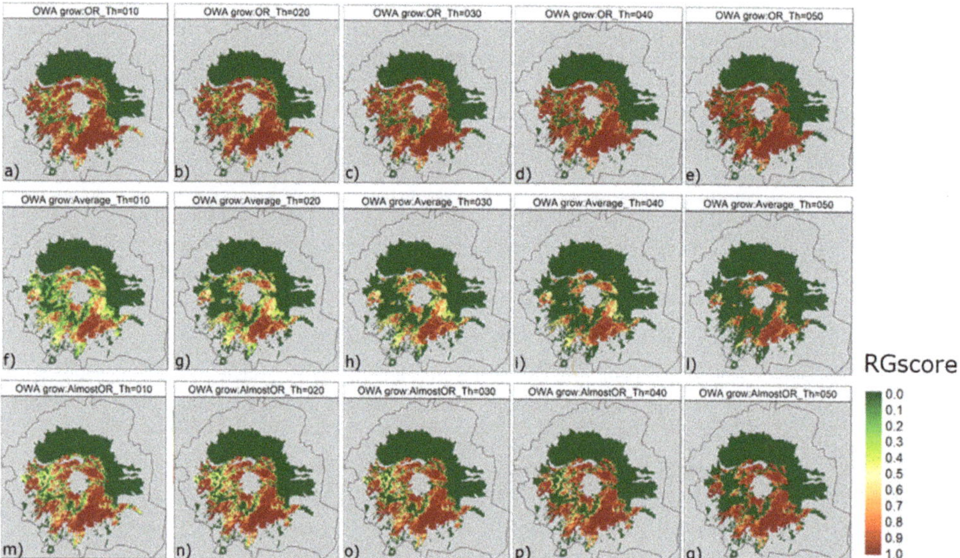

Figure A2. RG output score [0, 1] estimated over the Vesuvius training site OWA$_{AlmostOR}$ (**a–e**), OWA$_{Average}$ (**f–l**) and OWA$_{AlmostOR}$ (**m–q**) as growing boundaries and different threshold for identifying growing boundaries (Th) in [0.1–0.5] (left to right columns). Masked areas are in gray. In all cases the seed layer is OWA$_{AND}$ > 0.9.

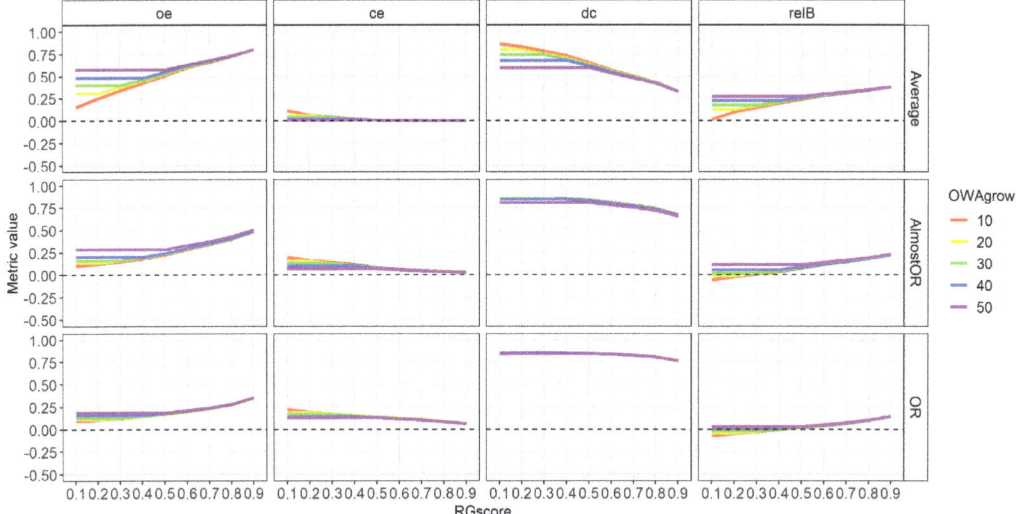

Figure A3. Accuracy metrics (omission error = oe, commission error = ce and Dice coefficient = dc) estimates for burned area maps obtained with the region growing (RG) algorithm over the Vesuvius site. Seed layer is the OWA_{AND} score map and growing layers are: $OWA_{Average}$, $OWA_{AlmostOR}$ and OWA_{OR}. Accuracy metrics are estimated for combinations of the threshold applied to the growing layers (OWA_{grow}, legend colors) and to the RG_{score} (x-axis).

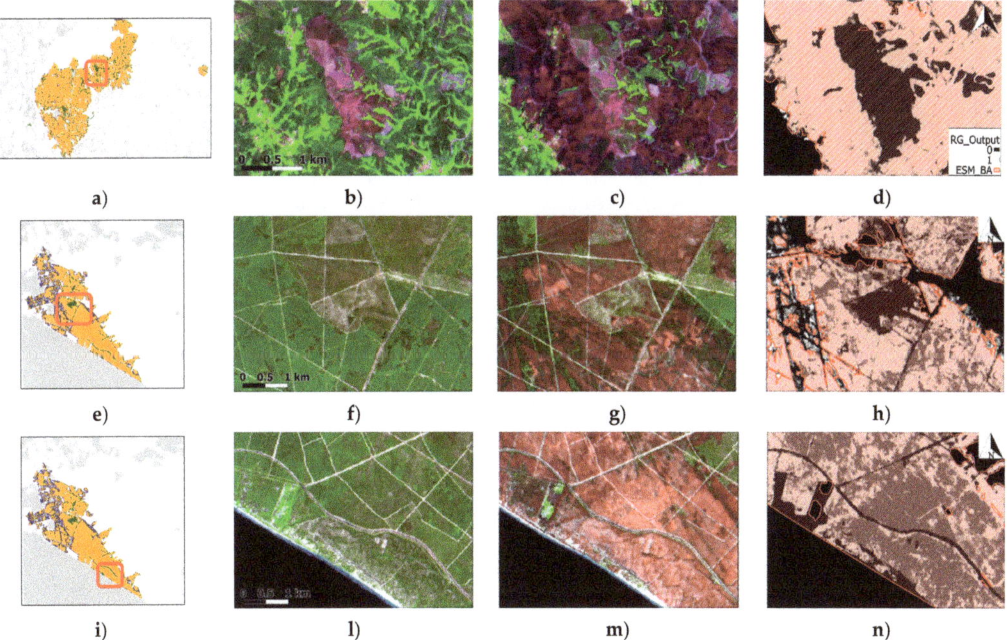

Figure A4. Example of apparent commission errors for the Leiria site, Portugal (**a–d**): pre-fire S2 image (4 June 2017, **b**), post-fire S2 image (4 July 2017, **c**) and burned areas from RG and EMS (**d**). Examples of commission errors in the Huelva site, Spain (**e–n**): pre-fire S2 image (11 June 2017, **f,l**), post-fire S2 image (1 July 2017, **g,m**) and burned areas from RG and EMS (**c**). S2 RGB are false color composites SWIR-NIR-Red. First column shows the location of the zoom areas and the last column RG classification (white to black background) and EMS reference perimeters (red line patterns).

References

1. Carvalho, A.; Monteiro, A.; Flannigan, M.; Solman, S.; Miranda, A.I.; Borrego, C. Forest fires in a changing climate and their impacts on air quality. *Atmos. Environ.* **2011**, *45*, 5545–5553. [CrossRef]
2. Knorr, W.; Jiang, L.; Arneth, A. Climate, CO_2 and human population impacts on global wildfire emissions. *Biogeosciences* **2016**, *13*, 267–282. [CrossRef]
3. Jolly, W.M.; Cochrane, M.A.; Freeborn, P.H.; Holden, Z.A.; Brown, T.J.; Williamson, G.J.; Bowman, D.M.J.S. Climate-induced variations in global wildfire danger from 1979 to 2013. *Nat. Commun.* **2015**, *6*, 7537. [CrossRef] [PubMed]
4. Abatzoglou, J.T.; Williams, A.P.; Barbero, R. Global emergence of anthropogenic climate change in fire weather indices. *Geophys. Res. Lett.* **2019**, *46*, 326–336. [CrossRef]
5. Dupuy, J.L.; Fargeon, H.; Martin-St Paul, N.; Pimont, F.; Ruffault, J.; Guijarro, M.; Hernando, C.; Madrigal, J.; Fernandes, P. Climate change impact on future wildfire danger and activity in southern Europe: A review. *Ann. Forest Sci.* **2020**, *77*, 35. [CrossRef]
6. Van der Werf, G.R.; Randerson, J.T.; Giglio, L.; Collatz, G.; Mu, M.; Kasibhatla, P.S.; Morton, D.C.; DeFries, R.S.; Jin, Y.; van Leeuwen, T.T. Global fire emissions and the contribution of deforestation, savanna, forest, agricultural, and peat fires (1997–2009). *Atmos. Chem. Phys.* **2010**, *10*, 11707–11735. [CrossRef]
7. Yue, C.; Ciais, P.; Cadule, P.; Thonicke, K.; van Leeuwen, T.T. Modelling the role of fires in the terrestrial carbon balance by incorporating SPITFIRE into the global vegetation model ORCHIDEE—part 2: Carbon emissions and the role of fires in the global carbon balance. *Geosci. Model Dev.* **2015**, *8*, 1285–1297. [CrossRef]
8. Li, F.; Levis, S.; Ward, D.S. Quantifying the role of fire in the Earth system—Part 1: Improved global fire modeling in the Community Earth System Model (CESM1). *Biogeosciences* **2013**, *10*, 2293–2314. [CrossRef]
9. Li, F.; Bond-Lamberty, B.; Levis, S. Quantifying the role of fire in the Earth system—Part 2: Impact on the net carbon balance of global terrestrial ecosystems for the 20th century. *Biogeosciences* **2014**, *11*, 1345–1360. [CrossRef]
10. Alonso-Canas, I.; Chuvieco, E. Global burned area mapping from ENVISAT-MERIS and MODIS active fire data. *Remote Sens. Environ.* **2015**, *163*, 140–152. [CrossRef]
11. Libonati, R.; DaCamara, C.; Setzer, A.; Morelli, F.; Melchiori, A. An algorithm for burned area detection in the Brazilian cerrado using 4 µm MODIS imagery. *Remote Sens.* **2015**, *7*, 15782–15803. [CrossRef]
12. Chuvieco, E.; Mouillot, F.; van der Werf, G.R.; San Miguel, J.; Tanase, M.; Koutsias, N.; García, M.; Yebra, M.; Padilla, M.; Gitas, I.; et al. Historical background and current developments for mapping burned area from satellite earth observation. *Remote Sens. Environ.* **2019**, *225*, 45–64. [CrossRef]
13. Giglio, L.; Schroeder, W.; Justice, C.O. The collection 6 MODIS active fire detection algorithm and fire products. *Remote Sens. Environ.* **2016**, *178*, 31–41. [CrossRef] [PubMed]
14. Shimabukuro, Y.E.; Dutra, A.C.; Arai, E.; Duarte, V.; Cassol, H.L.G.; Pereira, G.; Cardozo, F.S. Mapping burned areas of mato grosso state Brazilian amazon using multisensor datasets. *Remote Sens.* **2020**, *12*, 3827. [CrossRef]
15. Chuvieco, E.; Yue, C.; Heil, A.; Mouillot, F.; Alonso-Canas, I.; Padilla, M.; Pereira, J.M.; Oom, D.; Tansey, K. A new global burned area product for climate assessment of fire impacts. *Glob. Ecol. Biogeogr.* **2016**, *25*, 619–629. [CrossRef]
16. Pereira, A.A.; Pereira, J.M.C.; Libonati, R.; Oom, D.; Setzer, A.W.; Morelli, F.; Machado-Silva, F.; de Carvalho, L.M.T. Burned area mapping in the Brazilian savanna using a one-class support vector machine trained by active fires. *Remote Sens.* **2017**, *9*, 1161. [CrossRef]
17. Randerson, J.T.; Chen, Y.; van der Werf, G.R.; Rogers, B.M.; Morton, D.C. Global burned area and biomass burning emissions from small fires. *Biogeoscience* **2012**, *117*. [CrossRef]
18. Filipponi, F. Exploitation of sentinel-2 time series to map burned areas at the national level: A case study on the 2017 Italy wildfires. *Remote Sens.* **2019**, *11*, 622. [CrossRef]
19. Roteta, E.; Bastarrika, A.; Padilla, M.; Storm, T.; Chuvieco, E. Development of a sentinel-2 burned area algorithm: Generation of a small fire database for sub-Saharan Africa. *Remote Sens. Environ.* **2019**, *222*, 1–17. [CrossRef]
20. Stavrakoudis, D.; Katagis, T.; Minakou, C.; Gitas, I.Z. Automated burned scar mapping using sentinel-2 imagery. *J. Geogr. Inf. Syst.* **2020**, *12*, 221–240. [CrossRef]
21. Roy, D.; Huang, H.; Boschetti, L.; Giglio, L.; Yan, L.; Zhang, H.; Li, Z. Landsat-8 and Sentinel-2 burned area mapping—a combined sensor multi-temporal change detection approach. *Remote Sens. Environ.* **2019**, *231*, 111254. [CrossRef]
22. Boschetti, L.; Roy, D.; Justice, C.O.; Michael, L.; Humber, M.L. MODIS–Landsat fusion for large area 30 m burned area mapping. *Remote Sens. Environ.* **2015**, *161*, 27–42.
23. Li, J.; Roy, D.P. A Global analysis of sentinel-2a, sentinel-2b andlandsat-8 data revisit intervals and implications for terrestrial monitoring. *Remote Sens.* **2017**, *9*, 902. [CrossRef]
24. Liu, S.; Zheng, Y.; Dalponte, M.; Tong, X. A novel fire index-based burned area change detection approach using Landsat-8 OLI data. *Eur. J. Remote Sens.* **2020**, *53*, 104–112. [CrossRef]
25. Bastarrika, A.; Chuvieco, E.; Martín, M.P. Mapping burned areas from Landsat TM/ETM+ data with a two-phase algorithm: Balancing omission and commission errors. *Remote Sens. Environ.* **2011**, *115*, 1003–1012. [CrossRef]
26. Stroppiana, D.; Bordogna, G.; Carrara, P.; Boschetti, M.; Boschetti, L.; Brivio, P.A. A method for extracting burned areas from Landsat TM/ETM+ images by soft aggregation of multiple Spectral Indices and a region growing algorithm. *ISPRS J. Photogramm. Remote Sens.* **2012**, *69*, 88–102. [CrossRef]

27. Goffi, A.; Stroppiana, D.; Brivio, P.A.; Bordogna, G.; Boschetti, M. Towards an automated approach to map flooded areas from Sentinel-2 MSI data and soft integration of water spectral features. *Int. J. Appl. Earth Obs.* **2020**, *84*, 101951. [CrossRef]
28. Yager, R.R. On ordered weighted averaging aggregation operators in multi-criteria decision making. *IEEE Trans. Syst. Man Cybern.* **1988**, *18*, 183–190. [CrossRef]
29. Sánchez-Benítez, A.; García-Herrera, R.; Barriopedro, D.; Sousa, P.M.; Trigo, R.M. June 2017: The earliest european summer mega-heatwave of reanalysis period. *Geophys. Res. Lett.* **2018**, *45*, 1955–1962. [CrossRef]
30. Turco, M.; Jerez, S.; Augusto, S.; Tarin-Carrasco, P.; Ratola, N.; Jimenez-Guerrero, P.; Trigo, R.M. Climate drivers of the 2017 devastating fires in Portugal. *Sci. Rep.* **2019**, *9*, 13886. [CrossRef]
31. Ranghetti, L.; Busetto, L. Sen2r: An R Toolbox to Find, Download and Preprocess Sentinel-2 Data. R Package Version 1.0.0. 2019. Available online: http://sen2r.ranghetti.info (accessed on 1 May 2021).
32. Main-Knorn, M.; Pflug, B.; Louis, J.; Debaecker, V.; Müller-Wilm, U.; Gascon, F. Sen2Cor for sentinel-2. In Proceedings of the SPIE 10427, Image and Signal Processing for Remote Sensing XXIII, Warsaw, Poland, 10–14 September 2017; p. 1042704.
33. Louis, J.; Charantonis, A.; Berthelot, B. Cloud detection for sentinel-2. In Proceedings of the ESA Living Planet Symposium, Bergen, Norway, 28 June–2 July 2010.
34. Planet Team. Planet Application Program Interface: In Space for Life on Earth. San Francisco, CA. 2017. Available online: https://api.planet.com (accessed on 1 May 2021).
35. Lemajic, S.; Vajsová, B.; Aastrand, P. *New Sensors Benchmark Report on PlanetScope: Geometric Benchmarking Test for Common Agricultural Policy (CAP) Purposes*; Publications Office of the European Union: Luxembourg, 2018; ISBN 978-92-79-92833-8. JRC111221. [CrossRef]
36. Belgiu, M.; Drăguţ, L. Random forest in remote sensing: A review of applications and future directions. *ISPRS J. Photogramm. Remote Sens.* **2016**, *114*, 24–31. [CrossRef]
37. Ramo, R.; Chuvieco, E. Developing a random forest algorithm for MODIS global burned area classification. *Remote Sens.* **2017**, *9*, 1193. [CrossRef]
38. Breiman, L. Random forest. *Mach. Learn.* **2001**, *45*, 5–32. [CrossRef]
39. Boschetti, M.; Nutini, F.; Manfron, G.; Brivio, P.A.; Nelson, A. Comparative analysis of normalised difference spectral indices derived from MODIS for detecting surface water in flooded rice cropping systems. *PLoS ONE* **2014**, *9*, e88741. [CrossRef] [PubMed]
40. Carrara, P.; Bordogna, G.; Boschetti, M.; Brivio, P.A.; Nelson, A.; Stroppiana, D. A flexible multi-source spatial-data fusion system for environmental status assessment at continental scale. *Int. J. Geogr. Inf. Sci.* **2008**, *22*, 781–799. [CrossRef]
41. Goffi, A.; Bordogna, G.; Stroppiana, D.; Boschetti, M.; Brivio, P.A. Knowledge and data-driven mapping of environmental status indicators from remote sensing and VGI. *Remote Sens.* **2020**, *12*, 495. [CrossRef]
42. Congalton, R.G.; Green, K. *Assessing the Accuracy of Remotely Sensed Data*, 2nd ed.; CRC Press: Boca Raton, FL, USA, 2008; p. 346.
43. Roy, D.P.; Jin, Y.; Lewis, P.E.; Justice, C.O. Prototyping a global algorithm for systematic fire-affected area mapping using MODIS time series data. *Remote Sens. Environ.* **2005**, *97*, 137–162. [CrossRef]
44. Clevers, J.G.; Gitelson, A.A. Remote estimation of crop and grass chlorophyll and nitrogen content using red-edge bands on Sentinel-2 and -3. *Int. J. Appl. Earth Obs. Geoinf.* **2013**, *23*, 344–351. [CrossRef]
45. Saulino, L.; Rita, A.; Migliozzi, A.; Maffei, C.; Allevato, E.; Garonna, A.P.; Saracino, A. Detecting burn severity across mediterranean forest types by coupling medium-spatial resolution satellite imagery and field data. *Remote Sens.* **2020**, *12*, 741. [CrossRef]
46. Boschetti, L.; Roy, D.P.; Giglio, L.; Huang, H.; Zubkova, M.; Humber, M.L. Global validation of the collection 6 MODIS burned area product. *Remote Sens. Environ.* **2019**, *235*, 111490. [CrossRef]
47. Huang, H.; Roy, D.P.; Boschetti, L.; Zhang, H.K.; Yan, L.; Kumar, S.S.; Gomez-Dans, J.; Li, J. Separability analysis of sentinel-2a multi-spectral instrument (MSI) data for burned area discrimination. *Remote Sens.* **2016**, *8*, 873. [CrossRef]
48. Pulvirenti, L.; Squicciarino, G.; Fiori, E.; Fiorucci, P.; Ferraris, L.; Negro, D.; Gollini, A.; Severino, M.; Puca, S. An automatic processing chain for near real-time mapping of burned forest areas using sentinel-2 data. *Remote Sens.* **2020**, *12*, 674. [CrossRef]
49. Smiraglia, D.; Filipponi, F.; Mandrone, S.; Tornato, A.; Taramelli, A. Agreement index for burned area mapping: Integration of multiple spectral indices using sentinel-2 satellite images. *Remote Sens.* **2020**, *12*, 1862. [CrossRef]
50. Seydi, S.T.; Akhoondzadeh, M.; Amani, M.; Mahdavi, S. Wildfire damage assessment over Australia using sentinel-2 imagery and MODIS land cover product within the google earth engine cloud platform. *Remote Sens.* **2021**, *13*, 220. [CrossRef]

remote sensing

Article

Generating a Baseline Map of Surface Fuel Loading Using Stratified Random Sampling Inventory Data through Cokriging and Multiple Linear Regression Methods

Chinsu Lin [1,*], Siao-En Ma [1], Li-Ping Huang [2], Chung-I Chen [3], Pei-Ting Lin [1], Zhih-Kai Yang [1] and Kuan-Ting Lin [1]

[1] Department of Forestry and Natural Resources, National Chiayi University, Chiayi 600355, Taiwan; s1020004@mail.ncyu.edu.tw (S.-E.M.); s1090172@mail.ncyu.edu.tw (P.-T.L.); s1090183@mail.ncyu.edu.tw (Z.-K.Y.); s1090184@mail.ncyu.edu.tw (K.-T.L.)
[2] Forest Administration Division, Forestry Bureau, Taipei 100024, Taiwan; a0048@forest.gov.tw
[3] Department of Forestry, National Chung Hsing University, Taichung 402202, Taiwan; cichen@nchu.edu.tw
* Correspondence: chinsu@mail.ncyu.edu.tw

Abstract: Surface fuel loading is a key factor in controlling wildfires and planning sustainable forest management. Spatially explicit maps of surface fuel loading can highlight the risks of a forest fire. Geospatial information is critical in enabling careful use of deliberate fire setting and also helps to minimize the possibility of heat conduction over forest lands. In contrast to lidar sensing and/or optical sensing based methods, an approach of integrating in-situ fuel inventory data, geospatial interpolation techniques, and multiple linear regression methods provides an alternative approach to surface fuel load estimation and mapping over mountainous forests. Using a stratified random sampling based inventory and cokriging analysis, surface fuel loading data of 120 plots distributed over four kinds of fuel types were collected in order to develop a total surface fuel loading model (lntSFL-BioTopo model) and a fine surface fuel model (lnfSFL-BioTopo model) for generating tSFL and fSFL maps. Results showed that the combination of topographic parameters such as slope, aspect, and their cross products and the fuel types such as pine stand, non-pine conifer stand, broadleaf stand, and conifer–broadleaf mixed stand was able to appropriately describe the changes in surface fuel loads over a forest with diverse terrain morphology. Based on a cross-validation method, the estimation of tSFL and fSFL of the study site had an RMSE of 3.476 tons/ha and 3.384 tons/ha, respectively. In contrast to the average loading of all inventory plots, the estimation for tSFL and fSFL had a relative error of 38% (PRMSE). The reciprocal of estimation bias of both SFL-BioTopo models tended to be an exponential growth function of the amount of surface fuel load, indicating that the estimation accuracy of the proposed method is likely to be improved with further study. In the regression modeling, a natural logarithm transformation of the surface fuel loading prevented the outcome of negative estimates and thus improved the estimation. Based on the results, this paper defined a minimum sampling unit (MSU) as the area for collecting surface fuels for interpolation using a cokriging model. Allocating the MSUs at the boundary and center of a plot improved surface fuel load prediction and mapping.

Keywords: wildfire fuel loadings; sampling-based inventory data; ordinary cokriging method; regression analysis; lidar remote sensing

Citation: Lin, C.; Ma, S.-E.; Huang, L.-P.; Chen, C.-I; Lin, P.-T.; Yang, Z.-K.; Lin, K.-T. Generating a Baseline Map of Surface Fuel Loading Using Stratified Random Sampling Inventory Data through Cokriging and Multiple Linear Regression Methods. *Remote Sens.* **2021**, *13*, 1561. https://doi.org/10.3390/rs13081561

Academic Editors: Elena Marcos, Leonor Calvo, Susana Suarez-Seoane and Víctor Fernández-García

Received: 30 March 2021
Accepted: 14 April 2021
Published: 17 April 2021

Publisher's Note: MDPI stays neutral with regard to jurisdictional claims in published maps and institutional affiliations.

1. Introduction

Wildfires are recognized as one of the major disturbances in terrestrial forest ecosystems. Fire can significantly change forest attributes and destroy the habitat of wildlife while at the same time, it can create another important habitat. Naturally occurring wildfires caused by lightning or extreme climate events (a long dry season or drought and high temperature) are generally inevitable. Fires are also frequently used to clear forest for large-scale plantation establishment [1] and small-scale agricultural uses [2–4]. Although rapid warming has recently resulted in more wildfires worldwide, human activities have been recognized as the major causes of wildfire [5]. Most human-caused wildfires can therefore be prevented by using fires responsibly and taking preventative measures [6].

During a long, dry heat-wave period, an original wildfire can occur naturally and may become uncontrolled when the weather conditions, topography, and fuels, the drivers of fire behavior, are suitable for fire spread or propagation [7,8]. From the viewpoint of ecosystem succession, a burnt forestland will recover over a long-term period of succession and consequently be suitable for the development of new communities of vegetation and wildlife [9]. However, large-scale and high-intensity fires are likely to occur simultaneously over a landscape and therefore cause profound societal impacts [10]. The most recent example of such extreme natural disturbance would be the 2019–2020 Black Summer megafires in Australia [11]. In contrast to prescribed fires, the biomass burning events release a significant amount of particular matter into the air [12]. Uncontrolled forest fires can be recurrent and cause a significant loss of aboveground biomass stocks. Observations of forest fires in northern Brazil showed the tropical rain forests lost more than 60% of the biomass stocks in 3–7 years after the last major fire in a series of recurrent fires; even for large trees with diameter > 50 cm, the biomass loss was about 54% in areas that burnt three times [13]. Similarly, the carbon loss of tropical forest in northern Australia caused by surface fires was equivalent to 46% of the annual net primary production (NPP) of the forest [14]. Although smaller carbon-stock reductions of about 6% of annual NPP were reported for temperate forests versus 11% for natural vegetation landscapes, wildfire-caused biomass consumption has become a significant source of carbon emissions globally [15]. Management of fuel loads therefore becomes an important issue for the safe use of fire, wildfire prevention, and REDD achievement conservation [16].

Fuel is the combustible biomass found in forests and can be divided into fine fuels such as leaves, grasses, and small twigs, and larger fuels such as shrubs, branches on the ground, downed trees, and logs [17]. From the standpoint of vertical dimension, the fuel can be classified as ground fuels, aerial fuels (trees, snags, and ladder fuels), and canopy fuels (green leaves and branches of crowns). Accordingly, a fire occurring in the humus layers, moving slowly, which can probably smolder for a long time, is called a ground fire; a fire burning only surface litter and duff is a surface fire, while a fire that burns trees over their entire height to the top is called a crown fire. Human-caused wildfire generally begins as a ground/surface fire at a small scale but occasionally it can become a crown fire and cause huge damage to the forest. Therefore, a map of fuel load distribution can highlight the prevalence of wildfire risk and guide people in more careful use of fire in order to minimize the possibility of heat transfer via conduction/radiation/convection in forests.

When considering the possibility of ignition of a forest fire, identifying fuels that are most likely to burn is related to the water content and the amount of surface fuels in a forest. In other words, the fuel loadings should include the quantities of duff, litter, fine-woody debris, and coarse woody debris or logs (fallen dead woods) distributed over the ground surface of the forest because these are the primary factors for predicting fire effects from on-site fuels [18]. The fuel-loading models (FLMs) designed by Sikkink et al. [18] emphasized the importance of fuel composition for ground and surface fires. When fires become uncontrollable, the amount of crown materials will be included in the fuels available for crown fires and the trunk may or may not be burnt in crown fires. Therefore, many studies have been conducted to explore methods for deriving the distribution of fuel or biomass (ton/ha) using a variety of data such as forest inventory data and airborne lidar scanning (ALS) data for aboveground biomass [19–22] and canopy biomass [23–25]. According to the remote sensing-based IPCC method, a canopy fuel map can be derived using an ALS canopy height model by segmenting every single tree, using, for example, mathematical morphology-based watershed segmentation [26–29], Multilevel Morphological Active Contour (MMAC) [30] or Multilevel Slicing And Coding (MSAC) techniques [31]. Moreover, the latest development of lidar sensing enables precise inventories of surface fuel and canopy fuel using mobile terrestrial lidar instruments [32–35]. To overcome the high cost of high-density point cloud ALS data, alternate methods for surface fuel loading (SFL) estimation can be based on mathematical/empirical models using inventory data such as vegetation/species maps and related environmental factors [18,36], satellite full-waveform lidar data [37,38], and photon lidar data [39]. More recently, the approach has been extended to integrate optical sensing images such as infrared orthophotos and QuickBird images with ALS data to improve fuel mapping accuracy [40–43]. The major strength of lidar technology in fuel estimation is the ability to retrieve fuel heights and discriminate between fuel types [36]. It is impossible to measure surface fuel loads directly from lidar data because lidar pulses can rarely penetrate the dense litter and duff layers on the bare earth surface and travel back to the sensor.

Surface fuel inventories involve the complicated task of the collection of live biomass such as grasses, forbs, and small woody plants and dead fuels such as duffs, debris, and fallen dead wood over a large area of forest floor. Lidar technology is considered to be an efficient method of gathering information of detailed biomaterials distributed on the land surface. However, there still are difficulties in the determination of surface fuel loads over a wide range of forests using this technology. In practice, the determination of surface fuel loads is a process related to the collection and analysis of diverse surface fuels, fuel bed depth, and bulk density in the field. Fuel depth varies widely, for example, from 0.3 m for grasses to 1.8 m for shrub fields and 0.06–0.30 m for timber litter in forests [44]. Therefore, the surface fuels lying on the fuel bed 0.3 m above the earth surface are generally collected to account for the bulk density [45]. In addition, the ground is generally covered with surface mass due to the weathering effect and accumulation of fine materials. In spite of the ability of laser pulses to penetrate canopy gaps and reach the ground, collecting sufficient numbers of point clouds that lie on the surface fuel and the earth surface remains difficult and therefore, characterizing surface fuel though lidar point clouds is still challenging.

Surface fuel loads and bulk density are primarily subject to forest structures related to species composition, phenology, and canopy height [41,46–50]. Distribution of surface fuel loads is therefore a consequence of the interaction of multiple factors such as forest type or overstorey/understorey species, topographic relief, and climate. A traditional forest fuel inventory is capable of collecting the loads and bed depth of surface fuel and can further differentiate and measure a variety of fuel sizes, for example, duff mass, litter mass, fine-woody debris, coarse-woody debris, and fallen dead wood (FDW). In practice, a field inventory of surface fuel loads within the whole area of sample plots has to collect and measure the masses of the litter layer and duff layer. The work is time consuming and labor intensive and significantly disturbs the surface stability and seed bank when the plot covers a large area, for example, 10 by 10 square meters or even larger. The distribution of fuel masses is most likely spatially dependent in a relatively local space. This kind of spatial autocorrelation can be described via geostatistics which incorporates the spatial coordinates of sample data to interpolate values for locations where samples were not taken. Geostatistical methods such as kriging and cokriging (refer to Section 2.2 for the details) capture observed spatial dependence among regional points. In addition, the distribution of forests is typically a consequence of natural processes in which terrain morphology such as the slope, aspect, and elevation may influence the amount of solar radiance, temperature, humidity, and water availability on the slope, and these further affect the weathering and accumulation of surface fuel mass on a slope. Thus, the integration of traditional inventory and geostatistical methods could be helpful for mapping fuel distribution. A geospatially explicit continuous map of surface fuel loads can be a fuel baseline for the use of fire protection scenario generation and forest management. Therefore, the objective of this study was to propose an algorithm for generating surface fuel load maps through the integration of forest types, topographic variables, and in-situ inventory mass data using geostatistical analysis and multiple linear regression methods. The surface fuel load was presented as two types: fSFL (the fine- and coarse surface fuel loads) and tSFL (the total surface fuel load, i.e., the sum of fSFL and the FDW mass). With detailed composition complexity of fSFL and tSFL distribution over the study site, the uncertainty in mapping the surface fuel load was further examined, and a strategy to generate the surface fuel load was suggested.

2. Materials and Methods

2.1. Study Site and Data Acquisition

The forest in the area of the Dajiaxi Working Circle, managed by the Taiwan Forestry Bureau, was selected for this study. The forest located at 121°07′42″–121°27′03″E and 24°08′09″–24°26′31″N in north central Taiwan (Figure 1) has been frequently disturbed by fires during the last few decades. According to the official records revealed on the website of the Forestry Bureau (https://forecast.forest.gov.tw/Forecast/# accessed on 10 February 2021), a total of 2171 fires occurred in the 37 working circles of national forests from 1963 to 2019, averaging 37 ± 39 fires per year, with a minimum and maximum of three and 252 fires, respectively, while averaging 55 ± 74 per working circle in which the minimum frequency was 2 and the maximum was 287. Historical records of the Dajiaxi Working Circle revealed that larger fires occurred frequently during the period from 1963 to 1990. The annual occurrence frequency ranged from 0 to 20, and the burnt area was on average 526 ± 717 ha per year. Due to the implementation of a fire-fighting approach that incorporates the Incident Command System and Government Flying Service, the fire frequency and burnt area, after 1990, was significantly reduced to 0 to 4 and 39 ± 84 ha per year, respectively. Most of the fires occurred in the dry season from winter to the early spring; during this period, fire is most likely caused by careless use of fire. The prevalence of forest fire in the Dajiaxi Working Circle was around 10% and ranked 3rd among the 37 forest management units. The Dajiaxi national forest is therefore officially considered as a hotspot area for fire.

The altitude of the Island of Taiwan ranges from 0 to 3950 m. The forests on the island vary dramatically along with the changes in altitude, temperature, and latitude. With respect to the altitudinal variation, the forest is divided into the foothill zone (tropical forest), submontane zone (subtropical forest), montane zone (warm-temperate and temperate forests), upper montane (cool-temperate forest), subalpine zone (cold-temperate forest), and alpine zone (subarctic forest) [51]. The elevation of this site ranges between 1115 and 3885 m across the subtropical–temperate–subarctic forest zones, and the temperate forest dominates this area. The forest types include conifers, mixed pine–conifer–broadleaf (hereafter mixed), and broadleaf forests. Specifically, this site has a lot of Taiwan red pine (*Pinus taiwanensis*) plantations which were originally managed for wood production and thus account for a major part of the coniferous forest even though it has been suffering from a high risk of forest fire for decades. The pine is therefore listed together with the conifer, mixed, and broadleaf forests as one of the forest types for fuel load inventory and SFL modeling. Correspondingly, each of the forest types was sequentially encoded as 1: Taiwan red pine, 2: conifer, 3: mixed, and 4: broadleaf in this study. The map of forest types was generated using high-resolution ortho-photos obtained from the Taiwan Forestry Bureau.

The airborne lidar scanner data were acquired on 14 December 2018 via Strong Engineering Consulting Company using a P68C-TC plane. A small-footprint, full-waveform lidar system (Riegl LMS-Q780) mounted on the aircraft provided high-accuracy point cloud data. Lidar data were collected at an operating flight altitude of 3400–4000 m (or 1970–2370 m above ground level) with a laser pulse repetition rate of 240–270 KHz. The resulting lidar dataset with ground and canopy point cloud density around 2.5 and 15 points per square meters, respectively, was used to produce a 1.0-m cell resolution of a rasterized digital elevation model (DEM) and digital surface model (DSM) using a linear interpolation technique. Both DSM and DEM were used to produce CHM data for aboveground biomass mapping in a previous study, while only DEM data were used to derive topographic parameters such as elevation, slope, and aspect for this study. The range of the DEM data, classified degree slope (CS), and classified degree aspect (CA), as well as the reciprocal of degree slope (RDS) are shown in Figure 2. Forest type and topographical variables are abbreviated as BioTopo variables hereafter.

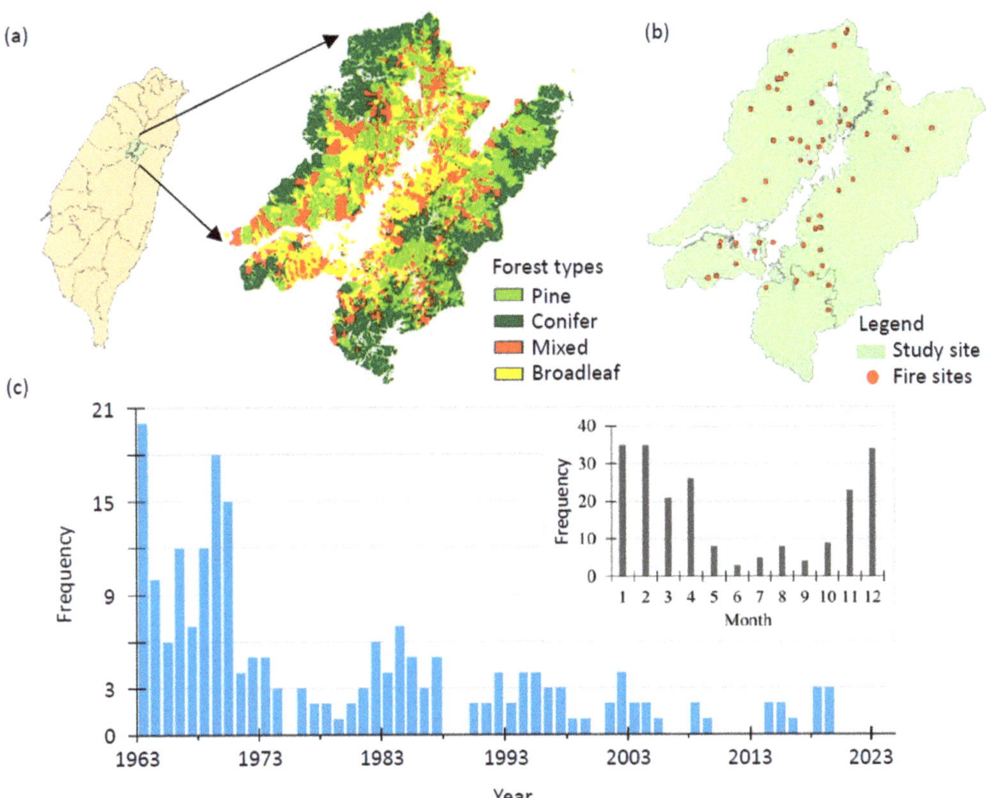

Figure 1. The biological data used in this study. The green polygon in (**a**) shows the geolocation of the forest with detailed forest types at the site. The total area of this study site is about 46,504 ha. Subfigure (**b**) depicts the sites of fires (red dots) occurring during the period from 1963 to 2019, and correspondingly the bar chart in subfigure (**c**) shows the fire frequency counted based on the year/month sequence. The white portion within the study site consists of rivers, inland lakes, and private agriculture land.

In implementing airborne laser scanning, aerial photographs with 80% endlap and 50% sidelap were acquired using a PhaseONE iXA 180 camera. The geometric distortion of the aerial photography had a value of $0.0169°$, $0.0209°$, and $0.0188°$ for the yaw, pitch, and roll error, respectively, which accounted for an overall error of $0.0328°$. The aerial photographs were used to generate a 0.2-m cell resolution orthophoto whose x-, y-, and z-coordinates had RMSE values of 5.05, 2.45, and 1.36 cm, respectively, accounting for an overall RMSE of 5.78 cm, based on the RTK-based ground control points. With the high-resolution DEM and orthophotos of the study site, a series of route planning was implemented in advance for in-situ inventory.

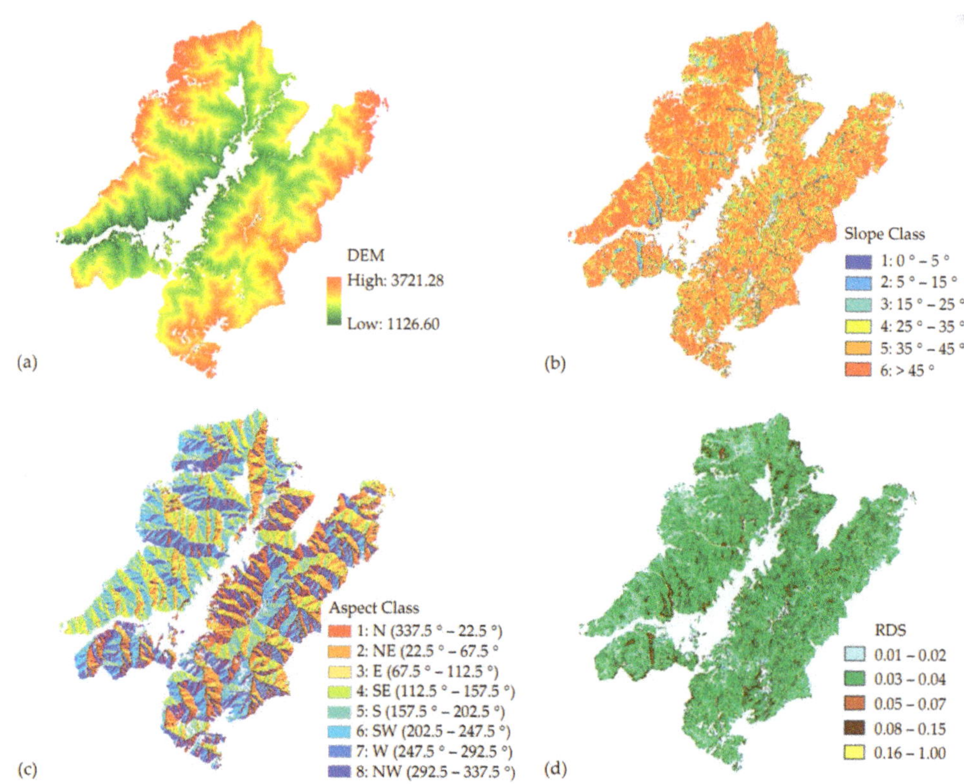

Figure 2. Topographical variables of the study site ((**a**): DEM, (**b**): slope class, (**c**): aspect class, and (**d**): reciprocal of the slope degree).

2.2. Surface Fuel Load Inventory Using Stratified Random Sampling and Cokriging Analysis

This study designed a three-level sampling strategy (Figure 3) to collect sufficient SFL sample data. The sampling process was first implemented using forest type as the strata and basemap-ID as the sampling unit. A series of random numbers indicating a sampling unit was selected in which a level-1 plot with an area of 20 m by 20 m was drawn for the pine, conifer, mixed, and broadleaf forestsover the study site (Figure 4). Second, a level-1 plot was divided into four subplots (level-2) with a size of 10 m by 10 m. Third, every subplot was evenly subdivided into 1-m-gridded microplots (the size is identical to topographical variables derived from ALS DEM data), and the particular microplots (level-3) located at the four corners and the center of each level-2 subplot whose surface fuels including dull, litter, fine-woody debris, coarse-woody debris, and fallen dead wood were investigated. Fourth, an ordinary cokriging method (Equation (1)) was applied to derive a distribution map of fSFL over the level-1 plot in the 1-m gridded cells, and finally, the fSFL of a level-2 subplot was determined by aggregating the values of all 1-m cells within the area of the corresponding subplot. After that, the tSFL of a subplot was determined by summing up the fSFL and the observed FDW mass within the corresponding area.

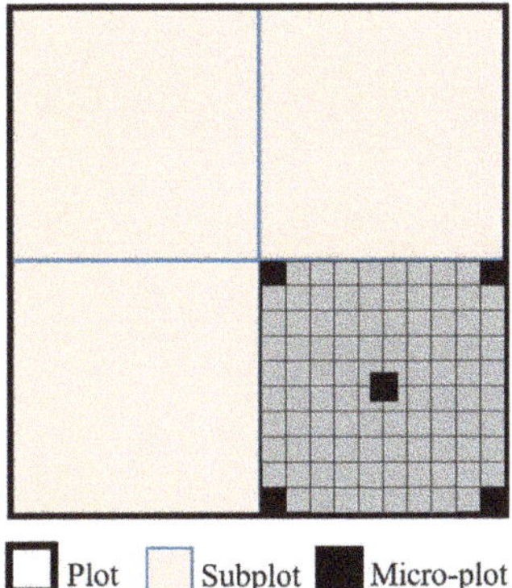

Figure 3. An illustration of the 3-level sampling scheme for the surface fuel load inventory.

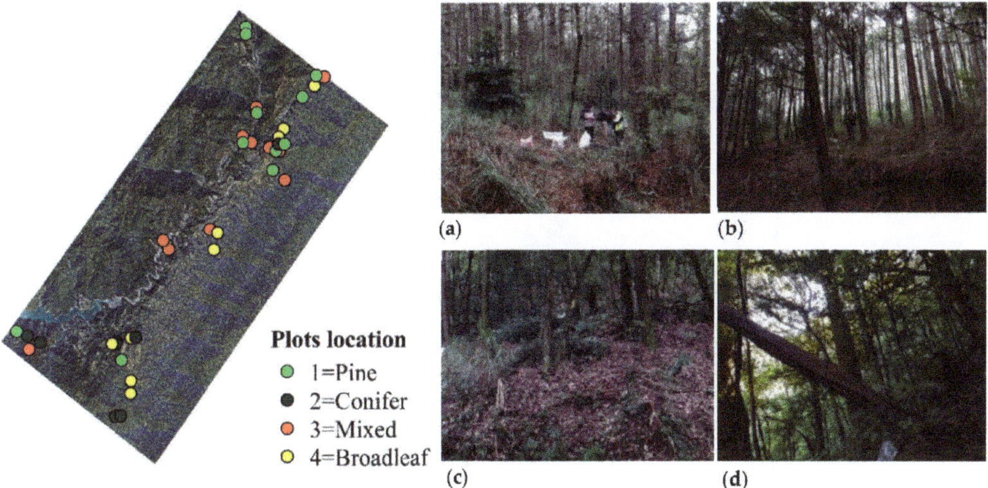

Figure 4. Location of plots for the fuel loading inventory. Subfigures (**a–d**) show the examples of the stands of pine, conifer, conifer-broadleaf mixed, and broadleaf forests.

The in-situ field inventory was carried out between February and July in 2019. Virtual base station real-time kinematic (VBS-RTK) technology was applied to locate the inventory plots in the forest. In addition, a GeoSLAM terrestrial mobile lidar scanner ZEB Horizon was used to capture 3D data for the plots for another aboveground biomass study (Figure 5). Live and dead surface fuels, including 1-hr, 10-hr, 100-hr, and 1000-hr timelag fuels, were collected separately [18,36,52]. For each fuel category, the fresh (or wet) weight and absolute dry weight were measured in-situ and in the laboratory, respectively. For each category, a fuel sample of 500 g was sent to be oven-dried at a temperature of 105 °C for measuring the absolute dry weight [53]. With the dry–fresh weight ratio of each subsample, the value of surface fuel loads of every plot was determined.

Figure 5. A site scene of a post-hoc field survey plot for the surface fuel model performance determination. The upper image displays surface fuel distributed over a pine stand mixed with a few broadleaf saplings and juveniles. The images to the lower left present a profile of litter/duff/soil in the plot, and those in the center to the lower middle show the cleared ground of the subplot area (10 m by 10 m) after the surface fuel collection. The image to the lower right shows the integrated control point for the VBS-RTK positioning and terrestrial mobile lidar scanning with a ZEB Horizon.

Under normal circumstances, surface fuel distribution is mainly governed by the location of vegetation/trees, which are distributed randomly/systematically over the forestland area in natural/planted forests, and is most likely related to topography, particularly the slope and aspect. This is because litterfall as well as fallen dead logs naturally move downslope due to gravity and collect at a position where the movement is blocked by topography or objects. In addition, when the fallen dead wood decomposes/decays, the rolling debris will accumulate more at the bottom of slopes over a limited local space. In other words, the amount of fine surface fuel or dead biomass can gradually change in terms of the physical characteristics over a slope, and fallen dead wood can occasionally interrupt the continuity in stands [54]. The possibility of discontinuity/continuity of surface fuel distribution increases with an increase in the area of interest of factors such as powerful typhoons or tropical cyclones that frequently bring high winds and rainfall, causing biomass to move to lowland areas, inland lakes or the ocean. A plot-based forest inventory is generally area limited, and attributes of the trees and the ground surface within a plot tend to be homogeneous; thus, the surface fuel properties of points in a plot are supposed to be spatially dependent. Because cokriging analysis conducts interpolation according to a semivariogram model, the technique can determine the spatial dependence among points [55,56]. In cokriging analysis, the fSFL in the level-3 1-m cells within a level-1 plot can therefore be presented as a function of the observed amount of surface fuel (the primary variable), with the slope degree, degree aspect, and fuel bed depth the three secondary variables. The cokriging system containing one primary and three secondary variables is defined as:

$$z_0^* = \sum_{i=1}^{n} \alpha_i z_i + \sum_{j=1}^{m} \beta_j x_j + \sum_{k=1}^{p} \gamma_k v_k + \sum_{l=1}^{q} \tau_l u_l \tag{1}$$

where z_0^* is the estimate of Z at location 0; $z_1, z_2, \ldots, z_n$ are the primary data at n locations; $x_1, x_2, \ldots, x_m, v_1, v_2, \ldots, v_p$, and $u_1, u_2, \ldots, u_l$ are the secondary data at m, p, and q locations, respectively. $\alpha_1, \alpha_2, \ldots, \alpha_n, \beta_1, \beta_2, \ldots, \beta_m, \gamma_1, \gamma_2, \ldots, \gamma_p$, and $l_1, l_2, \ldots, l_q$ are cokriging weights to be determined.

In the cokriging analysis, a variety of semivariogram models such as exponential, circular, spherical, and Gaussian models was tested and evaluated using the cross-validation method. Only the semivariogram model fitted with a smaller RMSE was used to generate the fSFL map of a plot. The microplot fSFLs over a level-1 plot area were further used to aggregate the level-2 subplot-based fSFL values and tSFL.

2.3. Modeling of Surface Fuel Loads Using Multiple Linear Regression

In this study, the regression coefficients (**B**) of a multiple linear regression (Equation (2)) were estimated by the method of ordinary least squares using Equation (3).

$$\mathbf{Y} = \mathbf{XB} + \in \tag{2}$$

$$\mathbf{\hat{B}} = \left(\mathbf{X'X}\right)^{-1}\mathbf{X'Y} \tag{3}$$

In Equation (3), **X** is the matrix of independent variables including CS, CA, RDS, and NFT (the normalized forest type, which is determined as the ratio of FT to the maximum value of FT) and **Y** is the dependent vector lnSFL. $\mathbf{X'X} = \mathbf{R}$ is the correlation matrix of independent variables as each of them is standardized by its own mean and standard deviation. In this study, 120 level-2 plots were collected and randomly divided into training and assessing sub-datasets. Based on leave-20%-of-the-plots-out (5-fold) cross-validation [38], all of the sample plots were evenly grouped into five assessing datasets. As a result, the average and standard deviation of performance measures were determined. In order to explore if estimation bias was related to fuel type, an additional evaluation was implemented based on leave-1-fuel-type-out cross-validation.

In the regression analysis, two surface fuel load models were generated, that is, the fSFL-BioTopo model and tSFL-BioTopo model in which fSFL and tSFL represent the fine surface fuel loads and total surface fuel load, respectively, and BioTopo is associated with the biological and topographic variables. The fSFL and tSFL models derived from the training dataset were further applied to derive a distribution map of the fSFL and tSFL for the Dajiaxi National Forest. Accuracy of the fSFL and tSFL maps was evaluated by a cross-validation method via the root-mean-square error (RMSE) and the percentage root-mean-square error (PRMSE). The RMSE is a scale-dependent accuracy measure and is presented on the same scale as the surface fuel load; in contrast, the PRMSE is scale-independent and measures the accuracy as an error percentage relative to the average of observations [22].

3. Results

3.1. The Derived fSFL Semivariogram Models and Their Performance in Estimating Level-1 Plot Surface Fuel Loads

Based on the rule of the smallest estimation bias in the cross-validation of a cokriging method in the level-3 microplot, a semivariogram model with the smallest RMSE was applied to generate a 1-m cell SFL map for the corresponding level-1 plot. Detailed information of the fitted models, prediction maps, RMSEs, and the percentage of RMSE related to the mean average of fSFL observations (PRMSEs) in the inventory data is shown in Table 1. As can be seen, the best prediction of fSFL for every single plot of the forest types was mostly achieved by an exponential semivariogram model. Out of 30 level-1 plots with the tested secondary variables, the variable slope was the most frequently used as the supplementary data to describe the spatial change in fSFL. Table 1 also shows 17 of the 30 models whose fSFL was predicted via the slope or simultaneously via the aspect and/or fuel bed depth. Fuel bed depth was another frequently selected variable which was used alone to account for the distribution of the amount of surface fuel in 11 models and another five models when combined with other secondary variables. In contrast, the aspect appeared to be additional supplementary data when accompanied with the slope.

Table 1. The ordinary cokriging method-derived maps of 1-m cell fine surface fuel loads (fSFL) and for the inventory plots of forest types.

Forest Type	Plot 1	Plot 2	Plot 3	Plot 4	Plot 5	Plot 6	Plot 7	Plot 8
Pine								
Model *	Gaus [1]	Exp [2]	Exp [3]	Exp [3]	Exp [4]	Spher [1]	Exp [4]	Exp [4]
RMSE	0.3036	0.3158	0.1604	0.1999	0.1762	0.1809	0.2667	0.0902
PRMSE	26.31	28.71	15.31	20.91	16.94	19.54	27.61	4.83
Conifer								
Model *	Exp [4]	Exp [4]	Exp [4]	Exp [1]	Exp [1]	Exp [1]	Exp [2]	Exp [4]
RMSE	0.2428	0.2594	0.2023	0.1687	0.3012	0.1423	0.2750	0.2387
PRMSE	32.81	31.18	18.46	29.91	43.15	22.80	39.51	35.42
Mixed								x
Model *	Exp [3]	Exp [5]	Exp [2]	Cir [4]	Exp [4]	Exp [3]	Exp [6]	x
RMSE	0.6082	0.7723	0.2239	0.2777	0.4112	0.2295	0.1988	x
PRMSE	39.14	50.95	36.58	35.06	31.29	30.68	48.88	x
Broadleaf								x
Model *	Exp [2]	Exp [4]	Exp [1]	Exp [5]	Exp [4]	Exp [4]	Exp [2]	x
RMSE	0.2229	0.1215	0.0934	0.2685	0.1332	0.1712	0.3411	x
PRMSE	24.44	17.21	21.13	45.05	23.13	29.12	43.61	x

*: Exp, Spher, Cir, and Gaus represent the exponential, spherical, circular, and Gaussian semivariogram models, respectively. The codes (1), (2), (3), (4), (5), and (6) after the type of semivariogram model indicate a combination of secondary variables used in deriving that semivariogram model. Correspondingly, the codes represent slope, slope–aspect, slope–fuel bed depth, fuel bed depth, slope–aspect–fuel bed depth, and aspect–fuel bed depth, respectively.

Recall the bias of the fSFL prediction map: the best accuracy had an RMSE of 0.0902 kg/100 m^2 or an equivalent error rate of PRMSE = 4.83% for a pine forest while the lowest accuracy had an RMSE of 0.7723 kg/100 m^2 and an PRMSE of 50.95% for a mixed forest. On average, the method of integrating 3-level stratified random sampling and cokriging analysis was able to derive the amount of surface fuels at an RMSE of 0.2533 $\pm$ 0.1390 kg/100 m^2 or a PRMSE of 29.66 $\pm$ 10.64%. In addition, the SFL distribution within the area of every single plot showed quite a different pattern among plots of the pine, conifer, mixed, and broadleaf forests. The natural variation of fSFL in a variety of forest types and topographic features revealed the spatial heterogeneity of fSFL distribution in forests, indicating that the method proposed makes sense for gathering plot-based surface fuel loads with a low cost of labor and time for forest inventories.

3.2. The Level-2 Subplot-Based fSFL-BioTopo Models and Their Performance in Generating the fSFL Map of the Whole Forest

The amount of fine surface fuels of the 120 level-2 subplots in an area of 10×10 m^2 is shown in Table 2 (hereafter a pixel) and was aggregated from the cokriging-derived level-1 SFL map. As can be seen, the pine stands had fSFL values that ranged from 73.71 kg/pixel to 149.75 kg/pixel, with an average of 107.32 ± 24.56 kg/pixel greater than 99.58 ± 56.11 kg/pixel, 81.34 ± 22.90 kg/pixel, and 73.54 ± 16.28 kg/pixel of the mixed stands, conifer stands, and broadleaf stands, respectively. The broadleaf stands on average had obviously smaller fSFL while the mixed stands, whose fSFL values displayed a significant and dramatic change indicated by their standard deviation, was almost three times larger than those of the pine, conifer, and broadleaf stands.

Table 2. The aggregated amount of fine surface fuel loads (fSFL) in level-2 subplots.

Forest Type	Subplot [¶]	Plot 1	Plot 2	Plot 3	Plot 4	Plot 5	Plot 6	Plot 7	Plot 8	AVG	STD
Pine (n1 = 32)	LL	149.75	132.98	75.89	161.72	121.82	63.54	92.59	79.22	107.32	24.56
	LR	122.12	120.37	125.91	126.11	139.38	101.47	87.41	83.19		
	UL	139.70	120.52	73.71	119.99	82.82	84.83	86.04	109.47		
	UR	120.34	110.68	128.56	78.36	97.3	112.78	76.40	109.42		
Conifer (n2 = 32)	LL	89.93	90.90	113.00	91.88	76.28	55.86	66.12	60.31	81.34	22.90
	LR	95.32	76.19	135.27	46.10	76.17	40.48	82.05	48.35		
	UL	93.78	85.37	110.91	68.92	99.70	82.24	90.00	92.86		
	UR	85.60	56.18	138.06	49.73	67.62	68.18	84.53	84.99		
Mixed (n3 = 28)	LL	86.04	223.01	104.74	87.04	225.84	73.07	36.48	x	99.58	56.11
	LR	80.86	227.40	48.80	94.39	174.59	89.64	41.86	x		
	UL	87.53	138.60	63.41	80.39	111.58	73.27	56.27	x		
	UR	91.69	180.47	39.15	87.23	104.04	46.44	34.30	x		
Broadleaf (n4 = 28)	LL	83.50	67.42	54.91	71.69	56.90	65.14	92.40	x	73.54	16.28
	LR	107.28	74.87	53.79	64.40	62.95	58.54	104.31	x		
	UL	74.98	92.06	49.62	85.58	63.65	73.2	84.48	x		
	UR	100.47	77.08	44.76	68.04	70.53	64.07	92.55	x		

[¶]: The abbreviations LL/LR/UL/UR indicate the level-2 subplot on the lower left/lower right/upper left/upper right of a level-1 plot listed in Table 1. The value of each entry has a unit of kg/pixel. A pixel has an area of 100 m^2. AVG and STD represent the respective average and standard deviation of the values for a forest type.

The ANOVA test showed that the derived fSFL-BioTopo model using the fSFL data of level-2 subplots displayed an R-squared value of 0.162 (F = 3.096, $p < 0.005$, $n = 120$). In Equation (4), the dependent variable is the natural log-transformed fSFL (lnfSFL, kg/pixel), and the independent variables are the original or first-order topographical variables (NFT, AC, and SC), their second-order interaction product (NFTxAC, NFTxSC, and ACxSC) and third-order interaction product (NFTxACxSC). Based on the cross-validation test, this model was able to achieve an accuracy of RMSE = 34.10 kg/pixel and PRMSE = 37.59%. In contrast, when the 6-class SC was replaced by four classes (that is, 1: <5°, 2: 5–10°, 3: 10–20°, 4: >20°) and the 8-class AC was regrouped as four classes (1: North, 2: East, 3: South, 4: West), the alternative lnfSFL-BioTopo model (Equation (5)) displayed an R-squared value of 0.173 (F = 8.063, $p < 0.001$, $n = 120$) and had an RMSE = 33.07 kg/pixel and PRMSE = 38.03%. The RMSE and PRMSE differences between Equations (4) and (5) were 1.03 kg/pixel and 0.44%, respectively, which accounts for a relative change rate of 3% and 1% in the RMSE and PRMSE. It was therefore concluded that the performance of the two models was almost identical.

$$\text{lnfSFL} = 3.785174 + 0.819635*\text{NFT} + 0.154689*\text{AC} + 0.416842*\text{SC} - 0.230969*\text{NFTxAC} - 0.671540*\text{NFTxSC} \\ -0.072628*\text{SCxAC} + 0.127985*\text{NFTxACxSC} \tag{4}$$

$$\text{lnfSFL} = 3.911091 + 1.209008*\text{NFT} + 0.076095*\text{AC} + 0.375311*\text{SC} - 0.237776*\text{NFTxAC} - 0.842360*\text{NFTxSC} \\ -0.040685*\text{SCxAC} + 0.130257*\text{NFTxACxSC} \tag{5}$$

3.3. The Level-2 Subplot-Based tSFL-BioTopo Model for Total Surface Fuel Loading Estimation

The detailed information of total surface fuel loads for the 120 subplots is shown in Table 3 in which the italic numbers indicate the fallen dead wood mass of that particular subplot. On average, the largest amount of FDW mass within the subplot area was found in the conifer stand, with an average of 27.57 ± 26.66 kg/pixel, followed in descending order by pine stands (13.42 ± 10.28 kg/pixel), broadleaf stands (12.04 ± 11.55 kg/pixel), and mixed forest stands (3.77 ± 3.44 kg/pixel). In contrast to the fSFL, the increasing amount of FDW mass in pine, conifer, mixed, and broadleaf forest stands was around 6.25%, 8.47%, 2.70%, and 4.68%, respectively; this indicates that the prevalence rate of FDW in the conifer and pine stands was significantly higher than that in the mixed and broadleaf stands.

Table 3. The aggregated amount of total surface fuel loads (tSFL) in level-2 subplots.

Forest Type	Subplot [1]	Plot 1	Plot 2	Plot 3	Plot 4	Plot 5	Plot 6	Plot 7	Plot 8	AVG	STD
Pine (n1 = 32)	LL	149.75	*145.32*	75.89	161.72	*125.38*	*70.93*	92.59	*109.61*	114.03	24.99
	LR	122.12	*132.71*	125.91	126.11	*142.93*	*108.86*	87.41	*113.58*		
	UL	139.70	*132.86*	73.71	119.99	*86.38*	*92.22*	86.04	*139.86*		
	UR	120.34	*123.02*	128.56	78.36	*100.86*	*120.17*	76.40	*139.81*		
Conifer (n2 = 32)	LL	89.93	90.90	113.00	91.88	*77.19*	55.86	66.12	*114.54*	88.23	26.70
	LR	95.32	76.19	135.27	46.10	*77.07*	40.48	82.05	*102.59*		
	UL	93.78	85.37	110.91	68.92	*100.61*	82.24	90.00	*147.10*		
	UR	85.60	56.18	138.06	49.73	*68.53*	68.18	84.53	*139.22*		
Mixed (n3 = 28)	LL	*90.18*	224.96	104.74	87.04	226.88	*83.38*	*37.88*	x	102.27	55.71
	LR	*85.00*	229.36	48.80	94.39	175.62	*99.94*	*43.26*	x		
	UL	*91.67*	140.55	63.41	80.39	112.61	*83.58*	*57.67*	x		
	UR	*95.83*	182.43	39.15	87.23	105.08	*56.75*	*35.70*	x		
Broadleaf (n4 = 28)	LL	83.50	67.42	54.91	*72.18*	56.90	88.72	*92.40*	x	76.98	16.63
	LR	107.28	74.87	53.79	*64.89*	62.95	82.12	*104.31*	x		
	UL	74.98	92.06	49.62	*86.06*	63.65	96.78	*84.48*	x		
	UR	100.47	77.08	44.76	*68.53*	70.53	87.65	*92.55*	x		

[1]: The same as in Table 2. The italics is used to highigh the partiular plots.

The inclusion of FDW mass did not change the relationship of the total surface fuel loads among the forest types, that is, the pine stand had the highest amount of tSFL, followed by the mixed, conifer, and broadleaf stands. The significant variation in tSFL in the mixed stands remained significantly larger than that in the other forest types. The R-squared value of the derived lntSFL-BioTopo model as shown in Equation (6) was 0.144 (F = 2.701, $p < 0.013$, $n = 120$). The performance of this model in predicting tSFL of the whole forest had an RMSE of 35.02 kg/pixel, corresponding to a PRMSE of 36.57%. Similarly, Equation (7) shows the alternative lntSFL-BioTopo model for tSFL estimation using NFT, 4-classes SC, 4-classes AC, and their interaction product variables. The R-squared value of this model was 0.168 (F = 7.836, $p < 0.001$, $n = 120$), and RMSE and PRMSE were 33.81 kg/pixel and 37.85%, respectively. The performance measures of the two models were also quite close, with a difference in RMSE and PRMSE of 1.21 kg/pixel and 1.28%, respectively. In contrast to Equation (6), the relative change in the two measures of Equation (7) was 3% and 4%.

$$\text{lntSFL} = 4.481948 + 0.171296*\text{NFT} + 0.082297*\text{AC} + 0.153252*\text{SC} - 0.170669*\text{NFTxAC} - 0.364650*\text{NFTxSC} \\ -0.041167*\text{SCxAC} + 0.092501*\text{NFTxACxSC} \tag{6}$$

$$\text{lntSFL} = 4.433072 + 0.635037*\text{NFT} - 0.012423*\text{AC} + 0.163312*\text{SC} - 0.145670*\text{NFTxAC} - 0.541648*\text{NFTxSC}$$
$$-0.000879*\text{SCxAC} + 0.077361*\text{NFTxACxSC} \tag{7}$$

For the 5-fold cross-validation, the respective average RMSE and PRMSE were 34.57 ± 8.76 kg/pixel and $37.58 \pm 6.11\%$ for the fSFL and 35.36 ± 9.09 kg/pixel and $36.38 \pm 5.98\%$ for the tSFL. However, for the leave-1-fuel-type-out cross-validation, a significant difference in estimation performance among the four trials was observed. When the pine, conifer, mixed, or broadleaf plots were sequentially excluded from modeling and then used for validation, the PRMSE for the fSFL estimation was 36%, 34%, 62%, and 109%, respectively, but 32%, 35%, 62%, and 102% for the tSFL estimation (Table 4). Obviously, the mixed and broadleaf surface fuel loads tended to be significantly over-estimated if they were not included in modeling. Both fSFL and tSFL of the broadleaf stands appeared to be significantly over-estimated because the surface fuel load in the stands was lower but under-estimated for the mixed stands due to the evidently diverse surface fuel load in the stands. In other words, a smaller PRMSE only occurred in the estimation when the plots of pine or conifer were not used for modeling, and the prediction bias appeared to be related to fuel type. Collecting appropriate numbers of plots for each fuel type for deriving a general model or having a sufficient number of plots for deriving fuel type-specific models should be able to prevent the significant over-estimation and/or under-estimation problem.

Table 4. Summary of the performance of fSFL and fSFL models based on cross-validation.

Model [¶]	fSFL Model R^2	RMSE (kg/m²)	PRMSE (%)	tSFL Model R^2	RMSE (kg/m²)	PRMSE (%)
Equation (4)/ Equation (6)	0.162 (F = 3.096, $p < 0.005$, $n = 120$)	34.10	37.59	0.144 (F = 2.701, $p < 0.013$, $n = 120$)	35.02	36.57
Equation (5)/ Equation (7)	0.173 (F = 8.063, $p < 0.001$, $n = 120$)	33.07	38.03	0.168 (F = 7.836, $p < 0.001$, $n = 120$)	33.81	37.85
DeGroup 1	0.154 (F = 2.295, $p = 0.034$, $n = 96$)	23.28	28.56	0.128 (F = 1.844, $p = 0.089$, $n = 96$)	24.10	28.84
DeGroup 2	0.167 (F = 2.526, $p = 0.020$, $n = 96$)	25.82	33.75	0.164 (F = 2.469, $p = 0.023$, $n = 96$)	25.73	31.44
DeGroup 3	0.182 (F = 2.801, $p = 0.011$, $n = 96$)	46.70	46.30	0.145 (F = 2.128, $p = 0.049$, $n = 96$)	47.96	45.59
DeGroup 4	0.136 (F = 1.986, $p = 0.066$, $n = 96$)	39.25	37.91	0.120 (F = 1.713, $p = 0.116$, $n = 96$)	39.05	36.11
DeGroup 5	0.193 (F = 3.016, $p = 0.007$, $n = 96$)	37.82	41.36	0.188 (F = 2.908, $p = 0.009$, $n = 96$)	39.94	39.90
DePine	0.106 (F = 1.352, $p = 0.237$, $n = 88$)	38.47	35.84	0.051 (F = 0.609, $p = 0.747$, $n = 88$)	37.01	32.45
DeConifer	0.232 (F = 3.455, $p = 0.003$, $n = 88$)	27.82	34.20	0.240 (F = 3.607, $p = 0.002$, $n = 88$)	30.70	34.79
DeMixed	0.257 (F = 4.160, $p = 0.001$, $n = 92$)	61.56	61.82	0.246 (F = 5.231, $p < 0.001$, $n = 92$)	63.02	61.62
DeBroadleaf	0.342 (F = 6.242, $p < 0.001$, $n = 92$)	80.11	108.94	0.258 (F = 4.167, $p = 0.001$, $n = 92$)	78.89	102.48

[¶]: The specific models (Equations (4) and (6)) used for generating surface fuel load maps of the whole study site. Performance measures of the models were determined based on all of the plots. DeGroup 1–5 represents the five evaluations of leave-20%-of-the-plots-out cross-validation; the respective average RMSE and PRMSE were 34.57 ± 8.76 kg/pixel and $37.58 \pm 6.11\%$ for the fSFL and 35.36 ± 9.09 kg/pixel and $36.38 \pm 5.98\%$ for the tSFL. DePine, DeConifer, DeMixed, and DeBroadleaf represent the four evaluations of leave-1-fuel-type-out cross-validation; the respective average RMSE and PRMSE were 51.99 ± 20.31 kg/pixel and $60.20 \pm 30.20\%$ for the fSFL and 52.40 ± 19.51 kg/pixel and $57.83 \pm 28.21\%$ for the tSFL. Both cross-validations included six slope classes and eight aspect classes.

4. Discussion

4.1. The Uncertainty of Surface Fuel Loading Estimation in fSFL and tSFL Models

Because the relative error in the two fSFL-BioTopo models (Equations (4) and (5)) was small, and that of the two tSFL-BioTopo models (Equations (6) and (7)) was almost identical, estimation performance of the paired models for both fSFL and tSFL can be considered equal. The uncertainty of surface fuel models is therefore discussed based primarily on the estimation of Equations (4) and (6).

The predicted values of surface fuel loading over the whole area of the study site are shown in Figure 6 and are summarized in Table 5. Based on the prediction maps of the whole forest, the fSFL mass of the pine stands ranged from 1.42 to 18.44 ton/ha and averaged 10.67 ± 1.72 ton/ha. Taking into account the forest areas, there was approximately 379,718.31 tons of fine surface fuel within the pine stands. This is the largest amount of fine fuel mass among the forest types over the whole forest. In contrast, the fine fuel mass of the conifer, mixed, and broadleaf stands averaged 9.29 ± 1.10 ton/ha, 8.22 ± 1.53 ton/ha, and 7.18 ± 2.40 ton/ha, respectively, and resulted in a total mass of 130,433.75 tons, 57,555.40

tons, and 52,283.57 tons. The relative amount of fine surface fuel mass among the four forest types derived from the lnfSFL-BioTopo model is quite similar to that observed in the subplot values. A similar trend appeared in the total surface fuel map derived from the lntSFL-BioTopo model. In addition, the amount of tSFL of the forest types showed the same sequence of pine > conifer > mixed > broadleaf, with an average value of 9.61 ± 1.01 ton/ha, 8.81 ± 1.03 ton/ha, 8.40 ± 1.49 ton/ha, and 7.71 ± 2.25 ton/ha, respectively. The results show that the lnfSFL-BioTopo and lntSFL-BioTopo models are capable of performing fSFL and tSFL estimations, and the estimates are generally consistent with the sampling inventory results. This reveals that the distribution map of surface fuel loading generated by each of the models is able to provide a baseline for accounting for the accumulation of surface fuel loads over time.

As noted by comparing the descriptive statistics of the two models in Table 5, the mixed and broadleaf fSFL estimate was generally smaller than the tSFL estimate by 0.18 ton/ha and 0.53 ton/ha while the pine and conifer stands' fSFL was generally greater than the tSFL by an amount of 1.05 ton/ha and 0.48 ton/ha. Mathematically, the situation of fSFL > tSFL in a forest stand should not happen according to their definitions as described in Section 2.2. Recall the descriptive statistics of inventory data shown in Tables 2 and 3: the fallen dead wood mass in the pine and conifer stands was almost three times higher than that in the mixed and broadleaf stands. In view of the range of surface fuel loading estimates of the models, the tSFL of pine and conifer stands was apparently smaller than the fSFL, with values of 10.66 vs. 17.02 and 10.60 vs. 12.29. In contrast to the mixed and broadleaf stands (11.75 vs. 11.87 and 13.56 vs. 13.30 for the range of fSFL and tSFL estimates), a significant uncertainty occurred in the estimation of the lntSFL-BioTopo model, and the source of uncertainty should be the inclusion of fallen dead wood mass. This kind of estimation uncertainty was also observed in the estimation bias of the subplots (Figure 7) in which an extra amount of bias in the estimates of tSFL was highlighted by an arrow with respect to those corresponding hollow bars.

Table 5. A summary of surface fuel loadings with respect to forest types in Dajiaxi National Forest.

Models	Forest Types	Areas (ha)	Minimum (ton/ha)	Maximum (ton/ha)	Average (ton/ha)	STD (ton/ha)	Total (tons)
lnfSFL-BioTopo	Pine	13,070	1.42	18.44	10.67	1.72	139,445.59
(Equation (4))	Conifer	14,039	1.04	13.33	9.29	1.10	130,433.75
	Mixed	7001	1.02	12.90	8.22	1.53	57,555.40
	Broadleaf	7280	0.66	13.96	7.18	2.40	52,283.57
	Sum	41,390	0.66	18.44	9.17	2.08	379,718.31
lntSFL-BioTopo	Pine	13,070	1.28	11.95	9.61	1.01	125,665.94
(Equation (6))	Conifer	14,039	1.03	11.62	8.81	1.03	123,659.56
	Mixed	7001	1.06	12.81	8.40	1.49	58,835.28
	Broadleaf	7280	0.76	14.32	7.71	2.25	56,147.02
	Sum	41,390	0.76	14.32	8.80	1.55	364,307.80
Difference	Pine	13,070	−6.50	1.34	−1.05	1.04	−13,779.65
(fSFL–tSFL)	Conifer	14,039	−3.34	0.83	−0.48	0.45	−6774.19
	Mix	7001	−2.52	0.79	0.18	0.26	1279.88
	Broadleaf	7280	−2.19	1.10	0.53	0.32	3863.45
	Sum	41,390	−6.50	1.34	−0.37	0.89	−15,410.51

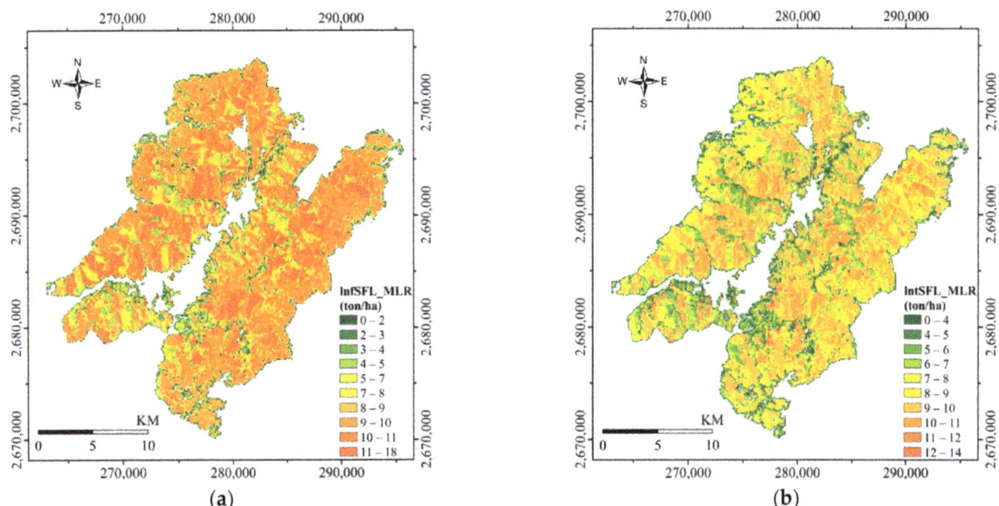

Figure 6. The surface fuel loading regression model-derived fSFL map (**a**) and tSFL map (**b**) of the Dajiaxi National Forest.

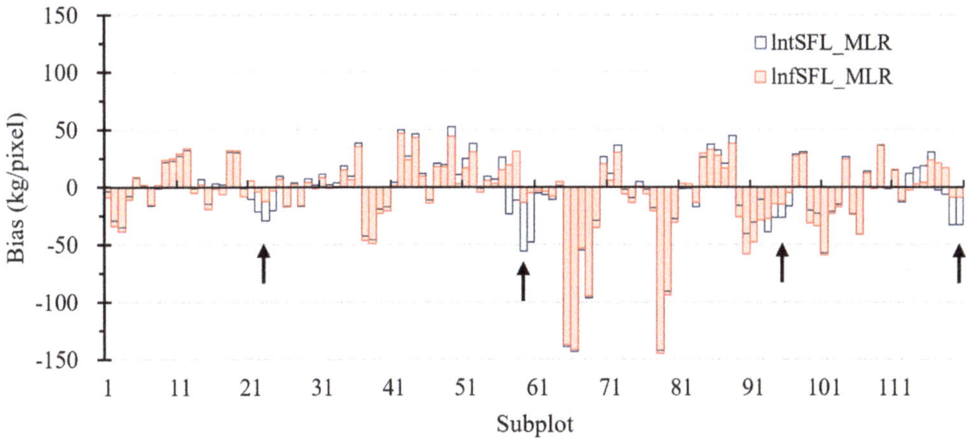

Figure 7. Prediction bias of surface fuel loadings in the lnfSFL–BioTopo and lntSFL–BioTopo models. The *x*–axis represents the identity of inventory samples, and the *y*–axis is the bias determined by the difference between estimated and inventory data. A negative value indicates an underestimation while a positive value represents an overestimation. The arrow below the hollow bars indicates that the corresponding inventory subplots (identity number: 21–24/57–60/93–96/119–120) had a significant FDW mass as well as a larger estimation bias in the tSFL estimates.

4.2. The Dependency of Estimation Bias on the Amount of Surface Fuel Loads

In Section 3, the cross-validation tests showed that fSFL and tSFL were predicted with a similar accuracy of RMSE (33.84 kg/pixel vs. 34.76 kg/pixel) and PRMSE (37.29% vs. 36.28%), indicating that the lnfSFL-BioTopo and lntSFL-BioTopo models have almost the same ability to predict surface fuel mass in the forest. The similarity was revealed through the bias of the inventory subplots as shown in the bar chart in Figure 7. However, the difference in fSFL and tSFL in a one-hectare areal basis over the whole forest of the study site showed more than 50% of the areas whose fSFL was larger than tSFL. This is particularly evident in the pine and coniferous forests (Figure 8).

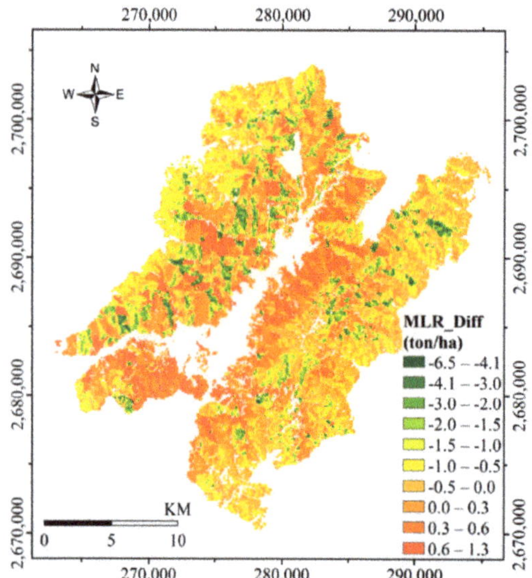

Summary of the classes of difference(tSFL – fSFL)

Classes (ton/ha)	Pine	Conifer	Mix	Broadleaf	SUM	%
-6.5 – -4.1	175	0	0	0	175	0.42
-4.1 – -3.0	567	1	0	0	568	1.37
-3.0 – -2.0	1446	23	6	3	1478	3.57
-2.0 – -1.5	1418	166	24	5	1613	3.90
-1.5 – -1.0	2001	1879	47	24	3951	9.55
-1.0 – -0.5	2855	4284	151	68	7358	17.78
-0.5 – 0.0	3184	5659	527	215	9585	23.16
0.0 – 0.3	1029	1876	4881	1093	8879	21.45
0.3 – 0.6	306	136	1340	2486	4268	10.31
0.6 – 1.3	89	15	25	3386	3515	8.49
SUM	13,070	14,039	7001	7280	41,390	100.00

Figure 8. Map of the difference between MLR-derived tSFL and fSFL estimates over the study site. The negative values are a result of larger fSFL and smaller tSFL, indicating prediction uncertainty in the lntSFL–BioTopo model.

Although the estimation appeared to be bias-independent and randomly based on the scatter plot of bias vs. estimates in Figure 9a,b, the prediction bias still revealed a linear dependency on the observed value. This is evident in Figure 9c,d, where the original variable was converted to the deviation from the mean of observed values, i.e., fSFL-fSFL$_{avg}$ or tSFL-tSFL$_{avg}$. The prediction bias can be presented as a negative linear function of the transformed variable, i.e., the bias is most likely to be compensated by the fuel mass itself, and the bias-adjustment value or compensatory value y$_{com}$ can be determined by the linear models shown in Figure 9c,d. The compensated fSFL and tSFL estimates can be retrieved by subtracting the y$_{com}$ from the original estimates of fSFL and tSFL using Equations (4) and (6), respectively. Accordingly, the compensated estimates of fSFL and tSFL through the bias-adjustment models in Figure 9c,d were significantly improved to 11.64 kg/pixel and 12.84% and 11.37 kg/pixel and 11.87% for RMSE and PRMSE, respectively. The bar chart shown in Figure 10a,b demonstrates the improvement of prediction bias for the original estimation and the compensated estimation of fSFL and tSFL with respect to each of the subplots.

4.3. A Possible Strategy for Improving Surface fuel Load Mapping

The prediction bias of the lnfSFL-BioTopo model and the lntSFL-BioTopo model (Equations (4) and (6)) can also be presented as a nonlinear function of the observed value of surface fuel loads. Estimates derived from the regression models can be over- or under-estimated and correspondingly generate a positive or negative prediction bias, determined as $\hat{y} - y$. Each bias can be adjusted to positive by introducing an additive component, c, to compensate for the negative values without changing the relationship between the bias and the observed values. Assuming that the compensated offset value c is $\geq$ the maximum observed value of tSFL, the reciprocal transformation of "fSFL$_{bias}$ + c" and "tSFL$_{bias}$ + c" can be presented as an exponential growth function of the observed value of fSFL or tSFL, respectively (Figure 11). The transformed bias is helpful to diagnose the prediction bias behavior with respect to the original scale of the fSFL and tSFL observed values. As shown in Figure 11a,b, the R-squared value of the two exponential growth models was 0.8872 and 0.8713 when a constant value of 250 was assigned to the compensative offset.

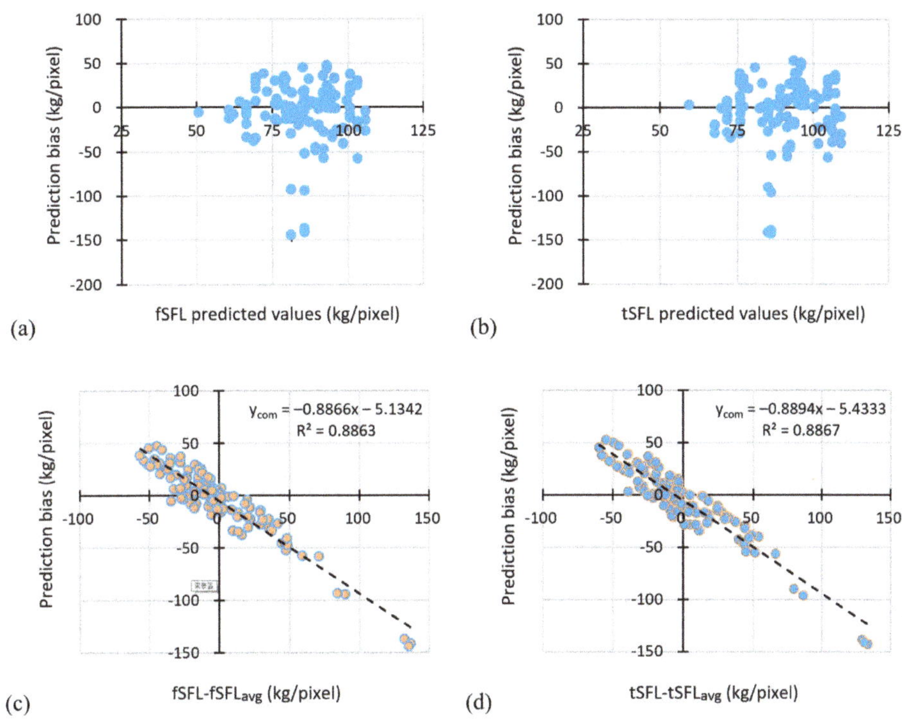

Figure 9. An examination of prediction bias of surface fuel loadings. The independency between the bias and predicted values of fSFL (**a**) and tSFL (**b**) via Equations (4) and (6). A linear dependency of the surface fuel mass estimation bias on the deviation from the mean of the observed values was formulated as a bias-adjustment model of Bias–adjfSFL–MLR for the estimate of fSFL shown in (**c**) and Bias–adjtSFL–MLR for the estimate of tSFL shown in (**d**). The compensated value (y_{com}) derived from the bias-adjustment model can be applied to appropriately restore surface fuel loading by subtracting y_{com} from the originally estimated value of fSFL or tSFL.

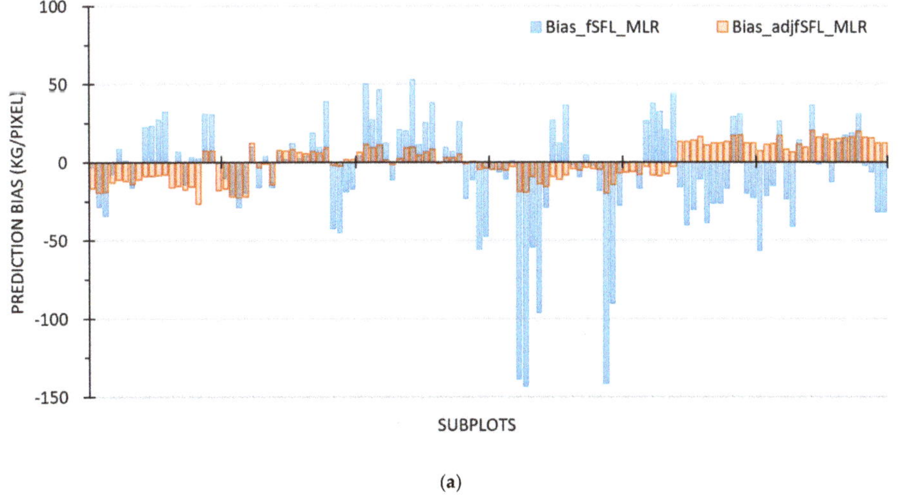

Figure 10. *Cont.*

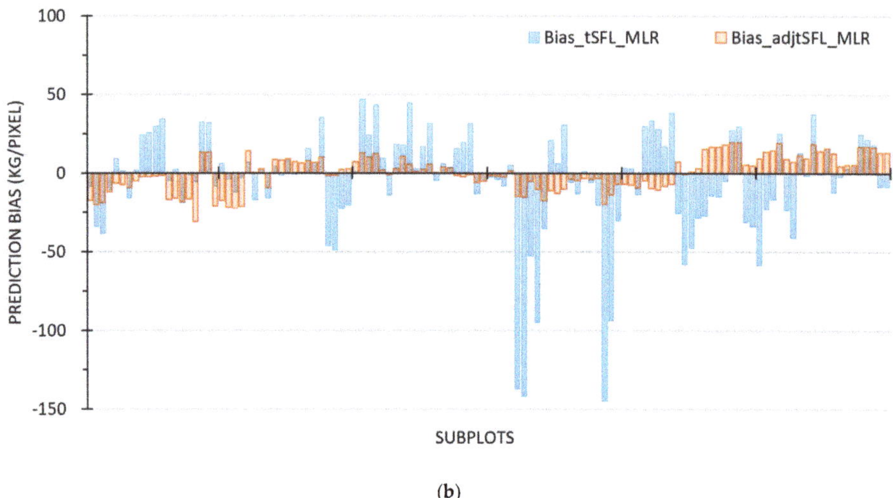

(**b**)

Figure 10. Changes in bias for the surface fuel loading estimation between the original lnfSFL–BioTopo model and its bias-adjustment model (**a**) and the lntSFL–BioTopo model and its bias-adjustment model (**b**).

In the multiple linear regression models (Equations (4) and (6)), a larger surface fuel load tended to be underestimated while a smaller fuel load was most likely overestimated, revealing that a smaller range of surface fuel load estimates was made by the lnfSFL-BioTopo model and the lntSFL-BioTopo model. Similarly, based on the reciprocal transformation, a smaller surface fuel load was generally found to have a smaller value of transformed bias and vice versa. The exponential growth function in Figure 11a,b shows that the greater the surface fuel load, the more significant the bias in the estimation. This is most likely induced by a shortage of samples of larger fuel loads in deriving the multiple linear regression models because the number of samples with respect to the diverse amounts of surface fuel loads is generally proportional to the size of the corresponding populations.

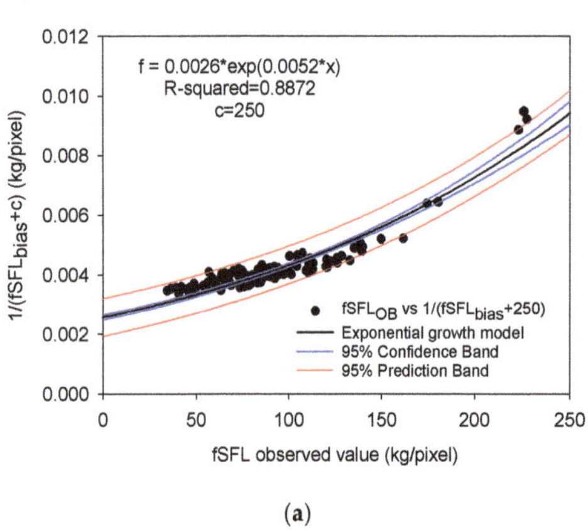

(**a**)

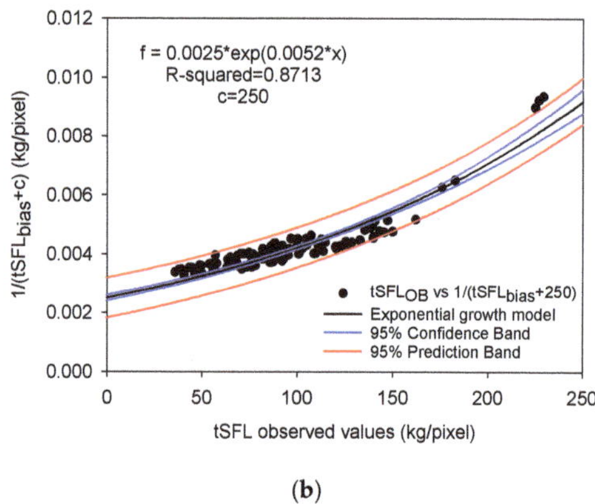

(**b**)

Figure 11. Nonlinear dependency of the surface fuel load estimation bias on the amount of observed surface fuel load. The transformed bias of the original lnfSFL-BioTopo model and the lntSFL-BioTopo model was positively related to the values of fSFL (**a**) and tSFL (**b**).

Figure 12 provides a generalized logistic distribution of the tSFL and fSFL observed values. The probability density function of this distribution is abbreviated as GL(x; α, β, γ for the shape, scale, and location parameters. Specifically, the distributions for tSFL and fSFL are GL(tSFL; 0.1804, 18.57, 90.04) and GL(fSFL; 0.2109, 17.38, 84.35), respectively. Figure 12a,b reveals a right skewed distribution, indicating the rareness of larger surface fuel load samples. In this study, the standard deviation and mean of the inventory data were 36.63 kg/pixel and 95.76 kg/pixel for the fSFL in the forests and 35.80 kg/pixel and 90.70 kg/pixel for the tSFL, which resulted in a coefficient of variation (CV) of around 0.38–0.39 for the forests. Statistically, the CV is used to examine the extent of data variability in relation to the arithmetic mean of a variable. The evidence of both the generalized logistic distributions of samples and CV suggests that increasing the number of inventory samples covering a wide range of a variable, particularly with a sufficient number of observations for diverse fuel loads, would be expected to upgrade a model's estimation performance.

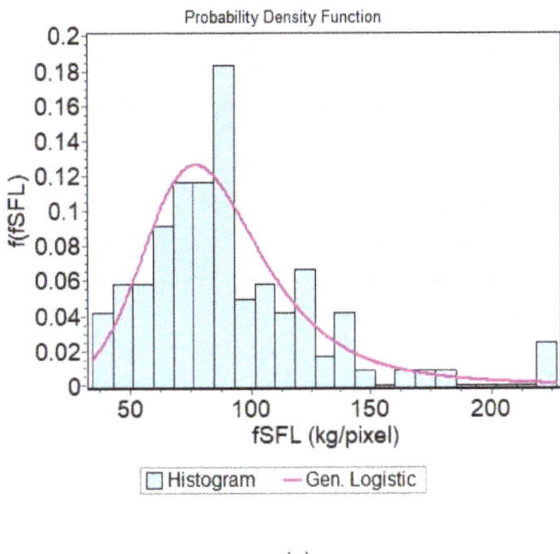

(a)

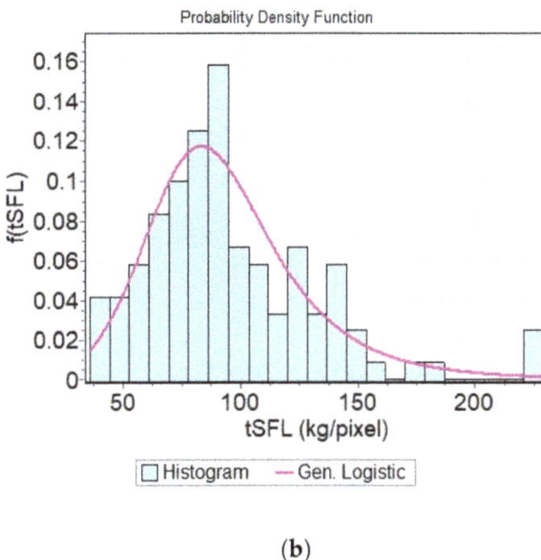

(b)

Figure 12. Probability density function of the observed values of surface fuel mass. The distribution of inventory fuel mass data is GL(tSFL; 0.1804, 18.57, 90.04) (**a**) and GL(fSFL; 0.2109, 17.38, 84.35) (**b**) where the *x*-axis is the observed value of fuel loads and the *y*-axis is the percentage frequency of individual fuel loads.

4.4. An Examination of the Appropriateness of the Cokriging-Based Surface Fuel Mapping Method

An additional independent inventory of surface fuel loads was carried out in an area of 10 × 10 square meters for examining the appropriateness of the sampling scheme in deriving the plot-based SFL map. The SFL of the whole plot was 100collected. As shown in Table 6, a few samples of 2-m size microplots located in the black cells of the locational template (template no. 2–5, with 5, 9, 13, and 25 samples), were used to derive a cokriging semivariogram model for generating the SFL prediction map. The RMSE of the four models was 0.65, 0.80, 0.69, and 0.40 kg/m^2, which is equivalent to a PRMSE of 29.71%, 39.05%, 34.34%, and 17.54%, respectively, indicating the best accuracy was achieved by a geospatial cokriging model that used all samples from the whole plot area at the scale of a 2-m size sampling scheme. The difference in PRMSE between the fourth case and the first case was approximately 12%, indicating the proposed method to collect surface fuel loads of inventory plots is a viable approach.

In contrast, templates 1 and 6 show alternative sampling schemes at a 1-m scale. The cokriging model derived using template number 6 had a prediction accuracy of RMSE 0.05 kg/m^2 and PRMSE 2.30%; the predicted values were quite close to the observed values. Compared with template 2, the smaller values of both RMSE and PRMSE achieved by template 1 also reveal the appropriateness of the proposed method in reducing labor costs while retaining accuracy for generating surface fuel load maps. The 1-m size cell is therefore defined as the minimum sampling unit (MSU). Table 7 further demonstrates the extrapolation map of surface fuel loads extended from a limited predefined map of the five sample microplots. The RMSE and PRMSE did not show significant differences among the four centralized sampling schemes (denoted as CenLL, CenLR, CenUL, CenUR), but the two measures showed an evident increase in the four edged sampling schemes. The results indicate that the extrapolation is not appropriate for deriving surface fuel loads of inventory plots, and the MSUs should be distributed along the boundary as well as in the center of a plot.

In published articles, much research demonstrates that the performance in the estimation of surface fuel loads using lidar data and optical images varies dramatically. From the perspective of PRMSE, the integrated approach of diverse remote sensing data was around 20–38% for a dense coniferous forest located in central Greece [33] and 37–98% for a bark beetle-affected forest in eastern Grand County in north-central Colorado [57]. A better accuracy of PRMSE around 5–47% was achieved for an upland oak-dominated forest in Kentucky using small footprint full-waveform lidar data [58]. According to Franke et al. [59], a PRMSE ranging from 21 to 41% was achieved when measuring diverse coarse woody debris in a forest savanna in the Brazilian Cerrado using only multi-temporal Landsat OLI images. In contrast, the 37% of PRMSE achieved in this study is quite close to the moderate accuracy revealed in previous studies, reflecting the potential of applying stratified random sampling to forest types, topographical variables, and inventory data in generating baseline information for surface fuel loading. The classification of forest types was determined based on the biological conditions of the study site. This system is quite similar to the fuel types of coniferous, deciduous, mixed wood, slash, and open grassland, as defined in the Canadian Fire Behavior Prediction System (FBPS) [36]. Various research has demonstrated the feasibility of integrating ALS and optical images to map the fuel types (alternatively fuel models), such as the ones defined by the Northern Forest Fire Laboratory (NFFL) [36,41,49,50,60,61]. The proposed algorithm for mapping the surface fuel load is therefore most likely able to substitute the fuel types of the FBPS, NFFL, and NFDRS classification systems to moderately improve the mapping performance for forests with undulating terrain morphology in mountainous area.

Table 6. A comparison of surface fuel load mapping using different numbers of samples within a spatial scale of a 10-m size plot.

Template No.	1	2	3	4	5	6
Locational template and the number of samples (NS) for deriving model [¶]	NS = 5@1 m	NS = 5@2 m	NS = 9@2 m	NS = 13@2 m	NS = 25@2 m	NS = 100@1 m
SFL Prediction map						
Semivariogram model	Exponential	Exponential	Gaussian	Gaussian	Gaussian	Exponential
Secondary variables	slope, aspect, fuel bed depth	slope, aspect, fuel bed depth	slope, aspect, fuel bed depth	slope, aspect, fuel bed depth	slope, aspect, fuel bed depth	slope, aspect, fuel bed depth
RMSE (kg/m^2)	0.59	0.65	0.80	0.69	0.40	0.05
PRMSE (%)	26.58	29.71	39.05	34.34	17.54	2.30

[¶] The black and gray boxes show the training and testing samples for cokriging model derivation and accuracy evaluation, respectively. The size value comes after the symbol "@" presenting the area of a microplot.

Table 7. A comparison of appropriateness of extended prediction based on the predefined cokriging surface fuel load map [¶].

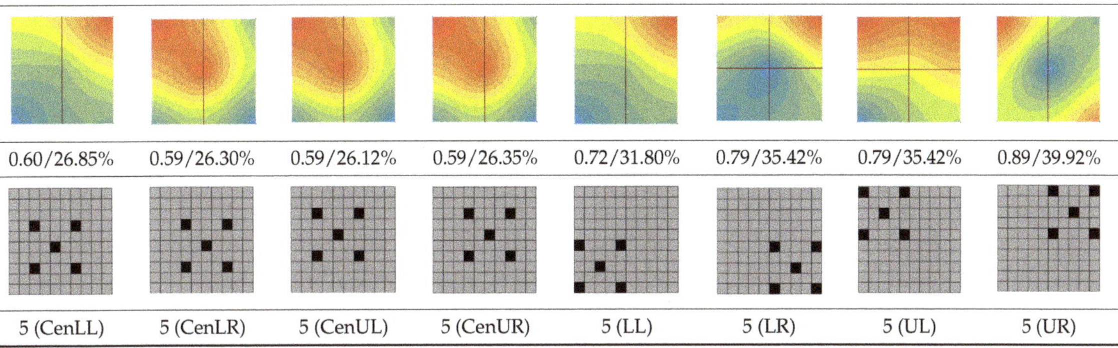

7	8	9	10	11	12	13	14
0.60/26.85%	0.59/26.30%	0.59/26.12%	0.59/26.35%	0.72/31.80%	0.79/35.42%	0.79/35.42%	0.89/39.92%
5 (CenLL)	5 (CenLR)	5 (CenUL)	5 (CenUR)	5 (LL)	5 (LR)	5 (UL)	5 (UR)

[¶] The black and gray boxes show the training and testing samples for cokriging model derivation and accuracy evaluation, respectively. The surface fuel load map for the particular area bounded by the training samples was first generated through the cokriging semivariogram model and then extended to the extent defined by templates no. 7–14. The xx/xx% values below each map specify the RMSE/PRMSE that was derived from the corresponding template.

5. Conclusions

Surface fuel load estimation and mapping are of particular importance, not merely for the origin of fire and better prediction of fire spread and intensity, but also for understanding the source of soil nutrients. Decomposed vegetative mass can easily enter into nutrient cycling and support the needs of vegetation growth and forest development. An in-situ field inventory is the direct method to collect real data of surface fuel load in a forest, but the method is only implementable over limited or small areas in the view of costs of labor, time efficiency, and finance. In general, the surface fuel load over a forest area is a result of the accumulation of litterfall and dried and short-lived vegetation mass as well as fallen dead wood, and wildfire consumes surface fuel. A fire-behavior model can be used to chart the post-fire fuel dynamics when the dynamics of fuel accumulation are established based on the historical records of fire regimes and initial surface fuel load [62]. Empirical models with predictors derived from lidar data and/or satellite images provide alternative methods of estimation at a certain accuracy or uncertainty. In fact, surface fuel masses generally form a dense cover on the ground surface. It is impossible for lidar pulses to penetrate the fuel bed and reach the bare ground. Consequently, measuring the surface fuel load directly with lidar point cloud data from the air and even the ground is obviously quite challenging.

The amount of surface fuel mass will change over time due to diverse influences induced by the interaction of biological, physical, and climatological factors; therefore, the information revealed through a map of surface fuel distribution is most likely valid or practical only for a limited period. Frequent updates of fuel maps are required for efficient management of forest fuel in order to control fire risks. In view of the economic cost of fuel mapping, a method of deriving accurate surface fuel load maps is needed that will be both easier to implement and at a lower cost. Considering the complexity of undulating terrain morphology and inaccessibility of vehicles in mountainous areas, the proposed method for estimation and mapping of surface fuel loads using topographic variables and classified fuel models (forest types) is highly appropriate to meet this need. To implement the 3-level stratified random sampling based approach for surface fuel load mapping, the user should apply a fuel type (also fuel models) classification as needed and then carry out inventories to collect data for generating a map of the surface fuel load.

For deriving a reliable prediction of surface fuel loads of an inventory plot, it is recommended that the minimum sampling unit for collecting surface fuel should include the four corners and the central position of an inventory plot in order to allow the cokriging method to achieve an accurate prediction. In the proposed method, the orthogonal decomposition design of the plots was mainly for the convenience of linking with raster-based remote sensing data. The size of a plot area can be determined based on the pixel size of the fuel type map and the topographic map. The plot design can be flexible in the geometries and sizes that allow compatibility with a variety of forest inventory systems, for example, the forest inventory and analysis (FIA) plot design of the USDA that establishes three additional plots next to a core plot at a fixed distance and in three directions [63]. The plot design of the Indonesia National Forest Inventory is a systematic cluster design. It is composed of clusters, temporary sample plots, and permanent sample plots. The fuel load data collected via the Cluster-TSP-PSP plot design [64,65] can directly adopt the proposed approach.

Biological variables are the primary leading factors of the surface fuel load. Some of the factors are likely to change over time due to growth, competition among trees, and disturbances. To address the temporal-related changes, a spatiotemporal dynamic model of biological mass transition would be a critical solution for better surface fuel load management.

Author Contributions: Conceptualization and methodology, C.L.; formal analysis, S.-E.M. and C.L.; investigation, S.-E.M., L.-P.H., C.-IC., P.-T.L., Z.-K.Y. and K.-T.L.; resources, C.L. and L.-P.H.; writing-original draft preparation, S.-E.M. and C.L.; writing-review and editing, C.L.; visualization, S.-E.M. All authors have read and agreed to the published version of the manuscript.

Funding: This research was completed with the financial support from the projects TFBC-1060113 sponsored by the Forestry Bureau, Taiwan, and MOST108-2621-M-415-001 and MOST109-2221-E-415-003, sponsored by the Ministry of Science and Technology, Taiwan.

Institutional Review Board Statement: Not applicable for studies not involving human or animals.

Informed Consent Statement: Not applicable for studies not involving human.

Data Availability Statement: Data are available from the authors upon reasonable request.

Acknowledgments: We sincerely thank the members of the Remote Sensing and Forest Biogeoscience Laboratory (RSFBioL) and the staff members at the Tungshih Forest District Office, Forestry Bureau, Taiwan, for their contributions to the fuel inventory.

Conflicts of Interest: The authors declare no conflict of interest.

References

1. Simorangkir, D. Fire use: Is it really the cheaper land preparation method for large-scale plantations? *Mitig. Adapt Strat. Glob. Chang.* **2007**, *12*, 147–164. [CrossRef]
2. Kurtulmuslu, M.; Yazici, E. Management of forest fires through the involvement of local communities in Turkey. In *Community-Based Fire Management: Case studies from China, the Gambia, Honduras, India, the Lao People's Democratic Republic and Turkey, Forest Resources Development Service*; Working Paper FFM/2; FAO-Regional Office for Asia and the Pacific: Bangkok, Thailand, 2003; Available online: http://www.fao.org/3/ad348e/ad348e0m.htm (accessed on 22 February 2021).
3. van der Werf, G.R.; Randerson, J.T.; Giglio, L.; Collatz, G.J.; Mu, M.; Kasibhatla, P.S.; Morton, D.C.; DeFries, R.S.; Jin, Y.; van Leeuwen, T.T. Global fire emissions and the contribution of deforestation, savanna, forest, agricultural, and peat fires (1997–2009). *Atmos. Chem. Phys.* **2010**, *10*, 11707–11735. [CrossRef]
4. Doyog, N.D.; Lin, C.; Lee, Y.J.; Lumbres, R.I.C.; Daipan, B.P.O.; Bayer, D.C.; Parian, C.P. Diagnosing pristine pine forest development through pansharpened-surface-reflectance Landsat images derived aboveground biomass productivity. *Forest Ecol. Manag.* **2021**, *487*, 119011. [CrossRef]
5. Shi, G.; Yan, H.; Zhang, W.; Dodson, J.; Heijnis, H.; Burrows, M. Rapid warming has resulted in more wildfires in northeastern Australia. *Sci. Total Environ.* **2021**, *771*, 144888. [CrossRef] [PubMed]
6. Short, K.C. *Spatial Wildfire Occurrence Data for the United States, 1992–2015. FPA_FOD_20170508*, 4th ed.; Forest Service Research Data Archive: Fort Collins, CO, USA, 2017.
7. Prichard, S.J.; Povak, N.A.; Kennedy, M.C.; Peterson, D.W. Fuel treatment effectiveness in the context of landform, vegetation, and large, wind-driven wildfires. *Ecol. Appl.* **2020**, *30*, e02104. [CrossRef] [PubMed]
8. Viegas, D. Overview of forest fire propagation research. *Fire Saf. Sci.* **2011**, *10*, 95–108. [CrossRef]
9. Burkle, L.A.; Myers, J.A.; Belote, R.T. Wildfire disturbance and productivity as drivers of plant species diversity across spatial scales. *Ecosphere* **2015**, *6*, 202. [CrossRef]

10. Fernandes, P.M.; Barros, A.M.G.; Pinto, A.; Santos, J.A. Characteristics and controls of extremely large wildfires in the western Mediterranean Basin. *J. Geophys. Res. Biogeosci.* **2016**, *121*, 2141–2157. [CrossRef]

11. Gonalez-Mathiesen, G.; Ruane, S.; March, A. Integrating wildfire risk management and spatial planning—A historical review of two Australian planning systems. *Int. J. Disaster Risk Reduct.* **2021**, *53*, 101984. [CrossRef]

12. Ryan, R.G.; Silver, J.D.; Schofield, R. Air quality and health impact of 2019-20 Black Summer megafires and COVID-19 lockdown in Melbourne and Sydney, Australia. *Environ. Pollut.* **2021**, *274*, 116498. [CrossRef]

13. Martins, F.; Xaud, H.; Dos Santos, J.; Galvão, L. Effects of fire on aboveground forest biomass in the northern Brazilian Amazon. *J. Trop. Ecol.* **2012**, *28*, 591–601. [CrossRef]

14. Murphy, B.P.; Prior, L.D.; Cochrane, M.A.; Williamson, G.J.; Bowman, D.M.J.S. Biomass consumption by surface fires across Earth's most fire prone continent. *Glob. Chang. Biol.* **2019**, *25*, 254–268. [CrossRef]

15. Keith, H.; Lindenmayer, D.B.; Mackey, B.G.; Blair, D.; Carter, L.; McBurney, L.; Okada, S.; Konishi-Nagano, T. Accounting for Biomass Carbon Stock Change Due to Wildfire in Temperate Forest Landscapes in Australia. *PLoS ONE* **2014**, *9*, e107126. [CrossRef] [PubMed]

16. Aragão, L.E.O.C.; Shimabukuro, Y.E. The incidence of fire in Amazonian forests with implications for REDD. *Science* **2010**, *328*, 1275. [CrossRef] [PubMed]

17. Fitzerald, S.; Berger, C.; Leavell, D. Fire FAQs: What is forest fuel, and what are fuel treatments? In *EM9230*; Oregon State University Extension Service: Corvallis, OR, USA, 2019.

18. Sikkink, P.G.; Lutes, D.E.; Keane, R.E. *Field Guide for Identifying Fuel Loading Models*; Gen. Tech. Rep. RMRS-GTR-225; Fort USDA, Forest Service, Rocky Mountain Research Station: Collins, CO, USA, 2009; 33p.

19. Skowronski, N.; Clark, K.; Nelson, R.; Hom, J.; Patterson, M. Remotely sensed measurements of forest structure and fuel loads in the Pinelands of New Jersey. *Remote Sens. Environ.* **2007**, *108*, 123–129. [CrossRef]

20. Lo, C.S.; Lin, C. Growth-competition-based stem diameter and volume modeling for tree-level forest inventory using airborne LiDAR Data. *IEEE Trans. Geosci. Remote Sens.* **2013**, *51*, 2216–2226. [CrossRef]

21. Anderson, R.S.; Bolstad, P.V. Estimating aboveground biomass and average annual wood biomass increment with airborne leaf-on and leaf-off lidar in great lakes forest types. *North. J. Appl. For.* **2013**, *30*, 16–22. [CrossRef]

22. Lin, C.; Thomson, G.; Popescu, S.C. An IPCC-compliant rechnique for forest carbon stock assessment using airborne lidar-derived tree metrics and competition index. *Remote Sens.* **2016**, *8*, 528. [CrossRef]

23. Erdody, T.L.; Moskal, L.M. Fusion of LiDAR and imagery for estimating forest canopy fuels. *Remote Sens. Environ.* **2009**, *114*, 725–737. [CrossRef]

24. Skowronski, N.S.; Clar, K.L.; Duveneck, M.; Hom, J. Three-dimensional canopy fuel loading predicted using upward and downward sensing LiDAR systems. *Remote Sens. Environ.* **2011**, *115*, 703–714. [CrossRef]

25. Pesonen, A.; Maltamo, M.; Eerikainen, K.; Packalen, P. Airborne laser scanning based prediction of coarse woody debris volumes in a conservation area. *Forest Ecol. Manag.* **2008**, *255*, 3288–3296. [CrossRef]

26. Beucher, S.; Meyer, F. The morphological approach to segmentation: The watershed transformation. In *Mathematical Morphology in Image Processing*; Marcel Dekker Inc.: New York, NY, USA, 1993; Volume 34, Chapter 12; pp. 452–464.

27. Wang, D. A multiscale gradient algorithm for image segmentation using watersheds. *Pattern Recogn.* **1997**, *30*, 2043–2052. [CrossRef]

28. Haris, K.; Efstradiadis, S.N.; Maglaveras, N.; Katsaggelos, A.K. Hybrid image segmentation using watersheds and fast region merging. *IEEE T. Image Process* **1998**, *7*, 1684–1699. [CrossRef]

29. Derivaux, S.; Lefevre, S.; Wemmert, C.; Korczak, J. On machine learning in watershed segmentation. In Proceedings of the IEEE International Workshop on Machine Learning in Signal Processing, Thessaloniki, Greece, 27–29 August 2007; pp. 187–192. [CrossRef]

30. Lin, C.; Thomson, G.; Lo, C.S.; Yang, M.S. A multi-level morphological active contour algorithm for delineating tree crowns in mountainous forest. *Photogram. Eng. Remote Sens.* **2011**, *77*, 241–249. [CrossRef]

31. Lin, C.Y.; Lin, C.; Chang, C.I. A multilevel slicing based coding method for tree detection. In Proceedings of the Geoscience and Remote Sensing Symposium (IGARSS), 2018 IEEE International, Valencia, Spain, 22–27 July 2018; pp. 7524–7527.

32. Alonso-Rego, C.; Arellano-Pérez, S.; Cabo, C.; Ordoñez, C.; Álvarez-González, J.G.; Díaz-Varela, R.A.; Ruiz-González, A.D. Estimating fuel loads and structural characteristics of shrub communities by using terrestrial laser scanning. *Remote Sens.* **2020**, *12*, 3704. [CrossRef]

33. Stefanidou, A.; Gitas, I.Z.; Korhonen, L.; Georgopoulos, N.; Stavrakoudis, D. Multispectral lidar-based estimation of surface fuel load in a dense coniferous forest. *Remote Sens.* **2020**, *12*, 3333. [CrossRef]

34. García, M.; Danson, F.M.; Riaño, D.; Chuvieco, E.; Ramirez, F.A.; Bandugula, V. Terrestrial laser scanning to estimate plot-level forest canopy fuel properties. *Int. J. Appl. Earth Obs. Geoinf.* **2011**, *13*, 636–645. [CrossRef]

35. Rowell, E.; Loudermilk, E.L.; Hawley, C.; Pokswinski, C.; Seielstad, C.; Queen, L.; Q'Brien, J.J.; Hudak, A.T.; Goodrick, S.; Hiers, J.K. Coupling terrestrial laser scanning with 3D fuel biomass sampling for advancing wildland fuels characterization. *Forest Ecol. Manag.* **2020**, *462*, 117945. [CrossRef]

36. Arroyo, L.A.; Pascual, C.; Manzanera, J.A. Fire models and methods to map fuel types: The role of remote sensing. *Forest Ecol. Manage.* **2008**, *256*, 1239–1252. [CrossRef]

37. Popescu, S.C.; Zhao, K.; Neuenschwander, A.; Lin, C. Satellite lidar vs small footprint airborne lidar: Comparing the accuracy of aboveground biomass stimates and forest structure metrics at footprint level. *Remote Sens. Environ.* **2011**, *115*, 2786–27987. [CrossRef]

38. Lin, C. Improved derivation of forest stand canopy height structure using harmonized metrics of full-waveform data. *Remote Sens. Environ.* **2019**, *235*, 111436. [CrossRef]

39. Narine, L.L.; Popescu, S.C.; Malambo, L. Using ICESat-2 to estimate and map forest aboveground biomass: A first example. *Remote Sens.* **2020**, *12*, 1824. [CrossRef]

40. Riaño, D.; Chuvieco, E.; Ustin, S.L.; Salas, F.J.; Rodriguez-Perez, J.R.; Ribeiro, L.M.; Viegas, D.X.; Moreno, J.F.; Fernandez, H. Estimation of shrub height for fuel-type mapping combining airborne LiDAR and simultaneous color infrared ortho imaging. *Int. J. Wildland Fire* **2007**, *16*, 341–348. [CrossRef]

41. Jakubowski, M.K.; Guo, Q.; Collins, B.; Stephens, S.; Kelly, M. Predicting surface fuel models and fuel metrics using lidar and CIR imagery in a dense, mountainous forest. *Photogram. Eng. Remote Sens.* **2013**, *79*, 37–49. [CrossRef]

42. Mutlu, M.; Popescu, S.C.; Stripling, C.; Spencer, T. Mapping surface fuel models using lidar and multispectral data fusion for fire behavior. *Remote Sens. Environ.* **2008**, *112*, 274–285. [CrossRef]

43. García, M.; Riaño, D.; Chuvieco, E.; Salas, J.; Danson, F.M. Multispectral and lidar data fusion for fuel type mapping using support vector machine and decision rules. *Remote Sens. Environ.* **2011**, *115*, 1369–1379. [CrossRef]

44. Cohen, J.D.; Deeming, J.E. *The National Fire Danger Rating System: Basic Equations*; Rep. No. PSW-82; Pacific Southwest Forest and Range Experiment Station: Berkeley, CA, USA, 1982.

45. Brown, J.K. Bulk densities of nonuniform surface fuels and their application to fire modeling. *For. Sci.* **1981**, *27*, 667–683. [CrossRef]

46. Lin, C.; Tsogt, K.; Zandraabal, T. A decompositional stand structure analysis for exploring stand dynamics of multiple attributes of a mixed-species forest. *Forest Ecol. Manag.* **2016**, *378*, 111–121. [CrossRef]

47. Stephens, S.L.; Finney, M.A.; Schantz, H. Bulk density and fuel loads of ponderosa pine and white fir forest floors: Impacts of leaf morphology. *Northwest Sci.* **2004**, *78*, 93–100.

48. Lin, C.; Dugarsuren, N. Deriving the spatiotemporal NPP pattern in terrestrial ecosystems of Mongolia using MODIS imagery. *Photogram. Eng. Remote Sens.* **2015**, *81*, 587–598. [CrossRef]

49. Dugarsuren, N.; Lin, C. Temporal variations in phenological events of forests, grasslands and desert steppe ecosystems in Mongolia: A remote sensing approach. *Ann. For. Res.* **2016**, *59*, 175–190.

50. Coates, T.A.; Johnson, A.; Aust, W.M.; Hagan, D.L.; Chow, A.T. Forest composition, fuel loading, and soil chemistry resulting from 50 years of forest management and natural disturbance in two southeastern Coastal Plain watersheds, USA. *Forest Ecol. Manag.* **2020**, *473*, 118337. [CrossRef]

51. Su, H.J. Studies on the climate and vegetation types of the natural forests in Taiwan (II)—Altitudinal vegetation zones in relation to temperature gradient. *Q. J. Chin. For.* **1984**, *17*, 57–73.

52. Battaglia, M.A.; Rocca, M.E.; Rhoades, C.C.; Ryan, M.G. Surface fuel loadings within mulching treatments in Colorado coniferous forests. *Forest Ecol. Manag.* **2010**, *260*, 1557–1566. [CrossRef]

53. Mathews, S. Effect of drying temperature on fuel moisture content measurements. *Int. J. Wildland Fire* **2010**, *19*, 800–802. [CrossRef]

54. Frandsen, W.H.; Andrews, P.L. Fire Behavior in Nonuniform Fuels, USDA Forest Service, Intermountain Forest and Range Experimental Station, Res. Paper INT-232, 1979.

55. Deutsch, C.V.; Journel, A.G. *Geostatistical Software Library and Users' Guide*; Oxford University Press: New York, NY, USA, 1992; 340p.

56. Berman, J.D.; Breysse, P.N.; White, R.H.; Waugh, D.W.; Curriero, F.C. Evaluating methods for spatial mapping: Applications for estimating ozone concentrations across the contiguous United States. *Environ. Technol. Innov.* **2015**, *3*, 1–10. [CrossRef]

57. Bright, C.B.; Tudak, A.T.; Meddens, A.J.H.; Hawbaker, T.J.; Briggs, J.S.; Kennedy, R.E. Predictin of forest canopy and surface fuels from lidar and satellite time series data in a bark beetle-affected forest. *Forests* **2017**, *8*, 322. [CrossRef]

58. van Aardt, J.A.; Arthur, M.; Sovkoplas, G.; Swetnam, T.L. Lidar-based estimation of forest floor fuel loads using a novel distributional approach. In Proceedings of the SilviLaser 2011, Tasmania, Australia, 16–20 October 2011; pp. 1–8.

59. Franke, J.; Barradas, A.C.S.; Borges, M.A.; Costa, M.M.; Dias, P.A.; Hoffmann, A.A.; Filho, J.C.O.; Melchiori, A.E.; Siegert, F. Fuel load mapping in the Brazilian Cerrado in support of fire management. *Remote Sens. Environ.* **2018**, *217*, 221–232. [CrossRef]

60. Prichard, S.J.; Sandberg, D.V.; Ottmar, R.D.; Eberhardt, E.; Andreu, A.; Eagle, P.; Swedin, K. *Fuel Characteristics Classification System version 3.0: Technical Documentation*; General Technicl Report PNW-GTR-887; Pacific Northwest Research Station: Corvallis, OR, USA, 2013; pp. 39–42.

61. Marino, E.; Ranz, P.; Tome, J.L.; Noriega, M.A.; Esteban, J.; Madrigal, J. Generation of high-resolution fuel model maps from discrete airborne laser scanner and Landsat-8 OLI: A low-cost and highly updated methodology for large areas. *Remote Sens. Environ.* **2016**, *187*, 267–280. [CrossRef]

62. Sah, J.P.; Ross, M.S.; Rnyder, J.R.; Koptur, S.; Cooley, H.C. Fuel loads, fire regimes, and post-fire fuel dynamics in Florida Keys pine forests. *Int. J. Wildland Fire* **2006**, *15*, 463–478. [CrossRef]

63. Bechtold, W.A.; Scott, C.T. *The Forest Inventory and Analysis Plot Design*; Gen. Tech. Rep. SRS-80; U.S. Department of Agriculture, Forest Service, Southern Research Station: Asheville, NC, USA, 2005; pp. 37–52.

64. Forest Resources Development Service, Brief on National Forest Inventory, Indonesia. MAR-SFM Working Paper 18. Forestry Department, UN-FAO. 2007; 1–14.
65. Lin, C.; Trianingsih, D. Identifying forest ecosystem regions for agricultural use and conservation. *Sci. Agric.* **2016**, *73*, 62–70. [CrossRef]

MDPI

St. Alban-Anlage 66

4052 Basel

Switzerland

www.mdpi.com

Remote Sensing Editorial Office

E-mail: remotesensing@mdpi.com

www.mdpi.com/journal/remotesensing